# Introduction to Aerosol Modelling

# Introduction to Aerosol Modelling

From Theory to Code

*Edited by*

*David Topping*
University of Manchester
Manchester, UK

*Michael Bane*
Manchester Metropolitan University &
High End Compute Ltd
Manchester, UK

*Registered Offices*
John Wiley & Sons, Inc., 111 River Street, Hoboken, NJ 07030, USA
John Wiley & Sons Ltd, The Atrium, Southern Gate, Chichester, West Sussex, PO19 8SQ, UK

*Editorial Office*
9600 Garsington Road, Oxford, OX4 2DQ, UK

For details of our global editorial offices, customer services, and more information about Wiley products visit us at www.wiley.com.

Wiley also publishes its books in a variety of electronic formats and by print-on-demand. Some content that appears in standard print versions of this book may not be available in other formats.

A catalogue record for this book is available from the Library of Congress

Paperback ISBN: 9781119625650; ePub ISBN: 9781119625711; ePDF ISBN: 9781119625667

Cover image: Cover illustration kindly provided by Harri Kokkola, Finnish Meteorological Institute. Photograph ©Tuukka Kokkola
Cover design by Wiley

Set in 9.5/12.5pt STIXTwoText by Integra Software Services Pvt. Ltd, Pondicherry, India
Printed and bound by CPI Group (UK) Ltd, Croydon, CR0 4YY

C9781119625650_030822

# Contents

# Contributors

**MICHAEL BANE** has spent several decades in optimizing codes and helping others understand how they can optimize codes. He is currently a Lecturer at Manchester Metropolitan University, UK, and Director of High End Compute LTD. Michael wrote Chapter 8 and gave supportive input to other chapters, particularly Chapter 1.

**JEFFREY CURTIS,** Postdoctoral Research Associate in the Department of Atmospheric Sciences at the University of Illinois at Urbana-Champaign. Jeffrey contributed to Chapter 5.

**HARRI KOKKOLA,** Group leader of the Atmospheric Modeling Group at the Atmospheric Research Centre of Eastern Finland. He has developed and applied models simulating atmospheric aerosol, clouds, and climate from process scale to global scale. Harri wrote Chapter 7.

**BENJAMIN MURPHY,** Physical Scientist in the United States Environmental Protection Agency Office of Research and Development at Research Triangle Park, North Carolina. Benjamin contributed to Chapters 1, 4, and 5.

**TINJA OLENIUS,** research scientist in the Air Quality Unit at Swedish Meteorological and Hydrological Institute (SMHI). Tinja's research interests include developing the chain of modeling tools from quantum mechanics to large-scale models for improving the representation of secondary particle formation from vapors. Tinja wrote Chapter 6.

**OLLI PAKARINEN,** university lecturer at the Institute for Atmospheric and Earth System Research (INAR), University of Helsinki, Finland. Olli contributed to Chapter 6.

**NICOLE RIEMER,** Professor in the Department of Atmospheric Sciences at the University of Illinois at Urbana-Champaign, IL, USA. Nicole's research interests include the development of aerosol models, from the process level to the global scale. Nicole wrote Chapter 5.

**PETROC SHELLEY,** PhD student in the Department of Earth and Environmental Science at the University of Manchester, UK. His current specialism is in the measurement and prediction of pure component properties. Petroc contributed to Chapter 4.

**DAVID TOPPING,** Professor in the Department of Earth and Environmental Science at the University of Manchester, UK. Since his PhD, David has developed and applied models of aerosol particles across a range of scales. David convened the writing team behind this book and wrote Chapters 1, 2, and 4.

**MATTHEW WEST,** Associate Professor in the Department of Mechanical Science and Engineering at the University of Illinois at Urbana-Champaign, IL, USA. His research interests include stochastic time integration methods and scientific computing. Matt wrote Chapter 5.

**ZHONGHUA ZHENG,** PhD in Environmental Engineering in Civil Engineering with a concentration in Computational Science and Engineering at the University of Illinois at Urbana-Champaign, IL, USA. Zhonghua contributed to Chapter 5.

**ANDREAS ZUEND,** Professor in the Department of Atmospheric and Oceanic Sciences at McGill University, Montreal, Quebec, Canada. His research interests include the development of predictive thermodynamic aerosol models for gas–particle partitioning, as well as reduced-complexity methods for mixture properties of aerosols and cloud droplets. Andreas wrote Chapter 3.

# Preface

Aerosol science is one that straddles many disciplines. There is a natural tendency for the aerosol scientist to therefore work at the interface of the traditional academic subjects of physics, chemistry, biology, mathematics, and computing. The impacts that aerosol particles have on the climate, air-quality, and thus human health are linked to their evolving chemical and physical characteristics. Likewise, the chemical and physical characteristic of aerosol particles reflect their sources and subsequent processes they have been subject to. Computational models are not only essential for constructing evidence-based understanding of important aerosol processes, but also to predict change and potential impact. Seminal publications provide an extensive overview on the history and basis of core theoretical frameworks that aerosol models are based on. However there is little on how we can translate such theory into code. While we focus on atmospheric aerosol in this book, the theory and tools developed are based on core aerosol physics that translate across multiple disciplines. Likewise, demonstrating a programming solution to common numerical operations is valuable to a large number of scientific disciplines. You may be reading this book as an undergraduate, postgraduate, seasoned researcher in the private/public sector or as someone who wishes to better understand the pathways to aerosol model development. Wherever you position yourself, it is hoped that the tools you will learn through this book will provide you with the basis to develop your own platforms and to ensure the next generation of aerosol modelers are equipped with foundational skills to address future challenges in aerosol science.

Manchester                                                                 *Professor David Topping*
July 2022

## Acknowledgments

The identified need for this book was inspired by the Aerosol and Droplet Science Centre for Doctoral Training [CDT], funded by the UK Engineering and Physical Sciences Research Council (EPSRC)(grant no. EP/S023593/1) (https://www.aerosol-cdt.ac.uk/).

The authors would also like to thank Hanna Vehkamäki, Ana Cristina Carvalho, Manu Thomas and Cecilia Bennet for useful discussions and comments. The Swedish Research Council VR (grant no. 2019-04853) and the Swedish Research Council for Sustainable Development FORMAS (grant no. 2019-01433) are acknowledged for financial support for Chapter 6.

## About the Companion Website

All of the code snippets and examples provided with the book can also be downloaded from the project Github repository:

https://github.com/aerosol-modelling/Book-Code.git

# 1

# Introduction and the Purpose of this Book

An aerosol particle is defined a solid or liquid particle suspended in a carrier gas. The term "aerosol" technically includes both the particle and carrier gas, though it is common to often hear this used when referring to just the particle. In this book, we will retain the use of the term "aerosol particle". Whilst we often treat scientific challenges in a siloed way, aerosol particles are of interest across many disciplines. For example, atmospheric aerosol particles are key determinants of air quality [1–3] and climate change [4–6]. Improving our understanding of sources, processes and sinks is important as we develop strategies to lesson the impacts we have on human health and environmental systems. Knowledge of aerosol physics and generation mechanisms is key to all factors of fuel delivery [7] and drug delivery to the lungs [8]. Likewise, various manufacturing processes require optimal generation, delivery and removal of aerosol particles in a range of conditions [9].

The purpose of this book is to provide you, the reader, with the tools to translate theory on which numerical aerosol models are based into working code. In following the content provided in this book, you will be able to reproduce models of key processes that can either be used in isolation or brought together to construct a demonstrator 0D box-model of a coupled gaseous-particulate system.

You may be reading this book as an undergraduate, postgraduate, seasoned researcher in the private/public sector or as someone who wishes to better understand the pathways to aerosol model development. Wherever you position yourself, the coupling between experimental and modeling infrastructure is important in any discipline. Whilst the driving factors that influence both can vary, Figure 1.1 presents an idealized workflow of model development and model scales both in response to and as a driving force behind aerosol experiments. Particular emphasis is given to atmospheric aerosol particles in this workflow where, as we move from left to right, we move from aerosol models at the molecular and single particle scale to aerosol models acting as an import component in regional and global scale models. The purpose of this figure is to represent a workflow that migrates our understanding of aerosol processes to a framework that may be used to predict impacts. In a perhaps controversial approach, we can imagine a scale at the bottom of the figure that assumes as we move from left to right we reduce the physical and chemical complexity of our aerosol models. This sets the scene for understanding the research landscape of much of the developments you will find in this book.

*Introduction to Aerosol Modelling: From Theory to Code.*
First Edition. Edited by David Topping and Michael Bane.
© 2022 John Wiley & Sons Ltd. Published 2022 by John Wiley & Sons Ltd.

If we start at the left-hand side of the figure, we use the term mechanistic model. In aerosol modeling parlance, a mechanistic model is one that is built around a numerical representation of an underlying physical theory. For example, this might include a set of coupled differential equations that describe the movement of mass between a gaseous and condensed phase, or between different compartments of a condensed phase. Parameters in these mechanistic models may describe chemical and physical properties that are included in these differential equations and have been derived from a series of experiments or provided through separate models. In a mechanistic model, our mathematical framework provides a clear numerical narrative and separation of the processes we wish to include. We can then choose an appropriate numerical method to provide, for example, a time-varying solution to a set of initial conditions or predict a point of equilibrium. Once we have constructed our mechanistic model, we can consider uncertainties associated with the model architecture itself and/or errors associated with the parameters we use in our simulation. Indeed, the next phase in our workflow in Figure 1.1 is to compare with targeted laboratory experiments that serve to quantify the accuracy of our model or identify uncertainties that need further reduction. You may find mechanistic frameworks used at the single particle level, or indeed in models that are designed to capture the evolution of a population of particles. Where mechanistic models cannot replicate observed behavior within a specific level of accuracy, or simply do not have an appropriate theoretical basis to build on, parameterizations can be developed. This can be used in combination with, or as an alternative to, the mechanistic model. We often state that a parameterization has a higher computational efficiency than an equivalent mechanistic model. Specifically, the time to solution is reduced. As we move further right in Figure 1.1, and typically start to consider populations of particles and multiple processes, we might refer to hybrid models that combine both mechanistic and parameterizations. We may also start to consider the computational resource available to conduct more complex simulations. At the global scale, an aerosol model is one of many components in a framework that attempts to capture the dynamics of multiple components of the earth system (e.g., ocean, biosphere, land-air interactions etc.). The level of physical and chemical complexity retained in our aerosol model is dictated by a number of factors. These include the computational resource available, the associated detail carried in components that drive and respond to aerosol processes (e.g., how many emissions that lead to aerosol formation are captured) and ongoing efforts to resolve how much detail is needed to resolve potential impacts on, for example, human health. Of course, this narrative is an ideal one but at least provides an insight into factors that dictate the methods we use to construct our aerosol models. In this manner, we start to appreciate the ecosystem of aerosols models and why they exist. The aerosol scientist may come across a range of "simple" and "complex" models that have been designed to provide benchmark simulations in isolation. You will find a description of these benchmark models in the chapters of this book. Once we begin to capture processes across multiple scales, an aerosol model developer starts to consider any approximations that may be needed according to the numerical methods and compute resource available.

Whilst we focus on atmospheric aerosol to define our composition space in this book, the theory and tools developed are based on core aerosol physics that translate

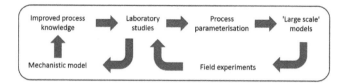

**Figure 1.1**   Ideal workflow of aerosol model development in environmental science.

across multiple disciplines. Likewise, demonstrating a programming solution to common numerical operations is valuable to a large number of scientific disciplines.

Research developments often move at a rapid pace and, as the global aerosol community develop new observational and modeling platforms, we continually hypothesize and verify new species and/or processes deemed important to improve our understanding. We do not provide a comprehensive coupling of all known and emerging chemical and process complexity in this book. Indeed, there are remaining challenges on how we actually do that from a programming and real-world validation perspective. The landscape of computing hardware and software also moves at a rapid pace. The choice of programming language to solve a particular problem, or provide a particular service, is influenced by a number of factors ranging from required time-to-solution and ability to share across multiple platforms. As with numerical representations of aerosol processes, we do not provide a comprehensive multilanguage demonstration in this book. It is anticipated that readers of this book will have a wide range of programming experience; from those who have no prior experience to those who regularly develop their own applications. We expect therefore that you will take away different lessons from working through the material provided, whether it be the solution to a set of common operations or learning how to develop your first numerical model in your first programming language. You will find there are often multiple ways to write a piece of code that performs a particular task. You will also find that as we often have our own style in writing, so too can we develop our own style of developing code. In this book, we provide complete demonstrations of how to develop working code around key concepts (highlighted in figure 1.2), but we do not force a particular style beyond requirements of the language syntax. We also however provide examples of how we can optimize the code we develop. By looking at a range of examples, this will help you start to more broadly consider how efficient your code is and perhaps embed these considerations as you start to develop mode applications. We also discuss best practice in sharing any code in the public domain and ensuring reproducibility.

Seminal publications [e.g., [2, 19]] provide an extensive overview on the history and basis of core theoretical frameworks that aerosol models are based on. We do not repeat that content in this book; rather we present the theoretical basis used in constructing a model and then focus on how we map this to code development.

The tools you will learn through this book are foundational. As the research community explore new hardware platforms and programming languages in an attempt to tackle growing complexity, these foundational skills will provide you with the basis to develop your own platforms. Will anyone tackle the entirety of aerosol modeling complexity? Maybe it will be you.

## 1.1  Aerosol Science and Chapter Synopses

Aerosol science is multidisciplinary by nature. This is reflected in the huge body of literature that now exists in peer-reviewed journals. There is a natural tendency for the aerosol scientist to therefore work at the interface of the traditional academic subjects of physics, chemistry, biology, mathematics, and computing. Of course, an aerosol scientist working in either medicine or climate change will find themselves focusing on distinct areas and the level of understanding in each will be dependent on the research challenge. However, chances are that both will, at some point, require training in key concepts of aerosol science that apply to both domains. Indeed, one benefit of becoming an aerosol scientist is that understanding, refinement and application of core concepts is transferable between disciplines.

In 2018, the Aerosol Society of the United Kingdom published the outcomes of an industrial engagement workshop [12], defining a pipeline of research, innovation

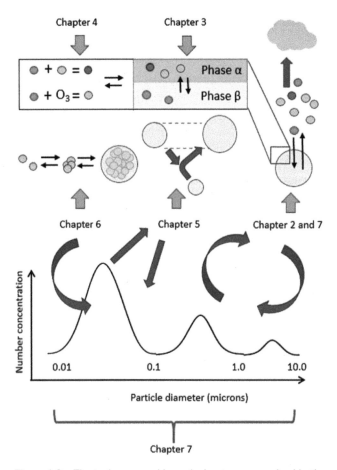

**Figure 1.2**   The topics covered in each chapter, summarized in the main body of text in this chapter, as a visual schematic that connects processes across the aerosol size spectrum. In this hypothetical example, the aerosol size distribution has three peaks represented as multiple log-normal contributions. We start to discuss log-normal distributions in Sections 1.3.2 and 1.3.3.

and technology development for aerosol science. In this they note that estimates of the global aerosol market size suggest it will reach \$84 billion per year by 2024 with products in the personal care, household, automotive, food, paints, and medical sectors. However, they also note that despite the growing interest into the macro-effects and industrial exploitation of aerosols, aerosol science is a relatively young discipline encompassing research topics which can concomitantly be understood as biological, chemical, engineering, environmental, material, medical, pharmaceutical, or physical science.

We focus on atmospheric aerosol particles for the remaining portions of this book. The impacts that aerosol particles have on the climate, air-quality and thus human health, are linked to their evolving chemical and physical characteristics [3, 5]. Likewise, the chemical and physical characteristic of aerosol particles reflect their sources and subsequent processes they have been subject to [2, 14]. Atmospheric aerosol particles can range in size from a few nanometers to hundreds of microns. They can be of primary or secondary origin. Primary particles are directly emitted into the atmosphere, whilst secondary particles are produced from gas-to-particle conversion processes. An aerosol particle may comprise inorganic and/or organic components which can be associated with both primary and secondary particles. Whilst the inorganic fraction may be comprised by a relatively small number of components [2], the organic fraction may comprise many thousands of compounds from multiple sources [15, 16]. As an aerosol particle resides in the atmosphere, we know that many processes taking place in/on atmospheric aerosol particles are accompanied by changes in the particles' morphology (size and shape) [17]. These processes also change the chemical composition of aerosol particles according to the availability of, for example, key gas phase oxidants and ambient conditions [16]. Likewise, particles of primary original (e.g. desert dust, volcanic ash, soot, pollen) can have widely varying morphological features [19–21]. Mechanisms are important in understanding lifetimes and potential impacts [22].

Let us begin with an idealized spherical representation of aerosol particles. Their size and composition will vary, but we wish to simulate how their concentration changes over time. Whilst controlled experiments may be able to isolate single levitated particles [23], under atmospheric conditions we expect a range of particle sizes and number densities. Let us take an isolated particle with diameter $d_p$ and density $\rho$ with units in m and $kg \cdot m^{-3}$, respectively. A single particle has a mass $M$ in kg calculated using Equation (1.1):

$$M = \frac{4}{3}\pi \left(\frac{d_p}{2}\right)^3 \rho \tag{1.1}$$

If we observe $N_L$ particles per cubic centimeter, we can use Equation (1.2) to calculate the total concentration of particles with size $d_p$, $M_{tot}$, using the common air quality metric of $\mu g.m^{-3}$.

$$M_{tot} = N_L M 10^{15} \tag{1.2}$$

where the factor $10^{15}$ is a product of converting $cm^{-3}$ to $m^{-3}$ ($10^6$) and kg to $\mu g$ ($10^9$). This formula has no information on the particle composition or morphology. Indeed, our idealized spherical representation may be wrong under certain conditions and/or for certain aerosol types. Nonetheless, we can use our particle representation, which has a volume $V$ (m) given by Equation (1.3), as a common particle reference.

$$V = \frac{4}{3}\pi\left(\frac{d_p}{2}\right)^3 \tag{1.3}$$

If we now specify that the concentration (number density) of particles with this specific volume, at a specific time $t$, can be represented by a variable $n_{v,t}$, we can start to formulate an expression that captures the evolution of particles with variable volumes into an algebraic form.

For example, in Figure 1.3 on the left-hand side we have a population of particles with a specific volume at a given point in time. The total concentration of these particles is $n_{v,t}$. On the right-hand side, this population has evolved after a time increment $\Delta t$ and we have particles at smaller and larger volumes, represented by a discrete change $\Delta v$. Specifically, $n_{v,t+\Delta t}$, $n_{v-\Delta v,t+\Delta t}$ and $n_{v+\Delta v,t+\Delta t}$ represent the concentration of particles that have the original volume, the concentration of particles with a smaller volume and the concentration of particles with a larger volume at a new time $t + \Delta t$, respectively. In reality these particles will have been created through primary and/or secondary mechanisms.

Equation (1.4) is the continuous general dynamic equation [24]. This ordinary differential equation (ODE) describes the rate of change of $n_{v,t}$ resulting from key processes which are identified as nucleation, coagulation and condensation. We also include a generic term that represents emission and removal mechanisms.

$$\frac{dn_{v,t}}{dt} = \left(\frac{dn_{v,t}}{dt}\right)_{nucleation} + \left(\frac{dn_{v,t}}{dt}\right)_{coagulation} + \left(\frac{dn_{v,t}}{dt}\right)_{condensation} +$$
$$\left(\frac{dn_{v,t}}{dt}\right)_{emission} - \left(\frac{dn_{v,t}}{dt}\right)_{removal} \tag{1.4}$$

Equation (1.4) is the first algebraic formulation of aerosol evolution in this book. Each contribution to this equation has its own theoretical basis and thus algebraic

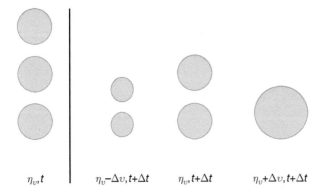

$\eta_v, t$ $\qquad$ $\eta_v-\Delta v, t+\Delta t$ $\qquad$ $\eta_v, t+\Delta t$ $\qquad$ $\eta_v+\Delta v, t+\Delta t$

**Figure 1.3** An initial population of particles with volume $V$ on the left-hand side of this figure evolve to a population with multiple volumes after a time increment $\Delta t$.

formulation. These are presented and discussed in specific chapters, and also highlighted in figure 1.2 and discussed shortly. We also need to translate our algebraic formulations into code. You will find a range of approaches are presented to perform this translation that provide you with a grounding in common approaches used across aerosol science. Figure 1.5 provides a schematic that maps processes on our book structure by highlighting the relevant chapters that focus on each in turn. Whilst we isolate processes including nucleation and coagulation in distinct chapters, we also cover properties and bridging technologies that are not explicitly highlighted in Equation (1.4) but are nonetheless important to providing a solution to it. These include developing models that capture condensed phase thermodynamics and simulate diffusion within a particle, simulating a reactive gas phase and presenting the structure required to build a simple model of aerosol to cloud droplet activation. A synopsis of each chapter is provided below. At the beginning of each chapter you will find an overview of the language chosen and use of any functions provided within that environment. You will also find that, in some chapters, we combine a breakdown of translating theory to code with instructions on how to run existing community models which have been written in a variety of programming languages.

- Chapter 1: Following on from an introduction to aerosol science, here we discuss general concepts around representing aerosol particles as a numerical model. This encourages you to consider what information an aerosol model may need to carry and how we then map that into a numerical model through the programming language constructs. For example, at the single particle level we may wish to construct a model that simulates the partitioning of mass between a gaseous phase and homogeneous droplet. If we have $n$ compounds in the gas phase and one particle, then it is reasonable to design a model built around a one dimensional array of size $n + 1$. Of course, once we consider populations of particles with varying sizes and, perhaps, interparticle morphology, then our choice becomes more complex. In this chapter, we therefore present some general considerations around these issues which help contextualize the design choices you will come across in the proceeding chapters. We follow this by introducing our first theory to code demonstrations where we implement two different approaches to represent a population of particles in Python. Known as the modal and sectional methods, this provides a basis for subsequent chapters.
- Chapter 2: In this chapter, we present and implement theories that allow us to simulate movement of mass between a gas and condensed phase. We first introduce equations that provide the basis for predicting the composition of both the gas and condensed phase at equilibrium. This is split between the processes of adsorption and absorption and considers both the particulate and gaseous phase to be nonreactive. Up to this point we assume our particulate phase to have no size, considering a total mass that compounds in the gas phase can adsorb or absorb to. Following on from this we then move to simulating dynamic absorptive partitioning by introducing and solving the droplet growth equation. Whilst both our gas and particulate phase remains nonreactive, we use Python to simulate the growth of both mono- and polydisperse populations as a function of time. The size and composition of the simulated aerosol particles influence the partitioning process through a change in equilibrium pressure above the condensed phase, but the condensed phase components do not interact with each other. We use both the modal and sectional distributions covered

in Chapter 1, where we compare Python and Julia models for the sectional approach. In each case, when designing our model structure, we also need to be aware of how we utilize any specific solver routines. In this case, we use existing ordinary differential equation (ODE) solvers within popular Python and Julia packages. Attention is then given to predicting aerosol water uptake through equilibrium frameworks, at the single particle level, below and above 100% relative humidity.

- Chapter 3: In this chapter, we introduce theoretical frameworks that underpin thermodynamics and nonideal mixing. Whilst in Chapter 2 we assumed our aerosol particles were constructed of a homogeneous ideal mixture, here we account for nonideal mixing. We specifically move to treating nonideal interactions that dictate the predicted equilibrium state and, in some cases, lead to phase separation in the particulate phase. You will see the use of Python again in constructing simulations of simple mixtures and the required structure of our code. You will also find information of how to use an existing community model to simulate phase separation in complex mixtures, written in Fortran. Once again, translating the relevant theory into code here requires some consideration of an appropriate model structure. In Chapter 2 we discuss separation of aerosol particle size as we design the arrays that will track information on aerosol composition through the simulation; likewise here we need to consider how we track the composition of our particle.

- Chapter 4: In all previous chapters, we have used a nonreactive "static" gas phase. As aerosol particles reside in a gaseous medium, they are subject to processes that are driven by the availability of compounds in said medium. Likewise, the availability of gaseous compounds is driven by the complex chemistry that unfolds as compounds react with each and a range of oxidants. In this chapter, we begin with an example of how to simulate, and thus track, the variable concentration of compounds in a gas phase. We discuss the concept of a chemical mechanism in the context of a file that holds information about the interaction of compounds in the gas phase. We then use tools in Python to extract information in these files and create a code structure that allows us to simulate the concentration of each compound as a function of time. Once again we consider an appropriate structure that is driven by the information we wish to track and our chosen ODE solvers. With an evolving gas phase, we can also consider the properties of individual compounds that dictate gas-to-particle partitioning. Predicting those properties requires us to map an algebraic form of a predictive technique to a chemical structure. You will therefore find the use of Python and existing informatics packages to extract and automate the prediction of properties for many thousands of compounds. With all of the previous work on required code structures and solvers for simulating condensational growth and a reactive gas phase, we also provide a demonstration of how to use existing community-driven models that have been designed to automate the process from reading a chemical mechanism file and then creating a model that will simulate the evolution of a coupled particulate and gas phase, written in both Python and Julia.

- Chapter 5: If the population of our particles have different velocities, there is a chance they will collide. Two particles colliding to produce a larger particle reduce the total number concentration over time. This process is called coagulation and is influenced by a number of factors including ambient conditions, the concentrations of aerosol particles and their phase state. In this chapter, we develop stochastic and deterministic representations of the coagulation process in Python. In all other chapters, we have built deterministic models. As part of this chapter's development of the stochastic and deterministic representations, we discuss the impact of design choices on computational efficiency, including a comparison between a modal and sectional approach. Following on from this we start to consider more complex coagulation scenarios including, for example, the treatment of nonspherical particles.

- Chapter 6: Whilst Chapter 2 focuses on gas-to-particle partitioning to an existing condensed phase, in this chapter we present underlying theories that are used to predict new particle formation. This is referred to as aerosol nucleation. As Chapter 5 treats the interaction between aerosol particles, here we move our focus to the molecular level. We contextualize this work in terms of aerosol size ranges we have met in all other chapters and now consider clusters of molecules as discrete units. The boundary between a molecule and an aerosol particle becomes blurred, but the challenge to design an appropriate numerical model remains. We build a Matlab solution to a discrete form of the "birth-death" equation for molecular clusters. You will find a discussion on how models we develop have a place and dependency in a wider ecosystem of numerical models. In this instance, that specifically includes those conducting molecular dynamics and quantum chemistry simulations. Moving beyond the Matlab based examples we build here, you will also find an introduction and tutorial on using the Atmospheric Cluster Dynamics Code (ACDC) which has components written in the languages Perl, Matlab and Fortran.

- Chapter 7: Chapters 1–6 can be looked at in isolation and the code examples allow you the reader to further develop them or integrate them into other software. Of course, we know from the general dynamic equation that we wish to connect these process descriptions such that we can simulate the life-cycle of an aerosol particle, or population of particles. In this chapter we therefore introduce the concept of a box-model and present a numerical and code design strategy to integrating nucleation, coagulation and condensational growth. Focusing on a sectional approach, you will find that the mechanism for tracking particle size presented in Chapter 2 has limitations when we wish to include nucleation and coagulation. We present methods for restructuring our numerical arrays in Python. We finish this chapter by assuming our aerosol distribution is within a rising parcel of air which leads to increasing relative humidity above saturation and leads to formation of cloud droplets. In this manner you will learn how to create a cloud parcel model and lay the foundation for another large area of research in capturing and predicting cloud micro-physics from a population of aerosol particles. In the end of the chapter, we will present a box aerosol-cloud model SALSA which comprises atmospherically relevant micro-physical processes which interact with aerosol particles, cloud droplets, precipitation droplets and ice nuclei.

Translating theory into code requires us to consider how we can represent information on the properties of the aerosol system we are interested in. This is influenced by choice of programming language, which in turn may be driven by the availability of numerical methods or your reliance on existing legacy code that may have been developed in your community. This also fundamentally requires consideration of how we represent aerosol particles in numerical models. Whilst each chapter provides you with a grounding in common approaches to solving the relevant process, with this in mind, in Section 1.2 we provide you with a brief overview of computer programming languages used in this book, should this be needed. Following this, in Section 1.3 we then present a general consideration of how we represent aerosol particles and their physical and chemical characteristics within numerical constructs in our software. This is designed to provide extra context to the content you will find in each chapter.

## 1.2 Computers and Programming Languages

A computational model is built around a set of rules; a single of set of algorithms. The interface between the aerosol scientist, who has drawn up these rules, and the computer is provided by a programming language. There are many programming languages available, each with its own set of advantages and disadvantages depending on what the problem is you wish to solve. There are general "rules of thumb" that can be followed in order to select the most appropriate language, and the more time you spend coding and exploring the multitude of applications available, the clearer this choice will become. One should avoid the pitfalls of snobbery that may surround software development, especially if you are starting your journey on becoming an aerosol model developer. If you select relatively simple examples to start practicing translating theory into code then you are free to pick a range of languages. True that some are faster, and others have a much broader ecosystem (e.g. for visualization, connecting numerical simulations to machine learning libraries etc). However, it is important to enjoy and explore the world of programming as best you can. When you start to require that element of speed, or specific dependencies, then you can revisit which language you wish to develop in. In this book you will indeed come across a range of languages. Python is used throughout the book, but we also present examples in Fortran, Julia and Matlab. These are influenced in part by the existing language in which aerosol community models are currently based, but it also provides us with an opportunity for you to explore these varying languages.

In order to appreciate the differences in computer languages, we need to understand a little about the hardware we wish to run our aerosol models on.

Let us start from the top. Imagine we have received a blue print of a high performance computing (HPC) center with plans for a cluster, nodes and individual processors. As illustrated in Figure 1.4, a **cluster** will have several nodes (represented by each yellow box); a **node** can be thought of as a motherboard with one of more sockets; a **socket** contains a CPU processor (represented by each blue box); and a **CPU processor** may have one or more **cores** as represented by each black box in the diagram. Much closer to home, you may have a laptop or desktop computer at your disposal that likewise contains a motherboard with (probably) one socket; the **socket** contains a CPU processor; and the **CPU processor** may have one or more **cores**. Your desktop will likely have fewer cores than the CPU processor in the HPC center. The CPU and other components will have been manufactured by a set of popular vendors that include Intel, AMD, ARM and IBM. As Section 8.3.2 discusses, in a cluster the nodes are connected by some form of **interconnect** as illustrated in orange in the figure.

We also need to consider how a high-level programming language that humans can read and understand ends up getting run on a computer.

Traditional computers (as opposed to the likes of quantum computing discussed in Section 8.3.3) have a very limited set of instructions they recognize. This is known

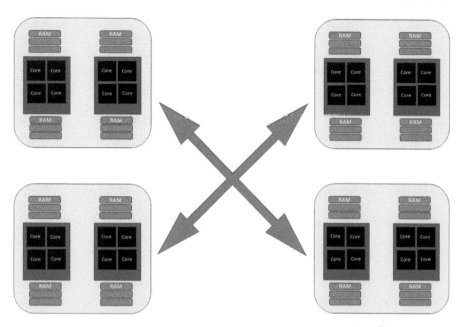

**Figure 1.4** Illustration of hardware components of an HPC cluster.

formally as the **Instruction Set Architecture** (ISA) and it is necessary to translate the high-level programming language in to **machine code** that adheres to a given CPU processor (since we may presume all cores on a given CPU processor have the same ISA).

There are two approaches to translating from the chosen programming language to machine code. **Compiled languages** do this translation in a separate step prior to running any of the code. Having compiled the source codes to an **executable** (shorthand for "executable file") for a given CPU processor, the executable can be run again and again on that CPU processor architecture, with further compilation only required when you make changes to the source codes. This is in contrast to **interpreted languages** which translate each line of code as it is required during the actual running of the source code. Many popular languages are "interpreted," which can be thought of as a basic "compile each line of code as we want to run it." From a functional viewpoint there is nothing wrong with this. Many interpreted languages are highly popular, due to their ease of learning, including Python.

Table 1.1 illustrates and compares these two approaches.

The main compiled languages are C, C++, Fortran and popular interpreted languages that include Python and Julia. You may be wondering why there is a choice and perhaps which you should use. In terms of which can be used to implement the algorithms discussed in this book, it doesn't matter—you can code each algorithm in any or all of these languages. But the ease of doing so, and the resulting performance, may vary. There are some pros and cons relating to compiled versus interpreted languages, as listed in Table 1.1. Programming languages themselves are based on different approaches, and most evolve over time. For example, Java was designed with the "object oriented" (OO) style in mind, where the focus of programming is the consideration of the hierarchical classes of objects and the various "methods" operating upon them.

**Table 1.1** Comparison of compiled and interpreted languages.

| Comparison of approaches | |
| --- | --- |
| **Interpreted languages** | **Compiled languages** |
| – | Requires a compiler |
| Translated line by line at run time | All of code is compiled in separate stage before running |
| A line may be translated many times | – |
| No opportunity to see bigger picture | Compiler can analyze full code to make deep optimizations |
| Generally felt to be easier to learn | Generally felt to give better performance |
| Can use same source on different platforms | Need to compile for each ISA |

Other languages such as C++, Julia, and Python (and to some extent modern Fortran) also now support OO programming.

Most programming languages scientists encounter are **imperative languages**, where the control is implemented by a series of statements that manipulate the data of the simulation. Fortran, C, C++, Julia, Python, and MATLAB are all imperative programming languages. The alternative is where it is the flow of data that is described by the programmer, leaving the implementation of how this is achieved to a compiler. These are known as **functional languages** and include Haskell, Lisp, Erlang, and Clojure. Data flow is key to programming concurrently and there is growing interest in use of functional programming for FPGAs (see Section 8.3.3 in Chapter 8) and highly concurrent systems.

Let us take a quick look at the languages used in this book in terms of their provenance and use. We reiterate that the best way to learn and become more familiar with each is to solve a particular problem, which you can do across the chapters in this book.

● Python [Official documentation: https://www.python.org/]: First released in 1991, Python is a universally popular language finding use across not just scientific domains, but from database design to web development. At of the time of writing, Python is released as version 3.8 with support for packages built in version 2 discontinued. Python is an interpreted language and is often regarded as the best choice to start programming. This is in part related to the readability of the code, but also the huge ecosystem of tools and facilities that you can now integrate with your own personal developments. For example, the ability to integrate with chemical informatics tools [25], machine learning [26], or powerful visualizations [27] is an attractive prospect for the multidisciplinary scientist with a number of examples used across aerosol science. Whilst speed has been a concern when developing relatively large numerical models using Python alone, Python can also act as the "glue" to connect libraries built in other languages whilst the Numba "High Performance Python Compiler" [28] translates Python functions to optimized machine code at runtime with minimal effort. Even if a researcher does not work directly with any form of numerical models, Python modules such as Pandas offer tremendous flexibility in ingesting and manipulating data. Similarly, Python provides an interface to a range of popular machine-learning modules such as Scikit-learn [26] and Keras [29]. Perhaps the most common route to installing Python on a personal machine is to use the Anaconda distribution (https://www.anaconda.com/products/individual), which comes with a flexible package manager called `Conda`. The Conda package manager can be very helpful for those who wish to build an environment with a number of modules that need to be installed separately or are not included in the default Anaconda distribution. This is because Conda manages the required dependencies and will install additional modules where required. Alongside Conda there is also the `pip` package installer which can be useful where there is no Conda channel for a particular module. There are multiple ways you can interact with and use Python for your own projects, but they all require us to write some code! For a graphical user interface (GUI) experience, the Spyder Interactive Development Environment (IDE) combines a text editor with interactive console and variable explorer (https://www.spyder-ide.org/). Jupyter notebooks are hugely popular environment for teaching Python. A Jupyter Notebook is an open-source web application that allows you to create and share documents that contain live code, equations, visualizations and narrative text (https://jupyter.org/) and is not limited to Python. Throughout this book we will provide you with Python examples written as individual text files that have a *.py* extension. Imagine we have created a file called *test.py*. Once you have a distribution of Python installed, we would open up a terminal window, or Anaconda prompt, and ask the Python Interpreter to run our file as follows:

```
python test.py
```

- Fortran [Official documentation: https://fortran-lang.org/]: The Fortran language has a rich history of, to date, more than six decades as a high-level language, with special emphasis on being a structured, compiled language offering good optimization potential for performance-critical numerical programs and libraries (e.g. 26). In particular, the FORTRAN 77 language (name at those times written in capital letters) and the substantially revised and improved Fortran 90 standard (released in the year 1991; perhaps the origin of "modern" Fortran) have found wide-spread applications in many fields of science and engineering and have been a popular choice for decades in those disciplines. The Fortran language also keeps evolving, with a new standard typically being released every five to ten years. The more recent versions of Fortran, described by the Fortran 2003, 2008, and 2018 standards, have added improved support of dynamic data types, object-oriented programming, new intrinsic array functions, native parallel programming, standardized interoperability with the C language, and clarified/revised language definitions (see https://wg5-fortran.org for official language standard publications and descriptions of added features). The standardized interoperability with the C language also supports improved interoperability of Fortran modules with Python (e.g., via the "f2py" method within Python's NumPy library).

  One valuable feature of the Fortran language development and revision process is that it remains backward-compatible with older standards and programs (so long as they were standard-conforming at their time). In the geoscience disciplines, Fortran programs have been at the core of numerous applications; many of the computationally expensive, highly parallelized programs for operational weather forecasting, atmospheric chemistry and transport modeling, and Earth system or climate simulations have been written in Fortran.

As Fortran is a compiled language we need a compiler. The choice of compiler may depend on the operating system and your personal preference. For example, Fortran code presented in this book has been developed on both the Microsoft Windows and the Linux platforms. Under Windows we have chosen to use the Intel® Fortran compiler, known as "ifort", via integration into Microsoft Visual Studio Community 2019 (which provides an edit-compile-run IDE (see below)). For students, the Intel® Fortran compiler is available for free as part of the Intel® oneAPI HPC Toolkit. We also compile and execute example programs using the free GNU "gfortran" compiler on a Linux environment.[1] In this case we create a text file, say `test.f90`. If we were to use the gfortran compiler, then within a terminal or command prompt we would create an executable file `example.out` by entering the following command:

```
gfortran  -o example.out test.f90
```

Upon successful compilation, we can run this executable within the Linux terminal by entering the command `./example.out`. In this very simple example we have omitted a number of additional compiler options that can tell the compiler what level of optimization is required, which we cover in more detail in Section 8.2.1.

- Julia [Official documentation: https://julialang.org/]: At the time of writing, Julia is a relatively new programming language positioned at version 1.6. There is growing interest in Julia as it was created with the understanding that *Scientific computing has traditionally required the highest performance, yet domain experts have largely moved to slower dynamic languages for daily work* (https://julia-doc.readthedocs.io/en/latest/manual/introduction/). Indeed, once you start creating code in Julia, you may find a number of similarities to the Python syntax though the speed of the code, thus time-to-solution, can approach that of Fortran [31]. Julia achieves this using just-in-time (JIT) compilation, where the code we write is compiled to machine code during execution of a program rather than before execution. There are a number of very useful features provided in Julia, including the ability to implement automatic differentiation of the code you write. Likewise, whilst the ecosystem of tools in Julia may not match that of Python, for numerical computing there are over one hundred differential equations solvers available. Like Python, Julia also has a package manager to help integrate a range of modules into your workflow. You can install Julia following the instructions on the official Julia web page. Once installed, you can open a Julia console and type ] to enter Package management state, also known as `Pkg`. Please refer to the documentation for installing packages (https://docs.julialang.org/en/v1/stdlib/Pkg/). Again, we can create a text file that contains Julia code which we then wish to run. Imagine we create a file called `test.jl`. Once Julia is installed on your machine, open up a terminal [Max/Linux] or command prompt and we can run our files as follows:

```
julia test.jl
```

---

1 Note that on some Linux distributions, "f95" is a synonym for "gfortran".

- Matlab [official web site: https://uk.mathworks.com/products/matlab.html]: Matlab is a proprietary software product that provides a user with an interactive development environment, through a GUI. Developed by Mathworks, Matlab has been used across academia and engineering widely for a number of decades and is particularly useful when prototyping new ideas. When working within the Matlab GUI you will have access to a range of tools packaged in to a series of `toolboxes`, depending on which license you have access to. For example, according to the official documentation, the *Partial Differential Equation Toolbox$^{TM}$ provides functions for solving structural mechanics, heat transfer, and general partial differential equations (PDEs) using finite element analysis.* We can write Matlab code in a text file that has the `.m` extension. In the most recent versions, the user can execute Matlab code from a number of languages, including C/C++, Fortran, Java, and Python. Likewise, Matlab code can call functions developed in other languages. Whilst you can run Matlab code from the command line, it is more common to run code from within the Matlab GUI using the menu options provided (clicking on the run icon provided).

Each language has its own unique syntax. As you move between different languages, you will find some are similar in style and this can make it easier to translate the same theory into multiple languages or convert an existing code base into another language. Products such as Matlab come with their own integrated development environment (IDE), providing a user with a code editor that is able to highlight the variable syntax you use to construct a code file. Alternatively, you may be using the Spyder IDE to develop Python or Microsoft Visual Studio to develop Fortran code that likewise provide syntax highlighting. However, there are a number of alternative text editors that can be used in isolation. **Syntax highlighting** refers to the keywords of the given programming language being highlighted in bold, or as another color, often with auto-repeat and hints on usage. Using an IDE as you code up the many code snippets provided in this book, you will notice that some of the text commands and words used are colored in a particular way. This allows us to more easily identify different structural components to our code. In terms of available text editors, Atom (https://atom.io/) is a cross-platform, free to use text editor that also integrates with Git and GitHub directly (see Section 8.4.3). For Windows users, Notepad++ likewise provides a flexible syntax highlighting environment (https://notepad-plus-plus.org/downloads/). Try a few different text editors; often the choice can be somewhat personal depending on the machine and how your project evolves.

## 1.3 Representing Aerosol Particles as Model Frameworks

In the previous section, we provide a brief overview on the physical and chemical characteristics of atmospheric aerosol particles and the subsequent processes covered in this book. As you read each chapter, you will find a specific method for translating the theory that underline these processes into code.

There are multiple approaches to represent an aerosol particle, and population of particles, within a numerical model. The information that we represent and therefore track throughout a simulation brings together layers of chemical complexity and metrics that represent the physical state of the condensed phase. Likewise, we have to make a decision on how the gaseous and condensed phase are connected. This needs a consideration of, for example, what information is important for a particular impact we might be interested in; what data is available; what computing resource is available; and so on. There can also be a trade-off between computational efficiency and chemical and/or process complexity to be represented.

We know that particles can range in size from a few nanometers to hundreds of microns. Likewise, we know that aerosol particles can exist in multiple physical states from a solid to an amorphous and nonviscous liquid. There may also be inhomogeneous mixing inside each particle which may coincide with phase separated regions.

With this in mind, we can start to consider how we might represent the chemical and physical state of individual particles and populations of particles. Do we begin at a molecular scale, a homogeneous single particle scale or perhaps by discretizing regions within each particle? Throughout this book you will find we can take either approach depending on our need.

Let us start with a general narrative. It is common to visualize and analyze aerosol populations in terms of number distributions or mass distributions as a function of particle size [2]. For atmospheric aerosol particles, we know size is important when understanding impacts on health and the climate. We also know that the composition is important when considering differential toxicity [2, 32, 33], for example, and the potential to form cloud droplets [4, 15, 34]. Likewise, for characterizing an aerosol, size distributions are limiting since the same size distribution can correspond to many different mixing states. We use the term "mixing state" as reference to the ways the aerosol chemical species are distributed amongst the particles in the population.

We can start to map this narrative to both algebraic formula and visual references. We can state that an aerosol particle contains mass $\mu_a \geq 0$ of species $a$, for $a = 1, ..., A$, so that the particle composition is described by the $A$-dimensional composition vector $\vec{\mu} \in \mathbb{R}^A$. The symbol $\in$ denotes a set membership and thus can be read as "belongs to," whilst $\mathbb{R}$ means a set of real numbers, and $\mathbb{R}^A$ thus an "A" dimensional set of real numbers. For example, if we have a single particle comprising two components "1" and "2," we could write our composition vector as $\vec{\mu} = (\mu_1, \mu_2)$ where both $\mu_1$ and $\mu_2$ represent a set of concentration axes. Each aerosol particle represents therefore a point in the $A$-dimensional *composition space*. Figure 1.5 illustrates this concept for an aerosol population that has a number distribution as shown in Figure 1.5(a), and consists of two species ($A = 2$) with mass distribution as shown in Figure 1.5(b). Figure 1.5(c)–(e) shows how particle samples from three possible populations populate the composition space. The composition space is in this case two-dimensional, so each particle is again represented by a vector $\vec{\mu} = (\mu_1, \mu_2)$. The broken grey lines are lines of constant diameter (assuming spherical particles). These populations are all consistent with the number and mass distribution above, but they differ in their mixing state. Figure 1.5(c)–(e) also shows corresponding pictorial representations of each population. At this point we are only concerned with the chemical composition, so the fact that the particles can have different shapes and structures is not represented, and the particles are shown as generic pie-charts with the size and colors representing particle size and composition respectively, rather than any particular morphology.

Figure 1.5(c) depicts the classic "external mixture," where each particle only contains one of the two species. In the composition space graph, the particles are arranged only

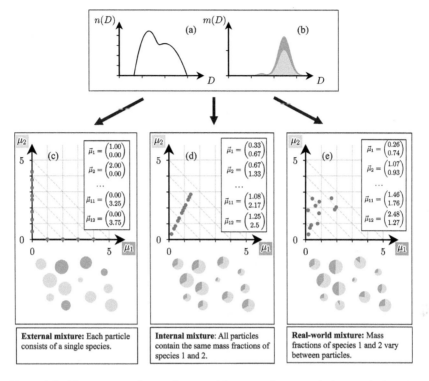

**Figure 1.5** The concept of aerosol composition space for a population consisting of two chemical species. (a) Example of a number size distribution, (b) corresponding mass size distribution, (c), (d), and (e) are composition space depictions of example particles from populations that are consistent with the distributions shown in (a) and (b).

along the two axes, and one of the two components of their composition vector is zero, as represented by points along either axis. In contrast, Figure 1.5(d) shows the classic "fully internal mixture," where each particle contains the same mixture of species 1 and 2. These particles are therefore arranged along a line through the origin in the composition space graph, with slope 3/2, since the mass ratio of the two species is 3:2 in each particle. Figure 1.5(e) shows a state in between, neither fully internally nor externally mixed. Such a population is more akin to ambient aerosol populations than the extreme cases mentioned above.

We have already noted that aerosol particles experience continuous transformations during their lifetime in the atmosphere. As the size and composition of the particles change, they move in composition space due to aerosol processes acting on them. For example, if species 2 in Figure 1.5 was semivolatile, we would observe the particles moving parallel to the $\mu_2$-axis as species 2 condenses on the particles or evaporates from the particles. Hence, condensation and evaporation (Chapter 2) can be seen as an advection process in composition space. In contrast, coagulation (Chapter 5) is a stochastic, discontinuous jump process that produces a particle that will be placed according to the sum of the vector components of its parent particles. Likewise, nucleation (Chapter 6) produces particles at the smallest end of the size spectrum whose composition vectors are formed initially through discrete summations of parent molecules.

In summary, aerosol mixing state is a dynamic quantity that changes continuously as the particles' compositions change, but also as particles are added to the population by new particle formation, emissions or transport, or removed from the population by dry or wet deposition. The impact of these changes on mixing state can be quantified more precisely by the use of mixing state metrics as explained in Ref. [35]. In reality the composition space of an aerosol population is high-dimensional, including tens or even hundreds of different dimensions (one for each chemical species). Common aerosol modeling approaches typically work with low-dimensional projections of this high-dimensional space.

So far we have discussed composition space in the context of chemical composition where each entry in a particle's composition vector is the amount of a chemical species. We can extend this concept by adding more components to the composition vector to characterize shape and structure. Candidate quantities are the particle's fractal dimension, the particle viscosity, surface tension, or information about the quantity and location of inclusions within the particle.

The above narrative helps to contextualize how we might now consider mapping this into numerical models. In this book each chapter focuses on a particular process and ways to represent that process and thus change in aerosol particle number, composition, size, and perhaps state. Specifically, in each chapter we will provide a narrative of the theoretical basis on which our process of interest, and thus models, are based. We then discuss how we approximate the particle size distribution, thus composition space, in a numerical representation of this process.

### 1.3.1 Size Distributions

When modeling aerosol particle evolution across a range of sizes, we can represent the variation in particle size and composition using one of the following approaches, illustrated in Figure 1.6:

- Track each particle.
- Assume a continuous distribution [modal].
- Place particles into distinct size ranges, or "bins" [sectional].

The first, whilst possible [36] and explicitly discussed in Chapter 5, requires significant computational resource for complex systems. We define complexity here in terms

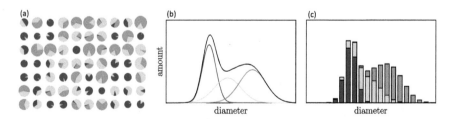

**Figure 1.6** Illustration of the three different aerosol representations: (a) particle-resolved: the aerosol population is represented by a sample of computational particles; (b) modal: the aerosol population is represented by several overlapping continuous distribution functions; (c) sectional: the aerosol population is represented by discrete size bins.

of both chemical and physical process descriptions, some of which we discussed in individual chapters. Likewise we use a level of subjectivity in defining a "significant" cost which will depend on the primary use of the constructed model. The third approach of assuming a continuous distribution often requires the least computational resource. Whilst still used in global schemes, and referred to as the modal approach, it has known limitations when we come to tracking individual components in particles. However, the trade-off between computational performance and accuracy will depend on the application in question. The second is perhaps the most popular across a range of scales and is known as the sectional approach. This approach requires us to split the continuous size distribution into discrete size bins and is, however, more computationally expensive than the modal approach.

If a population of particles has only one size, then we call that population mono-disperse. If the population has more than one size, we say this is poly-disperse. Imagine we observe a poly-disperse population through some empirical means, counting the number of particles of each size. Once we count the number of particles in our sampled volume, what would a size distribution look like? In other words, in a 2D plot of number versus size what shape would this plot resemble?

We can first start with common distributions. The probability density of many phenomena in the natural world often follows a normal distribution [37]. In a normal distribution, also referred to as a Gaussian distribution, data is symmetric about a mean value. Algebraically, the probability density for the normal (Gaussian) distribution is given by

$$p_{norm}(x) = \frac{1}{\sqrt{2\pi\sigma_p^2}}e^{-\frac{(x-\mu_p)^2}{2\sigma_p^2}}, \tag{1.5}$$

where $p_{norm}(x)$ is the probability density, $\sigma_p$ is the standard deviation, $\mu_p$ is the mean, and $x$ is the variable we are sampling from. A normal distribution can be visualized as per Figure 1.7, with the Python code used to generate the figure provided in code Listing 1.1:

In some cases you may hear reference to $\mu_p$ as the location parameter and $\sigma_p$ as the scale parameter. Indeed, if you were to import the Python package Scipy and use the `log-norm` function:

```
import scipy.stats.lognorm
```

you would find such reference in the documentation. The standard deviation gives us information about the shape since 68% of the data lies within one standard deviation of the mean. The definite integral of an arbitrary normal function is given in 1.6, confirming that the integral of Equation (1.5) is 1.0.

$$\int_{-\infty}^{\infty} e^{-a(x+b)^2} dx = \sqrt{\frac{\pi}{a}} \tag{1.6}$$

In many instances, data appears to be normally distributed on a logarithmic scale. In other words, when we sample from the logarithm of variable $x$, we might find a curve that is normally distributed. In this case our probability density function changes to Equation (1.7) and illustrated in Figure 1.8.

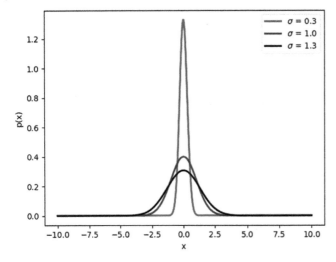

**Figure 1.7** Normal distributions, each with a mean ($\mu$) of 0 but three different standard deviations ($\sigma$).

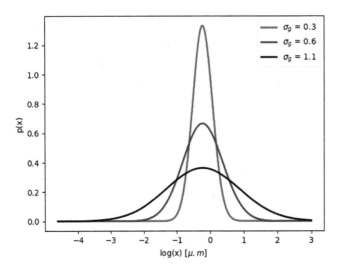

**Figure 1.8** Log-normal distributions, each with a mean ($\mu$) of 800 nm but three different geometric standard deviations ($\sigma_g$) of 0.3, 0.6 and 1.1.

$$p_{lognorm}(x) = \frac{1}{\sqrt{2\pi ln(\sigma_g)^2}} e^{-\frac{(ln(x)-ln(\mu_g))^2}{2ln(\sigma_g)^2}} . \tag{1.7}$$

Distinct from a normal distribution, a log-normal distribution has geometric mean $\mu_g$ (as opposed to an arithmetic mean) and a geometric standard deviation $\sigma_g$. Again, 68% of the area under the log-normal distribution lies within $log(\sigma_g)$ of the geometric mean.

As a first demonstration of translating equations to code, we are going to create our own normal and log-normal distributions before mapping these onto representative distributions of aerosol concentrations. Using the definition of a normal distribution, in the Python example provided in code listing 1.1 we assume that three normal distributions have the same mean ($\mu_p$) of "0" but 3 different standard deviations ($\sigma_p$) of 0.3, 1.0, and 1.3. We are going to use the `Numpy` and `Matplotlib` libraries to create our arrays and plot results accordingly. Notice that in this example we interchangeably refer to our *x* variable as a size distribution. In this instance we create nonphysical lower values that are negative but nonetheless should help to start considering how we map this to a physically relevant distribution. Once complete you should produce a plot identical to Figure 1.7:

```python
# Import the relevant libraries
import numpy as np
import matplotlib.pyplot as plt

# Define lower and upper limits
lower_size = -10.0
upper_size = 10.0

# Create 400 linearly seperated values for our 'x' variable
x = np.linspace(lower_size, upper_size, num=400) # values for x-axis

# Define the mean and standard deviation for each normal
    distribution
sigma_1 = 0.3
mean_1 = 0.0

sigma_2 = 1.0
mean_2 = 0.0

sigma_3 = 1.3
mean_3 = 0.0

# Implement equation 1.5
distribution_1 = (np.exp(-(x - mean_1)**2 / (2 * sigma_1**2)) /
    (sigma_1 * np.sqrt(2 * np.pi)))
distribution_2 = (np.exp(-(x - mean_2)**2 / (2 * sigma_2**2)) /
    (sigma_2 * np.sqrt(2 * np.pi)))
distribution_3 = (np.exp(-(x - mean_3)**2 / (2 * sigma_3**2)) /
    (sigma_3 * np.sqrt(2 * np.pi)))

# Plot the results
line1 = plt.plot(x, distribution_1 , linewidth=2, color='r',
    label='$\sigma$ = 0.3')
line2 = plt.plot(x, distribution_2 , linewidth=2, color='b',
    label='$\sigma$ = 1.0')
line3 = plt.plot(x, distribution_3 , linewidth=2, color='k',
    label='$\sigma$ = 1.3')
ax = plt.gca()
ax.legend()
ax.set_xlabel('x')
ax.set_ylabel('p(x)')
plt.show()
```

Listing 1.1 Code snippet for creating and plotting three normal distributions in Python.

In the next code example (Listing 1.2), we now initialize a log-normal distribution following Equation (1.7). In this example we now refer to an upper and lower size limit which we may think of in terms of aerosol particles with a set diameter in microns. We also now keep our geometric mean value $\mu_{ln}$ to represent 800 nm and vary the geometric standard deviation between 0.3, 0.6, and 1.1.

```
# Here we plot 3 different log-normal distributions with a mean
# diameter of 800nm and geometric standard deviation of 1.1, 1.3 and
    2.0
# We assume our smallest size is 10nm and largest is 20 microns

# Import the relevant libraries
import numpy as np
import matplotlib.pyplot as plt

# Define lower and upper limit of the size distribution [microns]
lower_size = 0.01
upper_size = 20

# Create an array of values in log space
x_log = np.linspace(np.log(lower_size), np.log(upper_size), num=400)

# Define geomatric standard deviations and mean values
sigmag_1 = 0.3
mean_1 = np.log(0.8)
sigmag_2 = 0.6
mean_2 = np.log(0.8)
sigmag_3 = 1.1
mean_3 = np.log(0.8)

# Implement equation 1.7
distribution_1 = (np.exp(-(x_log - mean_1)**2 / (2 * sigmag_1**2)) /
    (sigmag_1 * np.sqrt(2 * np.pi)))
distribution_2 = (np.exp(-(x_log - mean_2)**2 / (2 * sigmag_2**2)) /
    (sigmag_2 * np.sqrt(2 * np.pi)))
distribution_3 = (np.exp(-(x_log - mean_3)**2 / (2 * sigmag_3**2)) /
    (sigmag_3 * np.sqrt(2 * np.pi)))

# Plot the results
plt.plot(x_log, distribution_1 , linewidth=2, color='r')
plt.plot(x_log, distribution_2 , linewidth=2, color='b')
plt.plot(x_log, distribution_3 , linewidth=2, color='k')
ax = plt.gca()
ax.legend()
ax.set_xlabel('x')
ax.set_ylabel('p(x)')
plt.show()
```

**Listing 1.2** Log-normal distributions in Python.

Up to this point, we have created probability density distributions and referred to aerosol size as our *x* variable. How do we use this information to create and manipulate distribution of aerosol particle number concentrations? In the proceeding sections, namely Sections 1.3.2 and 1.3.3, we introduce and deploy two common approaches known as the sectional and modal approach, respectively.

## 1.3.2 The Sectional Distribution

The continuous log-normal probability density distribution does not provide us with information on the composition and state of individual or groups of particles. In the sectional approach, we break the continuous distribution into discrete sections along the size axis. We refer to these discrete sections as size bins. Each discrete section, or bin, may then be used to store information about the composition of particles within a given size range. As we reduce the total number of bins that represent our entire aerosol population, we force homogeneity of particulate composition across a wider size range. Likewise, we restrict the number of discrete particulate sizes that represent our distribution. In the following text, we discuss the numerical framework a discrete population can be built around. Limiting the number of bins in a model is often the easiest method of reducing computational burden. If we multiply Equation (1.7) by our total concentration of particles, $N_L$ (cm$^{-3}$), we can arrive at the log normal distribution given by 1.8:

$$n_{ln}(ln(x)) = \frac{N_T}{\sqrt{2\pi ln(\sigma_g)^2}} e^{-\frac{(ln(x)-ln(\mu_g))^2}{2ln(\sigma_g)^2}}.$$ (1.8)

The integral of 1.8 with respect to $ln(x)$ gives us the total number of particles $N_L$. We can therefore rewrite $n_{ln}(x)$ as a derivative function:

$$\frac{dN}{dln(x)} = n_{ln}(ln(x))$$ (1.9)

where we can also state that $n_{ln}(ln(x))dln(x)$ is the number of particles per unit volume in the range $ln(x)$ to $ln(x) + dln(x)$. Let us now use the variable $d_p$ (m) to represent a particle diameter such that $\frac{dN}{dln(d_p)} = n_{ln}(ln(d_p))$ and $n_{ln}(ln(d_p))dln(d_p)$ is the number of particles per unit volume in the size range $ln(d_p)$ to $ln(d_p) + dln(d_p)$. Of course, calculating a value for the concentration between these limits requires us to integrative the derivative. We may switch between using $d_p$ and $ln(d_p)$ as the size variable using 1.10 [2]:

$$n_{ln}(ln(d_p)) = d_p n_{ln}(d_p)$$ (1.10)

such that Equations (1.11) and (1.12) provide us with the final expressions for $n_{ln}(ln(d_p))$ and $n_{ln}(d_p)$, respectively:

$$n_{ln}(ln(d_p)) = \frac{dN}{dln(d_p)} = \frac{N_T}{\sqrt{2\pi ln(\sigma_g)^2}} e^{-\frac{(ln(d_p)-ln(\mu_g))^2}{2ln(\sigma_g)^2}},$$ (1.11)

$$n_{ln}(d_p) = \frac{dN}{d(d_p)} = \frac{N_T}{d_p\sqrt{2\pi ln(\sigma_g)^2}} e^{-\frac{(ln(d_p)-ln(\mu_g))^2}{2ln(\sigma_g)^2}}.$$ (1.12)

In a sectional approach, we wish to represent the size distribution as a collection of discrete size bins. How do we calculate the concentration of particles in a given bin? We could calculate the total number concentration $N_{1,2}$ (cm$^{-3}$) between two sizes, $d_p(1)$ and $d_p(2)$, through the following integral:

$$N_{1,2} = \int_{lnd_{p,1}}^{lnd_{p,2}} n_{ln}(lnd_p)dlnd_p$$ (1.13)

However, if we approximate and rewrite our derivative functions using a finite difference approach, Equation (1.12) becomes

$$\frac{dN}{d(d_p)} \approx \frac{\Delta N}{\Delta d_{p,i}} \approx \frac{N_T}{d_{p,i}\sqrt{2\pi ln(\sigma_g)^2}} e^{-\frac{(ln(d_{p,i})-ln(\mu_g))^2}{2ln(\sigma_g)^2}},$$ (1.14)

where we have introduced a new variable, $\Delta N$, to represent the concentration of particles between two generic diameters separated by $\Delta d_{p,i}$, which in turn we refer to as the concentration in a specific size bin number $i$ ($n_{sect,i}$) in a sectional framework such that the total concentration of particles can be calculated through 1.15:

$$N_L = \sum_{i}^{N_{sect,b}} n_{sect,i} \tag{1.15}$$

where $N_{sect,b}$ is the total number of size bins in a sectional model. This gives us an expression to calculate the number concentration in each size bin, $n_{sect,i}$, via Equation (1.16):

$$n_{sect,i} = \frac{\Delta d_{p,i} N_T}{d_{p,i}\sqrt{2\pi ln(\sigma_g)^2}} e^{-\frac{(ln(d_{p,i})-ln(\mu_g))^2}{2ln(\sigma_g)^2}} \tag{1.16}$$

We need to decide how we space our discrete bins and therefore arrive at a value for $\Delta d_{p,i}$. In a sectional approach, the bins should be spread geometrically and we can therefore use a volume-ratio distribution. In this distribution, a fixed volume ratio between bins defines the center of the bin and bin width.

As we increase the particle size, the volume of the current size bin $v_{sect,i}$ is determined by multiplying the volume of the previous size bin $v_{sect,i-1}$ by a fixed volume ratio $V_{rat}$, which is determined by the smallest and largest diameters, $d_{sect,1}$ and $d_{sect,Nb}$, and the total number of discrete bins desired $N_{sect,b}$. The ratio between discrete diameters $D_{rat}$ can be obtained from $V_{rat}$:

$$V_{rat} = \left(\frac{d_{sect,Nb}}{d_{sect,1}}\right)^{3/(N_{sect,b}-1)} \tag{1.17}$$

$$D_{rat} = V_{rat}^{1/3} \tag{1.18}$$

$$v_{sect,i} = V_{rat} v_{sect,i-1} \text{ for i=1 to} N_{sect,b} \tag{1.19}$$

Each bin has a finite width. The lowest volume limit $v_{sect,i,lo}$ of any given bin, and in turn the upper limit $v_{sect,i,hi}$, is related to the bin-center volume and $V_{rat}$ by

$$v_{sect,i} = \frac{1}{2}\left(v_{sect,i,hi} + v_{sect,i,lo}\right) \tag{1.20}$$

$$v_{sect,i,hi} = V_{rat} v_{sect,i,lo} \tag{1.21}$$

$$v_{sect,i,lo} = \frac{2v_{sect,i}}{1 + V_{rat}} \tag{1.22}$$

The diameter width of each bin $\Delta d_{sect,i}$ with diameter $d_{sect,i}$ can be obtained from the lower and upper diameter limits in each bin, $d_{sect,i,lo}$ and $d_{sect,i,hi}$, respectively:

$$\Delta d_{sect,i} = d_{sect,i,hi} - d_{sect,i,lo} = d_{sect,i} 2^{1/3} \frac{V_{rat}^{1/3} - 1}{(1 + V_{rat})^{1/3}} \tag{1.23}$$

Figure 1.9 illustrates how a subset of the variables map onto a hypothetical log-normal discrete distribution.

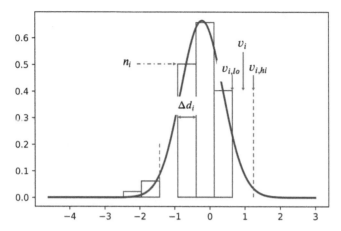

**Figure 1.9** Binned size distribution.

### 1.3.3 The Modal Distribution

Modal approaches are used widely in the atmospheric science community for representing aerosol size distributions and typically leverage the properties of the function to capture a wide range of aerosol sizes with a limited number of parameters. For a unimodal aerosol population, only one set of statistical parameters is required to describe the system. This simple scenario could be expected for a controlled laboratory chamber or flow reactor experiment. As the complexity of the aerosol population grows due to simultaneous sources or microphysical processes, numerical approximations of the distribution shape are necessary. Common examples that have been used in the past include log-normal and gamma functions.

With the log-normal function, we can diagnose the properties of the aerosol population across diameters that span several orders of magnitude. However, to represent the evolution of the population in response to microphysical processes or atmospheric transport, we calculate the integral moments of the log-normal function as these are conserved quantities. We define the $k^{th}$ moment of the size distribution as follows:

$$M_k = \int_{-\infty}^{\infty} d_p^k n_{ln}(\ln d_p)d(\ln d_p). \tag{1.24}$$

After substitution of Equation (1.11) for $d_p$ and integration, this expression becomes

$$M_k = N_T \mu^k \exp(\frac{k^2}{2} \ln^2 \sigma_g), \tag{1.25}$$

where $N_T$ is the total number of particles in the mode, $\mu$ is the geometric mean diameter, and $\sigma_g$ is the standard deviation of the log-normal distribution. Several of the moments calculated with this series have useful physical meaning for describing the aerosol population. The $0^{th}$ moment ($M_0$) is identical to $N_T$. An equation for the total mass of the mode ($M_{tot}$) expands the expression for the volume of a sphere ($V_T$) with the definition of the third moment ($M_3$) and scales by the aerosol density ($\rho$).

$$M_{tot} = \rho V_T = \rho \frac{\pi}{6} M_3 = \rho \frac{\pi}{6} N_T \mu^3 \exp(4.5 \ln^2 \sigma_g). \tag{1.26}$$

The total surface area may be calculated analogously as a function of $M_2$:

$$S_T = \pi M_2 = \pi N_T \mu^2 \exp(2 \ln^2 \sigma_g). \tag{1.27}$$

These integrated quantities ($N_T$, $S_T$, and $M_{tot}$) are useful for deriving perturbations on the aerosol mode due to processes that will be introduced in subsequent chapters. Once new quantities are calculated, they may be combined to arrive at new parameters describing the evolution of the log-normal function. Using the definitions of $M_0$, $M_2$, and $M_3$ from Equation (1.25), the geometric standard deviation may be calculated as

$$\sigma_g = e^{\left(\frac{1}{3} \ln M_0 - \ln M_2 + \frac{2}{3} \ln M_3\right)^{\frac{1}{2}}}, \tag{1.28}$$

and the geometric mean as

$$\mu = \left( \frac{M_3}{N_T \exp(\frac{9}{2} \ln^2 \sigma_g)} \right)^{\frac{1}{3}}. \tag{1.29}$$

General forms of Equations (1.28) and (1.29) were derived by [38] for the case where two arbitrary moments are known in addition to the zeroth moment ($N_T$):

$$\sigma_g = e^{\left( \frac{2}{k_1(k_1 - k_2)} \ln\left( \frac{M_{k_1}/N_T}{(M_{k_2}/N_T)^{k_1/k_2}} \right) \right)^{\frac{1}{2}}} \tag{1.30}$$

$$\mu = \left( \frac{M_{k_1}}{N_T} \right)^{1/[\frac{k_1}{k_2}(k_2 - k_1)]} \left( \frac{M_{k_2}}{N_T} \right)^{\frac{k_1/k_2}{(k_1 - k_2)}} \tag{1.31}$$

These transformations demonstrate the utility in describing aerosol populations and processes with moments of the log-normal function. With this approach, the impact of complex aerosol phenomena may be approximated in terms of quantities that are both intuitively understood and readily connected to common bulk particle measurements (e.g., mass, diameter, modal standard deviation, etc.). For example, as we will see in later chapters, a process like vapor condensation will add mass to the population but not change the total number, thus increasing $M_3$ and $M_2$ while not affecting $M_0$. The resulting changes to the system of moments will grow the geometric mean diameter and shrink the geometric standard deviation, thereby mimicking the well-known feature of condensational narrowing. In more atmospherically relevant applications, the full aerosol population is observed to contain a series of overlapping modes, or distinct aerosol sub-populations. Whitby et al. [38] provide extensive explanation of moment-based modal aerosol methods and derivation of the fundamental dynamic equations. In these cases, the aerosol distribution can be represented as the sum of independent log-normal functions, each with their own physical parameters. For example, the number of particles between diameters $dp_1$ and $dp_2$ can be calculated by integrating the log-normal number size distributions for all modes and then summing them.

$$n_{dp_1 - dp_2} = \sum_i^m \int_{dp_1}^{dp_2} n_{\ln,i}(\ln d_p) d(\ln d_p),$$

$$= \sum_i^m \frac{N_{T,i}}{2} \left( erf\left( \frac{\ln(dp_2/\mu_i)}{\sqrt{2} \ln \sigma_{g,i}} \right) - erf\left( \frac{\ln(dp_1/\mu_i)}{\sqrt{2} \ln \sigma_{g,i}} \right) \right). \tag{1.32}$$

Here, the total number of particles between diameters $d1$ and $d2$ are calculated as the sum of the integrated log-normal distribution of each mode (i) for the total number of modes (m). The error function, $erf$, is available in standard math libraries, and is defined as

$$erf\, z = \frac{2}{\sqrt{\pi}} \int_0^z e^{-\eta^2} d\eta \tag{1.33}$$

We can extend the number size distribution to determine distributions of other important properties of the population. Distributions of surface area and volume are written in terms of the number distribution as follows:

$$s_{ln}(\ln d_p) = \pi d_p^2 n_{ln}(\ln d_p) \tag{1.34}$$

$$v_{ln}(\ln d_p) = \frac{\pi}{6} d_p^3 n_{ln}(\ln d_p) \tag{1.35}$$

Ref. [2] shows that the surface area and volume distributions can be rearranged to a form analogous to the number distribution with the same geometric standard deviation and surface area median diameter $\mu_s$ and volume median diameter $\mu_v$ expressed as

$$\ln \mu_s = \ln \mu + 2 \ln^2 \sigma_g \tag{1.36}$$

$$\ln \mu_v = \ln \mu + 3 \ln^2 \sigma_g \tag{1.37}$$

These mode-dependent parameters can replace $\mu$ in Equation (1.32) to yield the surface area concentration and volume concentration between two arbitrary diameters.

We have chosen to introduce the concept of moment-based modal models using a specific example configuration with three moments and three adjustable log-normal parameters (N, $\mu$, and $\sigma_g$). Other configurations have been used in the past that differ in the selection and total number of moments used. Whitby et al. [38] and Binkowski and Shankar [39] selected the 0th, 3rd, and 6th moments for their aerosol models, with the 6th moment proportional to the radar reflectivity of the population. Many large-scale models that employ modal approaches use fewer moments and fix one or two of the log-normal parameters to representative values. A common configuration includes the 0th and 3rd moments, thereby allowing number and geometric mean diameter to vary, while fixing the geometric standard deviation. The simplest models only treat the 3rd moment and prescribe the geometric mean diameter and standard deviation. These implementations are highly efficient, but likely sacrifice accuracy, especially when representing processes that are nonlinearly dependent on particle size (e.g., dry deposition).

Moment-based modal approaches are attractive for their computational efficiency and intuitive formulation. They also have limitations when compared to more detailed size distribution approaches like sectional models. Most notably, numerical accuracy is likely sacrificed when fitting a realistic aerosol size distribution to a log-normal function. This approximation may manifest in discrepancies at the highest or lowest diameter size ranges or in biases in the height or location of the geometric mean diameter. Another important limitation emerges when representing size distributions during activation of aerosols to form cloud drops [40]. Under these conditions, modal models often erroneously adjust the log-normal to accommodate the change in moments and tend to underpredict the diameter of the remaining interstitial particle population. Finally, with few exceptions, moment-based modal models assume that chemical composition is independent of size within each mode, thus artificially diffusing some chemical components across size space, especially for wide modes with large standard deviation.

## 1.4 Code Availability

All of the code snippets and examples are provided with the book, or can be downloaded from the project Github repository https://github.com/aerosol-modelling/Book-Code.git. For the existing community codes referenced in each Chapter, they can be retrieved as follows:

- UManSysProp: UManSysProp is an open-source project, licensed under GPL v3.0. An archive of the UmanSysProp version used in this book can be found at https://doi.org/10.5281/zenodo.4110145. The project repository can be found at https://github.com/loftytopping/UmanSysProp_public.
- PyBox: PyBox is an open-source project, licensed under GNU General Public License v3.0. An archive of the PyBox version used in this book can be found at https://doi.org/10.5281/zenodo.1345005. The project repository can be found at https://github.com/loftytopping/PyBox.
- JlBox: JlBox is an open-source project, licensed under GNU General Public License v3.0. An archive of the JlBox version used in this book can be found at https://doi.org/10.5281/zenodo.4519192. The project repository can be found at https://github.com/huanglangwen/JlBox.
- SALSA: SALSA is an open-source project, licensed under the permissive MIT open-source licence. The project repository can be found at https://github.com/UCLALES-SALSA.
- PartMC: PartMC is an open-source project, licensed under the GNU General Public License version 2. The project repository can be found at https://github.com/compdyn/partmc.
- ACDC: ACDC is an open-source project, licensed under GNU General Public License v3.0. An archive of the version used in this book can be found at https://doi.org/10.5281/zenodo.5226490. The project repository can be found at https://github.com/tolenius/ACDC.

## Bibliography

[1] Hang Su, Yafang Cheng, and Ulrich Pöschl. New multiphase chemical processes influencing atmospheric aerosols, air quality, and climate in the anthropocene. *Accounts of Chemical Research*, 53:2034–2043, 10, 2020.

[2] Minhan Park, Hung Soo Joo, Kwangyul Lee, Myoseon Jang, Sang Don Kim, Injeong Kim et al. Differential toxicities of fine particulate matters from various sources. *Scientific Reports*, 8:1–11, 12, 2018.

**[3]** Carolina Molina, Richard Toro A., Carlos A. Manzano, Silvia Canepari, Lorenzo Massimi, and Manuel A. Leiva-Guzmán. Airborne aerosols and human health: leapfrogging from mass concentration to oxidative potential, 9, 2020.

**[4]** K. S. Carslaw, L. A. Lee, C. L. Reddington, K. J. Pringle, A. Rap, P. M. Forster et al. Large contribution of natural aerosols to uncertainty in indirect forcing. *Nature*, 503:67–71, 2013.

**[5]** Jlm Haywood, Chapter 30 - Atmospheric aerosols and their role in climate change, Editor(s): Trevor M. Letcher, Climate Change (Third Edition), Elsevier, 2021, pages 645–659, ISBN 9780128215753. https://www.sciencedirect.com/science/article/pii/B978012821575300030X?via%3Dihub.

**[6]** V. Faye McNeill. Atmospheric Aerosols: Clouds, Chemistry, and Climate. *Annual Review of Chemical and Biomolecular Engineering*, 8:427–444, 6, 2017. 10.1146/annurev-chembioeng-060816-101538.

**[7]** JoAnn Slama Lighty, John M. Veranth, and Adel F. Sarofim. Combustion aerosols: factors governing their size and composition and implications to human health. *Journal of the Air and Waste Management Association* 50:1565–1618, 2011.

**[8]** Pran Kishore Deb, Sara Nidal Abed, Hussam Maher, Amal Al-Aboudi, Anant Paradkar, Shantanu Bandopadhyay et al. Aerosols in pharmaceutical product development, Editor(s): Rakesh K. Tekade, In Advances in Pharmaceutical Product Development and Research, *Drug Delivery Systems*, Academic Press, 2020, 521–577, https://doi.org/10.1016/B978-0-12-814487-9.00011-9.

**[9]** Kihyon Hong, Se Hyun Kim, Ankit Mahajan, and C. Daniel Frisbie. Aerosol jet printed p- and n-type electrolyte-gated transistors with a variety of electrode materials: exploring practical routes to printed electronics. *ACS Applied Materials and Interfaces*, 6:18704–18711, 11, 2014.

**[10]** J. H. Seinfeld and S. N. Pandis. *Atmospheric Chemistry and Physics: From Air Pollution to Climate Change*. J. Wiley & Sons, New York, USA, 1998.

**[11]** M. Z. Jacobson. *Fundamentals of Atmospheric Modeling, Second Edition*. Cambridge University Press, New York, 2005.

**[12]** Darragh Murnane, Adam Boies, and Jonathan P. Reid. Building a UK pipeline of research, innovation and technology development for aerosol science a report supported by the UK Aerosol Society, March, 2018 (Final publication in July 2018). https://aerosol-soc.com/wp-content/uploads/2018/07/Building-a-UK-Research-and-Innovation-Pipeline-in-Aerosol-Science.pdf

**[13]** Charles E. Kolb and Douglas R. Worsnop. Chemistry and composition of atmospheric aerosol particles. *Annual Review of Physical Chemistry*, 63:471–491, 5, 2012.

**[14]** Kimberly A. Prather, Courtney D. Hatch, and Vicki H. Grassian. Analysis of atmospheric aerosols. *Annu Rev Anal Chem (Palo Alto Calif)*, 1:485–514, 2008.

**[15]** M. Hallquist, J. C. Wenger, U. Baltensperger, Y. Rudich, D. Simpson, M. Claeys et al. The formation, properties and impact of secondary organic aerosol: current and emerging issues. *Atmospheric Chemistry and Physics*, 9:5155–5236, 2009.

**[16]** Seinfeld, J. H., Pandis, S. N., Barnes, I., Dentener, F. J., Facchini, M. C., Van Dingenen, R. et al. Organic aerosol and global climate modelling: a review. *Atmospheric Chemistry and Physics*, 5:1053–1123, 2005. https://doi.org/10.5194/acp-5-1053-2005.

**[17]** Claudia Marcolli and Ulrich K. Krieger. Relevance of particle morphology for atmospheric aerosol processing. *Trends in Chemistry*, 2:1–3, 2020.

[18] N. M. Donahue, A. L. Robinson, C. O. Stanier, and S. N. Pandis. Coupled partitioning, dilution, and chemical aging of semivolatile organics. *Environmental Science and Technology*, 40:2635–2643, 5, 2006.

[19] Puneet Verma, Edmund Pickering, Mohammad Jafari, Yi Guo, Svetlana Stevanovic, Joseph F.S. Fernando et al. Influence of fuel-oxygen content on morphology and nanostructure of soot particles. *Combustion and Flame*, 205:206–219, 7, 2019.

[20] Yu Tian, Zhe Wang, Xiaole Pan, Jie Li, Ting Yang, Dawei Wang et al. Influence of the morphological change in natural Asian dust during transport: a modeling study for a typical dust event over northern china. *Science of the Total Environment*, 739:139791, 10, 2020.

[21] Eric Sauvageat, Yanick Zeder, Kevin Auderset, Bertrand Calpini, Bernard Clot, Benoit Crouzy et al. Real-time pollen monitoring using digital holography. *Atmospheric Measurement Techniques*, 13:1539–1550, 3, 2020.

[22] Delphine K. Farmer, Erin K. Boedicker, and Holly M. Debolt. Dry deposition of atmospheric aerosols: approaches, observations, and mechanisms. *Annual Review of Physical Chemistry*, 72:375–397, 1, 2021.

[23] Allen E. Haddrell, Rachael E. H. Miles, Bryan R. Bzdek, Jonathan P. Reid, Rebecca J. Hopkins, and Jim S. Walker. Coalescence sampling and analysis of aerosols using aerosol optical tweezers. *Analytical Chemistry*, 89:2345–2352, 2, 2017.

[24] Fred Gelbard and John H. Seinfeld. The general dynamic equation for aerosols. Theory and application to aerosol formation and growth. *Journal of Colloid and Interface Science*, 68:363–382, 1979.

[25] Noel M. O'Boyle, Michael Banck, Craig A. James, Chris Morley, Tim Vandermeersch, and Geoffrey R. Hutchison. Open Babel: an open chemical toolbox. *Journal of Cheminformatics*, 3(10):33, October, 2011.

[26] Fabian Pedregosa, Gael Varoquaux, Alexandre Gramfort, Vincent Michel, Bertrand Thirion, Olivier Grisel et al. Scikit-learn: Machine Learning in Python. Technical Report 85, 2011.

[27] Michael Waskom. seaborn: statistical data visualization. *Journal of Open Source Software*, 6(60):3021, April, 2021.

[28] Siu Kwan Lam, Antoine Pitrou, and Stanley Seibert. Numba: A LLVM-based Python JIT compiler. In *Proceedings of the Second Workshop on the LLVM Compiler Infrastructure in HPC - LLVM '15*, New York, New York, USA. ACM Press, 2015.

[29] François Chollet et al. Keras. https://keras.io, 2015.

[30] W. H. Press, S. A. Teukolsky, W. T. Vetterling, B. P. Flannery, and M. Metcalf. *Numerical Recipes in Fortran 90: Volume 2, Volume 2 of Fortran Numerical Recipes: The Art of Parallel Scientific Computing*. Cambridge, New York, NY: Cambridge University Press, 1996. https://pubmed.ncbi.nlm.nih.gov/31363196/.

[31] Jeffrey M. Perkel. Julia: come for the syntax, stay for the speed. *Nature* 2021 572:7767, July, 2019.

[32] Manabu Shiraiwa, Kayo Ueda, Andrea Pozzer, Gerhard Lammel, Christopher J. Kampf, Akihiro Fushimi et al. Aerosol health effects from molecular to global scales. *Environmental Science & Technology*, 51(23):13545–13567, 2017. PMID: 29111690.

[33] Janine Fröhlich-Nowoisky, Christopher J. Kampf, Bettina Weber, J. Alex Huffman, Christopher Pöhlker, Meinrat O. Andreae et al. Bioaerosols in the earth system: climate, health, and ecosystem interactions. *Atmospheric Research*, 182:346–376, 2016.

[34] John H. Seinfeld, Christopher Bretherton, Kenneth S. Carslaw, Hugh Coe, Paul J. DeMott, Edward J. Dunlea et al. Improving our fundamental understanding of the role of aerosol-cloud interactions in the climate system. *Proceedings of the National Academy of Sciences of the United States of America*, 113:5781–5790, 5, 2016.

[35] N. Riemer, A. P. Ault, M. West, R. L. Craig, and J. H. Curtis. Aerosol mixing state: measurements, modeling, and impacts. *Reviews of Geophysics*, 6:187–249, 2019.

[36] N. Riemer and M. West. Quantifying aerosol mixing state with entropy and diversity measures. *Atmospheric Chemistry and Physics*, 13:11423–11439, 11, 2013.

[37] S. A. Frank. The common patterns of nature, 8, 2009.

[38] Whitby, E. R., P. H. McMurry, U. Shankar, and F. S. Binkowski, Modal aerosol dynamics modeling, Rep. 600/3-91/020, Atmos. Res. and Exposure Assess. Lab., U.S. Environ. Prot. Agency, Research Triangle Park, N. C., 1991.

[39] F. Binkowski and U. Shankar. The regional particulate matter model 1. model description and preliminary results. *Journal of Geophysical Research*, 100:26191–26209, 1995.

[40] T. Korhola, H. Kokkola, H. Korhonen, A.-I. Partanen, A. Laaksonen, K. E. J. Lehtinen et al. Reallocation in modal aerosol models: impacts on predicting aerosol radiative effects. *Geoscientific Model Development*, 7(1):161–174, 2014.

## 2

## Gas-to-particle Partitioning

Atmospheric aerosol particles can be either primary or secondary in origin, with chemical and physical characteristics that can indicate varying contributions from both. We focus on new particle formation in Chapter 6, whereas in this chapter we discuss methods for simulating gas-to-particle partitioning. Simulating gas-to-particle partitioning is central to understanding how the properties of secondary particles change in response to ambient conditions. This is particularly important when trying to quantify the variable nature of ambient particulate matter in a changing environment. For example, understanding the interplay between inorganic and organic condensates has significant impact on aerosol-cloud interactions [1, 2]. Of course, the organic fraction of ambient particles comprises potentially many thousands of compounds [3, 4], each with varying properties. If we are to develop numerical models that capture this complexity, we need to identify and then translate the relevant physical theories into code.

As a particle is suspended in a gaseous medium, gas-to-particle partitioning may occur through adsorption to the particle surface and/or absorption into the particle surface and bulk. The forces that dictate this interaction are coupled to the state of the particulate surface and the heterogeneity of composition within the condensed phase. For example, there may be a gradient in composition as a function of distance from the center [5]. The onset of phase separation, caused by a change in ambient conditions, may change the morphology of the particle and thus interaction with the gas phase [6, 7]. Likewise, the surface layer may have a distinctly different composition from the bulk, both subject to varying chemical reactions which change the mass transfer process across the gaseous and condensed phase interface.

We can briefly summarize that the magnitude of material transferred during gas-to-particle partitioning therefore depends on a number of factors which can represented by a number of theoretical frameworks. We can treat these in isolation, or design frameworks that move to coupling processes together. With this in mind, in this chapter we provide the theoretical basis on which we can build multiprocess equilibrium and dynamic simulations of gas-to-particle partitioning.

Focusing on solutions that allow us to translate the relevant theory into code, we cross reference developments on condensed phase thermodynamics (Chapter 3), pure component and mixture properties (Chapter 4), and an evolving gas phase (Chapter 4) throughout. By doing so we aim to finish this chapter with a foundational understanding of equilibrium and dynamic gas-to-particle partitioning frameworks, whilst appreciating where additional complexity can be accounted for through inclusion of

*Introduction to Aerosol Modelling: From Theory to Code.*
First Edition. Edited by David Topping and Michael Bane.
© 2022 John Wiley & Sons Ltd. Published 2022 by John Wiley & Sons Ltd.

these additional processes. This contributes to expanding our knowledge base to better understand coupling of process during the presentation of box-models in Chapter 7.

In Section 2.1 we provide simple examples of implementing adsorption isotherms, using empirical parameters to predict the adsorption on particle surfaces. These provide us with introductory demonstrations of using function definitions and vector operations within the Python language. In Section 2.2 we implement equilibrium absorptive partitioning to a bulk condensed phase, which involves implementing our own root finding solver and calling those included within a standard package within the Scipy [8] set of tools available in Python.

In Section 2.4 we then demonstrate solution of the droplet growth equation and provide an example set of code listings where we calculate the condensational growth of both mono-disperse and poly-disperse populations. This initially relies on using Ordinary Differential Equation (ODE) solvers also provided within the Python Scipy environment, as an example. Here we discuss the required design of arrays and functions that capture the concentrations of both gas phase and condensed phase components using both a sectional and modal representation of particles sizes introduced in Chapter 1. Whilst we have focused on Python up to this point, we also provide a comparative example of how we implement the sectional approach using the Julia language. In Section 2.5 we consider cases for which water is the only compound to partitioning between the gaseous and condensed phases and build code around theories used to predict equilibrium water content in both sub- and supersaturated humid conditions. Here we demonstrate the use of developing our own functions within our own Python file which we then import into a separate script depending on which variation of Köhler theory we need.

## 2.1  Adsorption

Adsorption of a gas to the surface of an aerosol particle can lead to the formation of mono- and multilayer films. Depending on the strength of the interactions between the abdsorbed compounds and surface, the adsorption process can be classified as physisorption, driven by van der Waals forces, or chemisorption, with covalent-like interactions [9]. Once adsorbed, compounds may undergo further reactions and subsequently modify the impacts an aerosol particle can have. This has been the focus on studies assessing the change in water uptake potential of mineral particles [10], their role in ice-nucleation mechanisms [11] and improved understanding of heterogeneous processing to name a few.

Quantum chemical simulations have been used to study the interplay of adsorption mechanisms with surface-mediated reactions [9]. Including adsorption in complex models of the aerosol life-cycle relies on the use of semi-empirical adsorption isotherms. These relationships are based on molecular and macroscopic representations of the adsorption process, with parameters fit to experimental data for a range of surfaces and condensates. Tang et al. [11] provide a review of studies focusing on water adsorption to mineral dust particles, for example. Courtney et al. [12] note that many studies have shown that mineral dust can be active cloud condensation nuclei (CCN), even if it is only weakly hygroscopic as CCN activity is driven by pre-adsorbed water multilayers on the surface under subsaturated water vapor conditions.

There are a number of adsorption isotherms available, depending on the goodness of fit to experimental data. This, in turn, reflects the likelihood of the underlying theory representing the process taking place. For example, one of the best known and simplest equations is the Langmuir isotherm which presents the coverage as the partial pressure of the gas being adsorbed and constant specific to the surface material and temperature [13]. The Langmuir model assumes a mono-layer coverage, so it cannot be used when more than one layer is adsorbed. As Romakkaniemi et al. [13] note, the amount of adsorption is generally expressed in terms of surface coverage $\theta_{s,ads}$ (number of monolayers adsorbed), which can be rewritten as the ratio of the volume of gas adsorbed $V_{s,ads}$ at partial pressure $p$ to the volume of gas that will form a monolayer $V_{s,m}$ [2].

The Langmuir isotherm is given by Equation (2.1).

$$\theta_{s,ads} = \frac{V_{s,ads}}{V_{s,m}} = \frac{bp}{1+bp} \tag{2.1}$$

where $p$ is the partial pressure of the gas being adsorbed and $b$ is constant specific to the surface material and temperature. The BET (Brunauer, Emmett, and Teller) isotherm is given by Equation (2.2).

$$\theta_{s,ads} = \frac{V_{s,ads}}{V_{s,m}} = \frac{c_{sat,ads}}{\left(1 - _{sat,ads}\right)\left(1 - _{sat,ads} + c_{sat,ads}\right)} \tag{2.2}$$

and is based on the hypothesis that each separate layer obeys the Langmuir equation and the average heat of adsorption is equal to the heat of condensation for the second and higher layers. $S_{sat,ads}$ is a gas saturation ratio and $c$ is a constant. This theory assumes that the number of layers goes to infinity as the saturation ratio approaches unity. The Frenkel, Halsey, and Hill (FHH) isotherm is given by Equation (2.3)

$$\ln\left(\frac{1}{S_{sat,ads}}\right) = \frac{A}{\theta_{s,ads}^B} = \frac{A}{\frac{V_{s,ads}}{V_{s,m}}^B} \tag{2.3}$$

where $A$ and $B$ are constants. As already noted, the theoretical basis of a particular isotherm might not be appropriate for the system to be studied. For example, the Langmuir isotherm is restricted to systems where only one mono-layer is adsorbed. Laaksonen et al. [15] incorporate fractal dimension with Frenkel-Halsey-Hill (FHH)-adsorption-activation theory to improve closure between measured and modeled critical supersaturations of different types of water-insoluble particles. We discuss the critical supersaturation of water vapor in the context of cloud activation in Section 2.5. In this section, we focus on simple demonstrations of adsorptive partitioning. In the following code examples, we use three different approaches to populating an array of values for $\theta_{s,ads}$ as we change the saturation ratio $S_{sat,ads}$. We focus on the BET and FHH algorithms. The first embeds both approaches within a loop that cycles through each saturation ratio. The second places each algorithm within a separate function that is called from a loop, the third passes arrays of saturation ratios to each function and bypasses the need for a loop. We make sure all of our predictions are identical. Parameters used within both algorithms are taken from [13] assuming we are modeling adsorption of water on NaCl particles.

In Python, we first import the libraries we will be using and initialize constants (code Listing 2.1):

```python
# Import the libraries we will be using
import numpy as np
import matplotlib.pyplot as plt
import time

# Isotherm parameters for water on NaCl
A = 0.91
B = 0.67
c = 1.68
```

Listing 2.1 Import libraries and initialize constants for isotherm predictions.

Since we are comparing three different variations of the same predictions, we can also use three set of lists to store the values (code Listing 2.2):

```python
# Initialise a list of S values
# Arrays for 1st option
S_array1 = []
BET_theta_array1 = []
FHH_theta_array1 = []

# Arrays for 2nd option
S_array2 = []
BET_theta_array2 = []
FHH_theta_array2 = []

# Arrays for 3rd option
S_array3 = np.linspace(0.01, 0.9, 90) # Create a Numpy array of S
BET_theta_array3 = []
FHH_theta_array3 = []
```

Listing 2.2 Initialize empty lists for use during the different approaches outlined in the main body of text.

For our first variant (code Listing 2.3), we simply loop through a prescribed set of values for $S_{sat,ads}$, updating the lists that store values of $S_{sat,ads}$ and $\theta_{s,ads}$:

```python
# 1) Loop through values of S, the saturation ratio
for step in range(1,90):
    S = step / 100.0
    S_array1.append(S)
    BET_theta = c*S/((1.0-S)*(1.0-S+c*S))
    FHH_theta = np.power((A/(-1.0*np.log(S))),1.0/B)
    BET_theta_array1.append(BET_theta)
    FHH_theta_array1.append(FHH_theta)
```

Listing 2.3 Loop that sets the value for $S_{sat,ads}$ using a predefined range and calculates the values for $\theta_{s,ads}$ from BET and FHH isotherms before appending these values to two separate lists.

In the second variant (code Listing 2.4), we also loop through values of $S_{sat,ads}$, but this time we create specific functions for the BET and FHH models that are called from within our loop. This can be considered safer than the previous approach in that we do not need to redefine this function in any additional code and/or if we wish to make changes to the underlying structure of each theory. We can define two functions for the BET and FHH models as shown in code Listing 2.4:

```
# Define functions for both adsorption theories
def BET(c,S):
    theta = c*S/((1.0-S)*(1.0-S+c*S))
    return theta

def FHH(A,B,S):
    theta = np.power((A/(-1.0*np.log(S))),1.0/B)
    return theta
```

**Listing 2.4** Function definitions for BET and FHH isotherms.

where we need to pass in the constants for each method, $c$, $A$ and $B$. We can then modify our previous loop and call these methods within the loop as shown in code Listing 2.5:

```
# 2) Loop through values of S by calling the dedicated functions
for step in range(1,90):
    S = step / 100.0
    S_array2.append(S)
    BET_theta = BET(c,S)
    FHH_theta = FHH(A,B,S)
    BET_theta_array2.append(BET_theta)
    FHH_theta_array2.append(FHH_theta)
```

**Listing 2.5** Modified loop to call the separate BET and FHH isotherm functions.

Both methods produce values that we can plot as per Figure 2.1.

In the third and final option we bypass any loop dependency altogether and pass Numpy arrays to the BET and FHH functions directly as shown in code Listing 2.6.

```
# 3) Pass S as a Numpy array
BET_theta_array3 = BET(c,S_array3)
FHH_theta_array3 = FHH(A,B,S_array3)
```

**Listing 2.6** Calling the BET and FHH functions by passing Numpy arrays.

In this case we use vector operations. For an in depth discussion of optimization as it pertains to choice of code design, please refer to Chapter 8. Following these simple isotherms, we can also introduce the concept of a partitioning coefficient. Partitioning coefficients are also used in the next Section, Section 2.2, where we predict the mass of material partitioning from the gas to aerosol phase through the process of absorption. The definition of a partitioning coefficient [2] is shown in Equation (2.4)

$$\frac{C_{a,i}}{C_{g,i}} = K_p C_{t,i} \tag{2.4}$$

where $C_{a,i}$ is the concentration of compound $i$ in the condensed phase, $C_{g,i}$ the concentration in the gas phase, $C_{t,i}$ the total concentration, and $K_p$ the partitioning coefficient. The formulation of $K_p$, and subsequent units for gas and condensed phase concentrations, relates to the physical process leading to partitioning, which we build on in the next section.

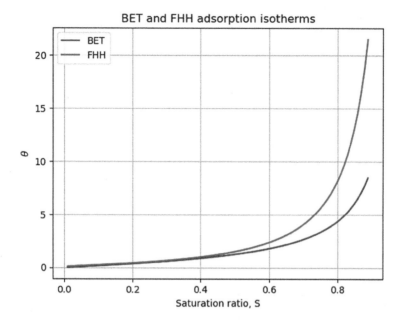

**Figure 2.1** Variation of $\theta_{s,ads}$ with $S_{sat,ads}$ for both the BET and FHH isotherms, focusing on water uptake onto NaCl. We do not consider any deliquescence behavior dictated by the solution thermodynamics of this particular system (see Chapter 3).

## 2.2 Equilibrium Absorptive Partitioning

Some implementations of gas-to-particle partitioning assume instantaneous partitioning between the condensed and gaseous phases. This has some limitations by virtue of the fact that we know gas-to-particle partitioning is a kinetic process, as we will see in Section 2.4. However, in some conditions this might be a reasonable assumption to make if the timescales for equilibrium are particularly fast. In any case, it can be useful to estimate how much condensed mass *might* result from partitioning if a gas phase abundance and volatility of this material can be estimated. A popular framework used in the atmospheric aerosol community is the mass-based equilibrium absorptive partitioning theory [16]. This is captured in the following equations:

$$C_{OA}^* = \sum_i C_{t,i}^* \varepsilon_i + core \tag{2.5}$$

$$\varepsilon_i = \left(1 + \frac{C_i^*}{C_{OA}^*}\right)^{-1} \tag{2.6}$$

$$C_i^* = \frac{10^6 M_i \gamma_i P_{sat,i}}{R_{gas}^* T} \tag{2.7}$$

where $C_{OA}^*$ is the total concentration of condensed mass comprising, usually, absorptive organic aerosol reported in $\mu g\,m^{-3}$, $C_{t,i}^*$ is the total concentration of specific

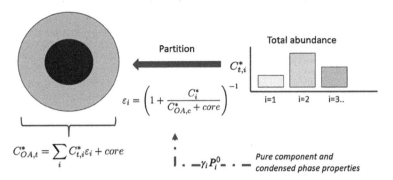

**Figure 2.2** Schematic illustration between a hypothetical aerosol particle with an in-volatile core and volatile condensed mass in a gaseous environment with a number of condensable products with total abundances $C_{t,i}^*$, volatility $\dot{C_i}$ and thus partitioning coefficients $\varepsilon_i$.

compound $i$ also given in $\mu g\,m^{-3}$, $\varepsilon_i$ is the partitioning coefficient of compound $i$ and has a value between 0 and 1. *core* is the concentration of core material. $C_i^*$ is a measure of component volatility (units $\mu g\,m^{-3}$) and, as shown in the third equation, relates to the compound molecular weight ($M_i$) and activity coefficient in the condensed phase $\gamma_i$. Unless the solution to which the compound condenses can be assumed ideal, $\gamma_i$ and thus $C_i^*$ changes with composition of the droplet, as we discuss in Chapter 3. The ideal gas constant $R_{gas}^*$ is given in units of $m^3\,atm\,K^{-1}\,mol^{-1}$ and the conversion factor $10^6$ is used to convert the molecular weight from $g\,mol^{-1}$ to $\mu g\,mol^{-1}$. The pure component vapor pressure $P_{sat,i}$, is given in atmospheres (atm). Methods for predicting $\gamma_i$ and $P_i$ are discussed in Chapters 3 and 4, respectively. Figure 2.2 illustrates the multiple dependencies which contribute to the partitioning of semivolatile organic compounds.

Whilst the units of $C_i^*$ ($\mu g\,m^{-3}$) can be confusing, it represents the condensed abundance when half of the compound is split between the gas and condensed phase according to its pure component volatility. Figure 2.3 displays the partitioning coefficients $\varepsilon_i$ for 10 representative compounds with linearly separated $Log10(C_i^*)$ values, thus orders of magnitude change in volatility, highlighted by the vertical dashed lines. From left to right, the curved solid lines represent the fraction of material condensed as a function of volatility [$x$ axis] for a fixed concentration of absorptive mass. For example, the first curved line on the left hand side of the figure represents the variation in $\varepsilon_i$ as a function of $C_i^*$ when there is $10^{-6}\mu g\,m^{-3}$ of existing condensed mass. For a compound with a $Log10(C_i^*)$ value of $10^{-6}\mu g\,m^{-3}$, half of the material will be in the particulate phase and half in the gaseous phase. For the same scenario, for a compound with a $Log10(C_i^*)$ value of $10^{-2}\mu g\,m^{-3}$ effectively 100% of the compound will be in the condensed phase. Each subsequent curved line increases the available absorptive mass by a factor of 10 such that the final line shows the variation in $\varepsilon_i$ for $10^3\mu g\,m^{-3}$ of absorptive mass. Code Listing 2.7 displays the Python function used to calculate the value of $\varepsilon_i$ as a function of $C_{OA}^*$ and $C_i^*$, which we represent as variables `COA_total` and `Cstar` in our code, respectively. In this example, we use the Numpy internal function **power** to implement Equation (2.6). The complete code used to create Figure 2.3 can be created using the file *chap2_alg2_Partitioning_coefficient.py*.

```
def partitioning_coefficient(Cstar,COA_total):
    # In the first instance we need to calculate mass that condenses
    from the core alone
    # Partitioning coefficient
    epsilon = np.power(1.0+(Cstar/(COA_total)),-1.0)
    return epsilon
```

**Listing 2.7** Highlighting the function used to populate an array with values for $\varepsilon_i$.

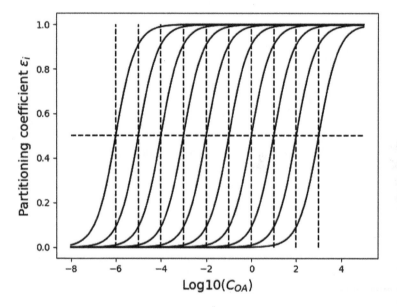

**Figure 2.3** Partitioning coefficients $\varepsilon_i$ for six representative compounds with linearly separated $\log 10(C_i^*)$ between −6 and 3. The horizontal dashed line highlights a value of $\varepsilon_i$ at 0.5, where each vertical line highlights the value of $\log 10(C_i^*)$ for each compound.

In Chapter 1, Section 1.3, we discussed how we might represent aerosol particles in a numerical framework. If you have arrived at this point in the book following Chapter 1 you will recall that in the sectional approach we populated a one-dimensional array of sizes, where each cell stored the concentration of particles of a given size. For example, the second cell in an array with 10 entries might store the concentration of all particles at 150 nm diameter. In this chapter, we want to modify the numerical structure of our code to track and simulate segregated composition of our particles. A natural inclination is to add another dimension to our existing size array for the sectional approach, or use a mechanism for extending that one-dimensional array to account for the composition of each size. In Section 2.4.1, we do indeed extend that array. However, in using the equilibrium framework presented in Equations (2.5)–(2.7), we have no explicit treatment of particle size, sizes, or even segregated composition. Rather, we are assuming our condensed phase is represented by a cumulative total mass and that mass comprises 10 components of varying volatility. This leaves us with a design choice to make with regards to the concentration of components in the gas and condensed phase. We are going to use two separate one-dimensional arrays to represent $C_{t,i}^*$ and $\varepsilon_i$ which will lead to an estimate for $C_{OA}^*$.

Our goal is to arrive at a value of $C^*_{OA}$ to calculate the condensed phase concentration at equilibrium. However, each time $C^*_{OA}$ is updated, $\varepsilon_i$ changes. We thus need to implement an iterative method for finding the root of our equation. We can use the Newton–Raphson method and rewrite our partitioning equation so that the root represents the equilibrium position. We have already presented the equations that describe the equilibrium absorptive partitioning approach. In Equations (2.8)–(2.10), we now separate out the *total* condensed and secondary absorptive mass through the variables $C^*_{OA,t}$ and $C^*_{OA,c}$, respectively.

$$C^*_{OA,t} = \sum_i C^*_{t,i} \varepsilon_i + core \tag{2.8}$$

$$C^*_{OA,c} = \sum_i C^*_{t,i} \varepsilon_i \tag{2.9}$$

$$\varepsilon_i = \left(1 + \frac{C^*_i}{C^*_{OA,c} + core}\right)^{-1} \tag{2.10}$$

How do we use this information to arrive at a prediction for total condensed secondary mass, or $C^*_{OA,c}$? We can simply use the Newton–Raphson method by reformulating the equations we wish to solve. Specifically, if we have a function $f$ of a variable $x$, each iteration gives us a new estimate for $x$ through Equation (2.11).

$$x_1 = x_0 - \frac{f(x_o)}{f'(x_o)} \tag{2.11}$$

where $f'(x_o)$ is, normally, the analytical expression for the gradient of $f$ with respect to $x$. In Equations (2.12)–(2.13), we rewrite our partitioning equation so we can use this method to find the value of $C^*_{OA,c}$ when $C^*_{OA,c} - \sum_i C^*_{t,i} \varepsilon_i$ is equal to zero.

$$f(C^*_{OA,c}) = C^*_{OA,c} - \left(\sum_i C^*_{t,i}\left(1 + \frac{C^*_i}{C^*_{OA,c} + core}\right)^{-1}\right) \tag{2.12}$$

$$f'(C^*_{OA,c}) = 1 - \sum_i C^*_{t,i}\left(1 + \frac{C^*_i}{C^*_{OA,c} + core}\right)^{-2} \frac{C^*_i}{(C^*_{OA,c} + core)^2} \tag{2.13}$$

Algorithm 2.1 outlines the steps in order to translate these equations into code from which we can estimate $C^*_{OA,c}$.

---

**Algorithm 2.1:** Newton–Raphson method for finding root of Equation (2.12)

---

1 **Result:** Secondary condensed mass $C^*_{OA,c}$
2 Initialization;
3 Initialize the number of condensing compounds, their volatility ($C^*_i$), and the concentration of preexisting core;
4 Prescribe an initial guess for $C^*_{OA,c}$, defined as $x_0$ in equations 2.12 and 2.13;
5 **while** *The absolute value of* $\frac{x_1 - x_0}{x_0}$ *is greater than a defined tolerance* **do**
6 $\quad$ Calculate $x_1$ using Equation (2.11);
7 **end**

---

Code Listing 2.8 presents the structure of our resulting code. In this example we have separated the contributions to both $f(C^*_{OA,c})$ and $f'(C^*_{OA,c})$ as individual functions **partitioning** and **partitioning_dash**, respectively. Within the function **Newtons_method** that calls these two functions, we implement a **while** loop that returns the value for $C^*_{OA,c}$ when a defined tolerance is met.

```python
import numpy as np

#Avogadros number
NA=6.02214076e+23

# Implement equation 2.8
def partitioning(Cstar,abundance,COA_c,core):
    # Partitioning coefficient
    epsilon = np.power(1.0+(Cstar/(COA_c+core)),-1.0)
    # Partitionined mass
    COA_c = np.sum(epsilon*abundance)
    return COA_c

# Implement equation 2.13
def partitioning_dash(Cstar,abundance,COA_c,core):
    epsilon = np.power(1.0+(Cstar/(COA_c+core)),-2.0)*(Cstar/((COA_c
    +core)**2.0))
    COA_dash = np.sum(epsilon*abundance)
    return COA_dash

# Implement Newtons method
def Newtons_method(Cstar,abundance,COA_init,core):

    COA_c = partitioning(Cstar,abundance,COA_init,core)
    f = COA_c - (partitioning(Cstar,abundance,COA_c,core))
    f_dash = 1.0 - partitioning_dash(Cstar,abundance,COA_c,core)
    COA_new = COA_c - f/f_dash

    # Iterate estimates of COA until tolerance met
    while (abs((COA_new-COA_c)/COA_new) > 1.0e-3):

        COA_c = COA_new
        f = COA_c - (partitioning(Cstar,abundance,COA_c,core))
        f_dash = 1.0 - partitioning_dash(Cstar,abundance,COA_c,core)
        COA_new = COA_c - f/f_dash

    return COA_c
```

**Listing 2.8** Defining a set of functions that allow us to implement the Newton–Raphson method for estimating $C^*_{OA,c}$.

With this structure in place, we can now initialize our simulation with the parameters listed in Table 2.1 for both $C^*_{t,i}$ and $C^*_i$.

This is shown in code Listing 2.9 that immediately follows code Listing 2.8 in the file *chap2_alg3_Absorptive_partitioning.py*. In this sample of code, we use the variables abundance to represent $C^*_{t,i}$ as a one-dimensional array and Cstar to represent $C^*_i$. Variable core, representing *core*, is set to 5 $\mu g\,m^{-3}$. When you run the file *chap2_alg3_Absorptive_partitioning.py*, we print to screen the initial concentration of the core absorptive mass, the total available secondary mass, and the final concentration estimated from our VBS simulation. With all the parameters defined earlier, you should find, we estimate a final concentration of secondary mass is ~2.25 $\mu g\,m^{-3}$.

**Table 2.1** Initializing the abundance of condensable material in each volatility bin.

| Partitioning parameters | |
| --- | --- |
| $Log10(C_i^*)$ | $C_{t,i}^*$ $(\mu g \, m^{-3})$ |
| −6 | 0.1 |
| −5 | 0.1 |
| −4 | 0.15 |
| −3 | 0.22 |
| −2 | 0.36 |
| −1 | 0.47 |
| 0 | 0.58 |
| 1 | 0.69 |
| 2 | 0.84 |
| 3 | 1.0 |

```
# Set an initial bulk core mass and condensed secondary mass
core=5.0
COA_first_guess = 1.0
print("Core absorptive mass = ", core)

# Populate volatility basis set with gas phase abundance
abundance = np.zeros((10), dtype = float)
abundance[0] = 0.1
abundance[1] = 0.1
abundance[2] = 0.15
abundance[3] = 0.22
abundance[4] = 0.36
abundance[5] = 0.47
abundance[6] = 0.58
abundance[7] = 0.69
abundance[8] = 0.84
abundance[9] = 1.0
print("Available secondary mass = ", np.sum(abundance))

# Define array of log10 C* values
log_c_star = np.linspace(-6, 3, 10)
Cstar = np.power(10.0,log_c_star)

# Call our function 'Newtons_method'
COA_final = Newtons_method(Cstar,abundance,COA_first_guess,core)
print("Secondary mass = ", COA_final)
```

**Listing 2.9** Defining a set of functions that allow us to implement the Newton-Raphson method for estimating $C_{OA,c}^*$.

Barley et al. [17] noted that the partitioning predictions of Donahue [16] and [18] are defined by two closely related expressions:

$$C_i^* = \frac{10^6 M_i \gamma_i P_i^o}{R_{gas}^* T} \tag{2.14}$$

$$K_{p,i} = \frac{R_{gas}^* T}{10^6 M_{om} \gamma_i P_i^o} \tag{2.15}$$

where we have already used the saturation concentration, $C_i^*$, in previous equations. $K_{p,i}$, the equilibrium coefficient, is not the reciprocal of $C_i^*$ as $M_{om}$ is calculated as the mass weighted molar mass of all condensates as given in Equation (2.16):

$$M_{om} = \frac{\sum_i C_{a,i}^* M_i}{C_{OA}} \tag{2.16}$$

They further redefine a molar based partitioning approach, where the saturation concentration is given in $\mu mol\, m^{-3}$ through removal of the molecular weight of the condensate:

$$C_{i,mol}^* = \frac{10^6 \gamma_i P_i^o}{R_{gas}^* T} \tag{2.17}$$

and $C_{OA}$ is also provided on a molar basis. In the above example, we manually set the value for the existing *core* mass. In the first exercise of this chapter, we ask you to set this initial value by initializing a monodisperse population of particles with a given size, mass, density, and number concentration.

## 2.3 Knudsen Regimes and the Kelvin Effect

In the previous section, we did not account for any size dependency in our partitioning calculations. We remind ourselves, in numerous locations in this book, that aerosol particles can span a wide range of physical sizes. Consider the definition of an aerosol particle, using the idealized spherical representation. When we discuss new particle formation in Chapter 6, we move away from macroscopic properties of our sphere and need to consider interactions at the molecular level. As you will find in that chapter, we consider aerosol particles not only as discrete entities, but also the molecular units that they are made of. In addition, as our particles decrease in size, we also need to consider the relevant theories that best describe the interaction they have with the surrounding medium. If the aerosol particle is large enough, we can consider it to reside in a continuous fluid. As the size decreases, this assumption starts to break down. Rather than a continuous fluid, the aerosol particle now enters a regime where physical interactions such as coagulation, and mass transfer through condensational growth, transitions into a stochastic nature. We refer to both the Kinetic and Continuum regimes [2] as those in which we treat the random motion of individual molecules striking an aerosol particle and the gas behaving as a continuous fluid, respectively. Transitioning between these two regimes is referred to as the transition regime. Indeed, in Section 2.4 we provide a number of steps that lead to the derivation of the droplet growth equation. This ordinary differential equation (ODE) allows us to calculate the mass flux across surface of our particle and thus predicts the growth and shrinkage as a function of time. This derivation implicitly assumes our particle resides in a continuum regime. However, as our particles decrease in size and enter a transition regime we can expect this flux to decrease and we need to introduce a correction factor to determine a more accurate flux estimate.

At what point do we need to account for these transitions and correct any predictive frameworks? In Equation (2.18) we find an expression that allows us to calculate the Knudsen number for a condensing gas that may partition to an aerosol particle.

$$Kn_i = \frac{2\lambda_i}{d_p} \tag{2.18}$$

where subscript $i$ is an index to simply point to a particular gaseous compound. $d_p$ is the diameter of our particle $(m)$, whilst $\lambda_i$ is the mean free path of our condensing gas, also in $m$. The mean free path represents the average distance a gaseous molecule travels before colliding and exchanging momentum with any other gas molecule [19]. Therefore, for large Knudsen numbers the aerosol particle of interest is small relative to the distance between molecular collisions in the surrounding medium and the particle is intercepted by gas molecules in a stochastic [19] rather than a continuous process. However, for smaller Knudsen numbers, the particle is large relative to the distance between molecular collisions and sees the fluid as a continuous medium.

$Kn_i$ is unitless, requiring both $d_p$ and $\lambda_i$ to have the same spatial units. We have not specified a range of values that would allow us to broadly classify predictions from 2.18 as "large" or "small" Knudsen numbers. However, we can now introduce a correction factor that will be used in solving the droplet growth equation in Section 2.4 to ensure we calculate appropriate fluxes across the transition regime.

Seinfeld and Pandis [2] provide an overview of the flux-matching theories and subsequent empirical studies that are used to derive a correction factor. Equation (2.19) presents the Suchs and Futugin [20] correction factor $f(Kn_i, \alpha_i)$ with the subsequent definition of the mean free path of the condensing gas in Equation (2.20).

$$f(Kn_i, \alpha_i) = \frac{0.75\alpha_i(1 + Kn_i)}{Kn_i(1 + Kn_i) + 0.283\alpha_i Kn_i + 0.75\alpha_i} \tag{2.19}$$

$$\lambda_i = \frac{3D_{g,i}}{c_{g,i}} \tag{2.20}$$

where $\alpha_i$ is the mass accommodation coefficient, $D_{g,i}$ is the gas phase diffusion coefficient of compound $i$ and $c_{g,i}$ is mean thermal velocity of compound $i$. The accommodation coefficient $\alpha_i$, which assumes values in the range of 0-1, describes the probability of a vapor molecule adhering to the surface of a liquid upon their collision [2].

If we assign units of cm$^2$s$^{-1}$ to $D_{g,i}$ and cm s$^{-1}$ to $c_{g,i}$, then $\lambda_i$ is reported in cm. Likewise, if we use units of m$^2$s$^{-1}$ and m s$^{-1}$ then $\lambda_i$ is reported in m requiring the matching definition of $d_p$. This may seem obvious, but when writing code, it is important to perform regular checks on units.

There are multiple variations of formulation (2.19) provided by [2], each with their own definition of $\lambda_i$ that, when used consistently, provide very similar profiles as a function of size.

$$c_{g,i} = \sqrt{\frac{8R_{gas}T}{\pi M_i}} \tag{2.21}$$

$$D_{g,i} = 1.9m_i^{-2/3} \tag{2.22}$$

Equations (2.21) and (2.22) allow us to calculate the mean thermal speed and diffusion coefficients of gas phase molecules as a function of their molecular weight, where $M_i$ and $m_i$ are given in $\mathrm{kg\,mol^{-1}}$ and $\mathrm{g\,mol^{-1}}$, respectively. In the current formulation, $c_{g,i}$ and $D_{g,i}$ are given in units of $\mathrm{m\,s^{-1}}$ and $\mathrm{cm^2\,s^{-1}}$, respectively, which would require a simple modification to (2.20) as follows:

$$\lambda_i = \frac{3D_{g,i} * 10^{-4}}{c_{g,i}} \tag{2.23}$$

to arrive at $\lambda_i$ in units of m. Alternatively we might write (2.20) as follows:

$$\lambda_i = \frac{3D_{g,i}}{c_{g,i}10^2} \tag{2.24}$$

to arrive at $\lambda_i$ in units of cm.

In the previous section we used the total concentration of condensed matter, thus absorptive mass $C_{OA}$, rather than treatment of individual particle size. Aside from the consideration of the transition regime, we also need to account for the influence of curvature on the equilibrium pressure above a droplet surface. Equation (2.25) presents Raoult's law, which dictates that the equilibrium vapor pressure of compound $i$, $p_i^{eq}$, above an ideal mixture of liquids (please see Chapter 3) is a product of the pure component saturation vapor pressure $p_i^o$ and the mole fraction of compound $i$ in solution, $x_i$.

$$p_i^{eq} = p_i^o x_i \tag{2.25}$$

Typically, $p_i^o$ is given in units of atm. The Kelvin effect leads to an increase in the equilibrium vapor pressure required to maintain a given composition in the condensed phase. The smaller the particle, the larger the increase required. Equation (2.25) can be modified to include the Kelvin effect via Equation (2.26).

$$p_i^{eq} = K_{surf} p_i^o x_i \tag{2.26}$$

where $K_{surf}$ is the factor that accounts for the increase and is calculated by Equation (2.27):

$$K_{surf} = exp\left(\frac{4v\bar{\sigma}_{sol}}{RTd_p}\right) \tag{2.27}$$

where $\sigma_{sol}$ is the surface tension of the air–surface interface of the particle ($\mathrm{N\,m^{-1}}$) and $v$ the volume occupied by compound $i$ in the liquid phase ($\mathrm{m^3\,mol^{-1}}$). In most instances, it is safe to assume that this can be calculated using the compound molecular weight $M_i$ ($\mathrm{kg\,mol^{-1}}$) and density of the solution $\rho_{sol}$ ($\mathrm{kg\,m^{-3}}$), leading to Equation (2.28).

$$K_{surf} = exp\left(\frac{4M_i\bar{\sigma}_{sol}}{RT\rho_{sol}d_p}\right) \tag{2.28}$$

Finally, we can rewrite Equation (2.26) in terms of a saturation ratio of compound $i$ via Equation (2.29).

$$S_i = exp\left(\frac{4M_i\bar{\sigma}_{sol}}{RT\rho_{sol}d_p}\right)x_i \tag{2.29}$$

In the following section, we use the Kelvin factor in solving the droplet growth equation, whilst also translating an equilibrium framework to estimate water update to aerosol particles as code.

## 2.4 Kinetic Absorptive Partitioning: The Droplet Growth Equation

Condensational growth is driven by a difference in the partial pressure above the surface of a particle and that in the gas phase. As concentrations of condensable products within the gas phase change, along with any changes in the composition, morphology, and phase state of the particle, the partitioning process redistributes species between the gaseous and condensed phases according to a combination of pure component and mixture properties. We discuss the prediction of both in Chapters 4 and 3, respectively.

For absorptive partitioning (see Section 2.2), the dynamic process of mass, and heat, transfer is described by a differential equation known as the droplet growth equation. The assumption of instantaneous equilibrium between the gaseous and condensed phases, discussed in Section 2.2, may be a valid approximation in certain conditions but this requires an assessment of timescales and required accuracy of particle size and/or composition.

If we follow Jacobson [19] by considering a single water droplet, we know that from Fick's first law of diffusion the rate of change of mass of a single homogeneous liquid water droplet is described by the following ODE:

$$\frac{dm}{dt} = 4\pi R_{wet}^2 D_{g,w} \frac{d\rho_{g,w}}{dR_{wet}} \tag{2.30}$$

where $m$ is the mass of the drop (g), $R_{wet}$ is the droplet radius (cm), $D_{g,w}$ is the molecular diffusion coefficient of water vapor (cm$^2$ s$^{-1}$), and $\rho_{g,w}$ is the density of water vapor (g cm$^{-3}$). Integrating (2.30) from the surface where $d\rho_{g,w} = d\rho_{g,w(R_{wet})}$, to infinity, gives

$$\frac{dm}{dt} = 4\pi R_{wet} D_{g,w} (\rho_{g,w} - \rho_{g,w(R_{wet})}) \tag{2.31}$$

If we assume there are $n_w$ water droplets per cubic centimeter, also referred to as the number density, the rate of change of condensed liquid water is given by

$$\frac{dm_{a,w}}{dt} = 4\pi R_{wet} D_{g,w} n_w (\rho_{g,w} - \rho_{g,w(R_{wet})}) \tag{2.32}$$

where $m_{a,w}$ is the mas of liquid water in the condensed phase. We can also rewrite this equation using the difference in molecular concentrations between the gas phase $C_{g,w}$ and that above the droplet surface $C_{a,w}^*$:

$$\frac{dC_{a,w}}{dt} = 4\pi R_{wet} D_{g,w} n_w (C_{g,w} - C_{a,w}^*) \tag{2.33}$$

where $C_{a,w}$ is the concentration of water in the condensed phase in molecules per cm$^{-3}$ of air.

As water vapor condenses latent heat is released, changing the temperature gradient between the surface of the droplet and surrounding air. An additional ODE that describes the cooling rate at the droplet surface due to conduction is given by (2.34):

$$\frac{dQ_r}{dt} = 4\pi R_{wet}^2 \kappa_{w,air} \frac{dT}{dR_{wet}} \tag{2.34}$$

where $\kappa_{w,air}$ is the thermal conductivity of moist air ($\mathrm{J\,cm^{-1}\,s^{-1}\,K^{-1}}$). When integrated from the droplet surface to infinity, we arrive at the following ODE for the cooling rate at the surface:

$$\frac{dQ_r}{dt} = 4\pi R_{wet} \kappa_{w,air} (Tr - T) \tag{2.35}$$

Combining the two equations leads to an expression for the temperature change at the droplet surface. By combining this equation with the ideal gas law, the Clausius–Clapeyron equation and solving for $\frac{dm}{dt}$ we arrive at the droplet growth equation for a single homogeneous droplet:

$$\frac{dm}{dt} = \frac{4\pi R_{wet} D_{g,w} \left(p_{g,w} - p_{g,w(R_{wet})}\right)}{\frac{D_{g,w} L_{e,w} p_{vw,r}}{\kappa_{w,air} T} \left(\frac{L_{e,w}}{R_{g,w} T} - 1\right) + R_{v,w} T} \tag{2.36}$$

where $p_{g,w}$ and $p_{g,w(R_{wet})}$ are the partial pressure and equilibrium vapor pressures of water vapor away from and at the droplet surface. $R_{g,w}$ is the gas constant for water vapor, given in ($\mathrm{J\,g^{-1}\,K^{-1}}$), and is obtained by dividing the ideal gas constant $R_{gas}$ by the molecular weight of water vapor $M_w$. $L_{e,w}$ is the latent heat of evaporation of water vapor ($\mathrm{J\,g^{-1}}$). The vapor pressure $p_{g,w}$ is related to both the gaseous density of water vapor $\rho_{g,w}$ and the number of moles of the vapor in the gas phase, $n_{mol,g,w}$, through (2.37).

$$p_{g,w} = \rho_{g,w} R_{g,w} T = \frac{n_{mol,v,i} M_w}{V} \tag{2.37}$$

When the gaseous density $\rho_{vw}$ is given in $\mathrm{g\,cm^{-3}}$ and the compound specific gas constant given as $\mathrm{J\,mol^{-1}\,K^{-1}}$, units of $p_{v,w}$ are equivalent to $10^4$ hPa. We can also relate $\rho_{vw,r}$, in units of $\mathrm{g\,cm^{-3}}$, to the concentration of water vapor $C_{g,w}$, given in molecules $\mathrm{cm^{-3}}$, as follows:

$$p_{g,w} = \frac{C_{g,w} M_w}{N_A} \tag{2.38}$$

where the molecular weight $M_w$ is given in units of $\mathrm{g\,mol^{-1}}$. Equation (2.36) is in units of $\mathrm{g\,s^{-1}}$, for one droplet, when using the above constant and variable definitions. If we assume there are $n_w$ droplets per cubic centimeter, we can calculate the net mass flux to that population of droplets, in units of $\mathrm{g\,s^{-1}\,cm^{-3}}$ via Equation (2.39):

$$\frac{dm}{dt} * n_w \tag{2.39}$$

Whilst equation 2.39 focuses on water vapor, Equation (2.36) can be written more generally for any condensing gas $i$. In the following we use the $a$ subscript to denote the particulate phase, where $dm_{a,i}$ therefore represents the change in mass of compound $i$ to the particulate phase:

$$\frac{dm_{a,i}}{dt} = \frac{4\pi r D_{g,i} \left(p_{g,i} - p_{g,i,r}\right)}{\frac{D_{g,i}L_{e,i}P_{g,i,r}}{\kappa_{i,air}T}\left(\frac{L_{e,i}}{R_{v,i}T} - 1\right) + R_{v,i}T} \tag{2.40}$$

where the equilibrium pressure above the droplet surface can be influenced a number of processes and mixture properties. Whilst the release of latent heat is treated for water vapor, for other gases this often ignored as the contribution to the denominator in 2.40 is negligible. Equation (2.40) then becomes

$$\frac{dm_{a,i}}{dt} = \frac{4\pi r D_{g,i} \left(p_{g,i} - p_{g,i,r}\right)}{R_{v,i}T} \tag{2.41}$$

Using the definition of vapor pressures $p_{g,i}$ and $p_{g,i,r}$, we then arrive at

$$\frac{dm_{a,i}}{dt} = 4\pi r D_{g,i} \left(\rho_{g,i} - \rho_{g,i,r}\right) \tag{2.42}$$

where $\frac{dm_{a,i}}{dt}$ is still given in $\mathrm{g\,s^{-1}}$. As we have been also using molecular abundances in our calculations so far, we can also rewrite (2.42) as the following:

$$\frac{dC_{a,i,m}}{dt} = 4\pi r D_{g,i} \left(C_{g,i} - C_{a,i,m}^{eq}\right) \tag{2.43}$$

where $C_{a,i,m}$ is the concentration of component $i$ in size bin $m$ of the condensed phase, $C_{g,i}$ is the concentration of $i$ in the gas phase, and $C_{a,i,m}^{eq}$ is the concentration of $i$ above the droplets in size bin $m$ in order to maintain the equilibrium pressure between the droplet surface and gas phase. The units of $\frac{dC_{a,i,m}}{dt}$ are in molecules $\mathrm{cm^{-3}\,s^{-1}}$ and the net mass flux to all particles in size bin $m$ is calculated by multiplying Equation (2.43) number of particles in size bin $m$, $n_m$. If we start to consider each relevant size bin, whether for a monodisperse or polydisperse population, we can start to consider the design of our code. In Chapter 7 we will revisit Equation (2.36) specifically for water vapor where a cloud parcel model is presented. From this point on we will assume that we can use Equation (2.43) and demonstrate an example body of code that solves the droplet growth equations for a generic condensate to a monodisperse and polydisperse population. We will also assume ideal homogeneous droplets for the remaining part of this chapter, whilst in Chapter 3 we present solutions to common problems associated with morphological and phase changes that would need to be accounted for in the calculation of $C_{a,i,m}^*$ in Equation (2.43). In addition, so far we have not considered any corrections for a non-continuum regime. Before we do that, as presented by Zaveri et al. [21], we often write the droplet growth equations as

$$\frac{dC_{a,i,m}}{dt} = k_{i,m} \left(C_{g,i} - C_{a,i,m}^{eq}\right) \tag{2.44}$$

where $k_{i,m}$ is the first order mass transfer coefficient for species $i$ and bin $m$ [21]. For condensation of each component to each size bin, we also need to account for the subsequent loss from the gas phase, given by

$$\frac{dC_{g,i}}{dt} = -\sum_m k_{i,m}\left(C_{g,i} - C^{eq}_{a,i,m}\right) \tag{2.45}$$

where the summation is over all size bins in the size distribution. Figure 2.4 provides a schematic of how this might look when updating values in a numerical 1D array of condensed phase concentrations $C_{a,i,m}$.

We are therefore going to "track" the concentration of every condensing component in the gas phase and the subsequent concentration in each size bin, where relevant. Before we design any code to solve the droplet growth equation, we also need formulae for all parameters in the droplet growth equation. We can calculate an equivalent gas phase concentration above the particle surface, $C^*_{a,i,m}$, as follows:

$$C^{eq}_{a,i,m} = p^{eq}_{i,m}\frac{N_A}{R^*_{gas}10^6 T} \tag{2.46}$$

$$P^{eq}_{i,m,a} = p^o_i x_{i,m}\gamma_{i,m}K_{surf,i,m} \tag{2.47}$$

$$K_{surf,i,m} = exp\left(\frac{4M_i\sigma_{sol}}{R_{gas}T2r\rho_{i,l}}\right) \tag{2.48}$$

where we express our equilibrium vapor pressure $p^{eq}_{i,m}$ in atm, since the pure component vapor pressure of component $i$ $p^o_i$ is often given in the same units as discussed in Chapter 4. Here the ideal gas constant $R^*_{gas}$, with a value of $8.205736 10^{-5}$, is given in units of $m^3$ atm $K^{-1}$ $mol^{-1}$ and the conversion factor $10^6$ is used to convert the volume concentration from $cm^{-3}$ to $m^{-3}$.

### 2.4.1 Solving the Droplet Growth Equation: A Sectional Approach

Implementing a strategy for solving the droplet growth equation will depend on the programming language you are using since the available libraries for solving an ODE will have specific interface requirements. If we use the ODE solvers exposed through

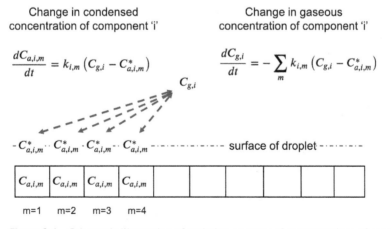

Change in condensed concentration of component 'i'

$$\frac{dC_{a,i,m}}{dt} = k_{i,m}\left(C_{g,i} - C^*_{a,i,m}\right)$$

Change in gaseous concentration of component 'i'

$$\frac{dC_{g,i}}{dt} = -\sum_m k_{i,m}\left(C_{g,i} - C^*_{a,i,m}\right)$$

$C_{g,i}$

$-C^*_{a,i,m}\cdots C^*_{a,i,m}\cdot C^*_{a,i,m}\cdots C^*_{a,i,m}$ ------- surface of droplet -------

| $C_{a,i,m}$ | $C_{a,i,m}$ | $C_{a,i,m}$ | $C_{a,i,m}$ | | | | | |
|---|---|---|---|---|---|---|---|---|

m=1    m=2    m=3    m=4

**Figure 2.4** Schematic illustration of updating an array of concentrations of species $j$ in the gas phase, $C_{g,i}$, through a difference between the concentrations above the surface of the droplet, for size bin $m$, $C^*_{a,i,m}$.

the Scipy Python package [22], which we do in this example, we might use a code structure illustrated in code snippet (2.10).

```
# Import modules
import numpy as np
from scipy.integrate import odeint


# -- define some initial conditions --
# .....
# ------------------------------------

def dydt(input):

    # -- perform some calculations --
    # .....
    # -------------------------------

    return dydt_array

# Define the time over which the simulation will take place
t = np.linspace(0, 10000, num=1000)

# Call the ODE solver with reference to our function, dydt, setting
    the
# absolute and relative tolerance, atol and rtol respectively.
solution = odeint(dy_dt, array, t, rtol=1.0e-6, atol=1.0e-4, tcrit=
    None)

   # Do something with the output contained within variable 'solution
    '..
```

Listing 2.10 An example layout of solving an ODE using the ODEint Scipy method.

The function **dydt** will contain the set of equations that dictate the loss of mass from the gas phase and subsequent growth of our aerosol size distribution. This can be further summarized following the pseudocode outlined in Algorithm (2.2).

---

**Algorithm 2.2:** Structure for implementing code to solve the droplet growth Equation (2.44).

---

1  **Result:** Solution of the droplet growth equation; particle size and condensed mass/composition
2  **Import** libraries;
3  **Initialize simulation:** define number of condensates, fundamental properties and properties of existing size distribution;
4  **Function** RHS function($y, t$):
5  |   Calculate size of droplets and composition
6  |   Calculate concentration gradients
7  |   Calculate rate change in gas and particulate phase
8  |   **return** Mass flux of every component in the gas phase and in each size bin
9
10 **Call an ODE solver** passing the function handle defined above

---

The complete code we will discuss can be found in the file *chap2_alg4_Droplet_growth_single_bin.py*. To start transferring the theory into code, we need to think about the structure of our numerical arrays.

For example, in Figure 2.5, we see an arrangement whereby two gaseous components $i$ and $j$ have concentrations that are stored in the first and second cells in the array. Following this, we track the concentration of each compound in each size bin $m$. We also need to calculate a first-order mass transfer coefficient which requires properties of the condensing gases and size of our droplets relative to the properties of the carrier gas. As we have already noted, so far we have not accounted for any noncontinuum corrections to our growth equation. However, using the expressions introduced in Section 2.3, we can introduce a correction factor $f\left(Kn_{i,m}, \alpha_i\right)$ to arrive at an expression for our first-order transfer coefficient $k_{i,m}$ given in Equation (2.49)

$$k_{i,m} = 4\pi R_{wet} D_{g,i} n_m f\left(Kn_{i,m}, \alpha_i\right) \tag{2.49}$$

where $Kn_{i,m}$ is the Knudsen number for component $i$ relative to size bin $m$, $\lambda_i$ is the mean free path of $i$, $c_{g,i}$ is the mean thermal velocity of component $i$, and $n_m$ is the number of particles in size bin $m$ using a sectional approach.

We also need to define some initial conditions. In our example, let us assume we have 10 volatility bins, using the volatility basis set (VBS) model described in Section 2.2, with the individual component volatility ($log10(C^*)$) ranging from $-6$ to 3. We also assume each condensing compound is inert and has a molecular weight of $200\,\mathrm{g\,mol^{-1}}$. The initial monodisperse population consists of an inert, involatile component with a density of $1400\,\mathrm{kg\,m^{-3}}$ and molecular weight of $200\,\mathrm{g\,mol^{-1}}$. We set the initial dry diameter to 150 nm and total number density of $300\,\mathrm{cm^{-3}}$. With the molecular weight set, through Equations (2.18)–(2.22), we can calculate the gas phase diffusion coefficient of $D_{g,i}$, the mean thermal velocity $c_{g,i}$, mean free path $\lambda_i$ of each condensate and thus Knudsen number $Kn_{i,m}$ and correction factor $f\left(Kn_{i,m}, \alpha_i\right)$ during the simulation. We also assume an accommodation coefficient, $\alpha_i$, of 0.1 for the purposes of this example. Under these conditions, we start with a preexisting absorptive mass of $0.74\,\mathrm{\mu g\,m^{-3}}$.

To start constructing our code, we can first define an array that will store concentrations of each condensate in the gas phase and condensed phase which, in this case, is represented using only 1 size bin. Note that, in this simulation, since we assume the core to be an involatile absorbing mass, we do not need to track any change

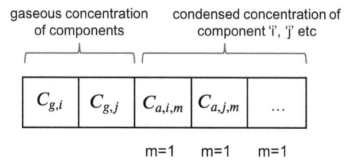

**Figure 2.5** Schematic illustration of a numerical array ordered in such a way to hold the concentrations of each specie in the gas phase ($C_{g,i}, C_{g,j}$, etc.) and then the concentration of each specie in each size bin in the preceding cells ($C_{a,i,m=1}, C_{a,i,m=2} \ldots C_{a,j,m=1}$, etc.).

in concentration but rather use it in the calculation of the droplet growth equation through the impact on equilibrium partial pressures and droplet size. In code snippet 2.11 we define the properties of our mono-disperse population and condensing compounds. At the end of the code listing, we are left with a *Numpy* array of gas phase and condensed phase concentrations to be passed into our ODE solver. These concentrations are stored in the units of molecules $cm^{-3}$.

```python
# Specify the temperature
Temp_K=298.15

# Number of condensing species from the gas phase
num_species = 10
# Number of size bins
num_bins = 1

# The molecular weight of each condensing specie [g/mol]
# Assuming a constant value for all components
mw_array=np.zeros((num_species), dtype=float)
mw_array[:]=200.0

# Define array of log10 C* values
log_c_star = np.linspace(-6, 3, num_species)
Cstar = np.power(10.0,log_c_star)
# Convert C* to a pure component saturation vapour pressure [atm]
P_sat = (Cstar*R_gas_other*Temp_K)/(1.0e6*mw_array)

# Initialise abundance in each volatility bin [micrograms/m3]
abundance = np.zeros((num_species), dtype = float)
abundance[0] = 0.1
abundance[1] = 0.1
abundance[2] = 0.15
abundance[3] = 0.22
abundance[4] = 0.36
abundance[5] = 0.47
abundance[6] = 0.58
abundance[7] = 0.69
abundance[8] = 0.84
abundance[9] = 1.0

# Unit conversion of gas abudance to molecules / cc
gas_abundance = ((abundance*1.0e-6)/(mw_array))*1.0e-6*NA

# Set accomodation coefficient
alpha_d_org=np.zeros((num_species), dtype=float)
alpha_d_org[:]=1.0
# Set density of condensing species [kg/m3]
density_org=np.zeros((num_species), dtype=float)
density_org[:]=1400.0

# Molecular diffusion coefficient in air (Equation 2.22). [cm2/s]
DStar_org = 1.9*np.power(mw_array,-2.0/3.0)
# Mean thermal velocity of each molecule (Equation 2.21). [m/s]
mean_them_vel=np.power((8.0*R_gas*Temp_K)/(np.pi*(mw_array*1.0e-3))
    ,0.5)
# Mean free path for each molecule (Equation 2.20). [m]
gamma_gas = ((3.0*DStar_org)/(mean_them_vel*1.0e2))*1.0e-2

# Define a monodisperse size distribution
# Assume each particle starts with an involatile core
# of absorptive organic with a mass of 200 g/mol and
# density 1400 km/m3. We store this information in an array
# as 'core'. This will ensure, in this example, that we do
# not get 100% evaporative loss
```

```
# Define total number of particles [per cc]
N_total = 300.0

# We carry an inert and involatile in each size bin
core = np.zeros((num_bins), dtype=float)
core_abundance = np.zeros((num_bins), dtype=float)
density_core = np.zeros((num_bins), dtype=float)
core_mw = np.zeros((num_bins), dtype=float)
density_core[:] = 1400.0
core_mw[:] = 200.0

# The number of particles is only carried in one bin
N_per_bin = np.zeros((num_bins), dtype=float)
N_per_bin[0] = N_total

# Define the diameter of our particles [microns]
size_array = np.zeros((num_bins), dtype=float)
size_array[0] = 0.150 #microns

# Use the size to calculate concentration of 'core' [molecules / cc]
# This aligns with the units used for volatile component
core_abundance[0] = (N_per_bin[0])*((4.0/3.0)*np.pi*
    np.power(size_array[0]*0.5e-6,3.0)*density_core[0]*1.0e3)
core_abundance[0] = (core_abundance[0] / core_mw[0])*NA

# What is our existing dry mass in micrograms per cubic metre?
dry_mass = np.sum((core_abundance/NA)*core_mw)*(1.0e12)
print("Initial dry mass = ", dry_mass)


# Define the gas and condensed concentration arrays
# Here we initialise the array used throughout the simulation and
# populate with initial concentrations in gas and condensed phase
array = np.zeros((num_species+num_species*num_bins), dtype=float)
# The first num_species cells hold the gas phase abundance
array[0:num_species] = gas_abundance
# The following cells hold the concentration of each specie in each
    size bin
array[num_species:num_species+num_species*num_bins] = 1.0e-20
```

Listing 2.11 Initializing a simulation based on solving the droplet growth equation for a monodisperse population.

As illustrated in code snippet 2.10, we now need to define the RHS of our differential equation. Let us recap on the structure of this approach. We have defined an array that holds the concentrations of each component in the gas and condensed phases, through one size bin, in the following line of our Python file:

```
array = np.zeros((num_species+num_species*num_bins), dtype=float)
# The first num_species cells hold the gas phase abundance
array[0:num_species] = gas_abundance
# The following cells hold the concentration of each specie in each
    size bin
array[num_species:num_species+num_species*num_bins] = 1.0e-20
```

where the variable `num_species` represents the number of secondary compounds in our simulation. In defining our function **dy_dt**, following code Listing 2.10, we need an expression that provides us with the rate of change of concentration of every cell in `array`. In this example, we assume there are no chemical reactions or production mechanisms taking place in the gas phase. Therefore, the gas phase concentrations will decrease at a rate defined by the total contributions to condensational loss given by Equation (2.45) and illustrated in figure 2.4. In code Listing 2.12, we further define our differential equation function, initialize an array of time periods to conduct the simulation over, call the ODE solver with user set tolerances, and extract the time varying concentration array.

```python
def dy_dt(array,t):

    # Retrieve the gas phase concentrations
    Cg_i_m_t = array[0:num_species]
    # Retrieve concentrations in our size bin as a slice
    temp_array=array[num_species:num_species+num_species*num_bins]
    # Sum total molecules in the condensed phase, plus core
    total_molecules=np.sum(temp_array)+core_abundance[0]
    # Calculate mole fractions in the, assumed, liquid phase for
    # calculating the equilibrium pressure above the droplet
    mole_fractions=temp_array/total_molecules
    # Calculate the density of the assumed liquid phase
    density_array = np.zeros((num_species+num_bins), dtype=float)
    density_array[0:num_species]=density_org[0:num_species]
    density_array[num_species]=density_core[0]
    # Create array that holds mass concentrations [g/cc] for
    # calculation of solution density [kg/m3]
    mass_array = np.zeros((num_species+1), dtype=float)
    mass_array[0:num_species] = (temp_array/NA)*mw_array
    mass_array[num_species] = (core_abundance[0]/NA)*core_mw[0]
    total_mass=np.sum(mass_array)
    mass_fractions_array=mass_array/total_mass
    density=1.0/(np.sum(mass_fractions_array/density_array))

    # Now calculate the size [cm] of our particles according
    # to the condensed phase abundance of material
    # We need to remember that mass is in [g/cc] whilst density
    # is in [kg/m3].
    # We convert mass and number concentrations to kg/m3 and /m3
    size=((3.0*((total_mass*1.0e3)/(N_per_bin[0]*1.0e6)))/
        (4.0*np.pi*density))**(1.0/3.0)

    # Calculate the Knudsen number for all condensing
    # molecules based on this new size
    # This relies on mean free path for each species [cm]
    # and particle radius [cm]
    Kn=gamma_gas/size
    # Calculate Non-continuum regime correction (Equation 2.19)
    # Calculate a correction factor according to the continuum
    versus
    # non-continuum regimes
    Inverse_Kn=1.0/Kn
    Correction_part1=(1.33e0+0.71e0*Inverse_Kn)/(1.0e0+Inverse_Kn)
    Correction_part2=(4.0e0*(1.0e0-alpha_d_org))/(3.0e0*alpha_d_org)
    Correction_part3=1.0e0+(Correction_part1+Correction_part2)*Kn
    Correction=1.0/Correction_part3
```

```
# Now calculate a kelvin factor for every semi-volatile compound
  in this
# size bin (Equation 2.28)
kelvin_factor=np.exp((4.0E0*mw_array*1.0e-3*sigma)/
    (R_gas*Temp_K*size_array[0]*2.0e0*density))
# Calculate the equilibrium pressure at the surface
Pressure_eq=kelvin_factor*mole_fractions*P_sat
# Calculate the equilibrium concentration [molecules/cc]
# equivalent of this pressure
Cstar_i_m_t=Pressure_eq*(NA/(R_gas_other*1.0e6*Temp_K))

# Implement the droplet growth equation (Equation 2.44)
# The equation relies on the following parameters
# radius [m]
# DStar_org [m2/s] - pass in [cm2/s] so needs to be converted by
  *1.0E-4
# Pressure_gas [Pascals]
# Pressure_eq [Pascals]
# R_gas [m3 Pascals /(K mol)]
# molw [g/mol]
# T [K]
# The units of the equation should therefore be g/s
k_i_m_t_part1 = DStar_org*Correction
k_i_m_t=4.0e0*np.pi*size*1.0e2*N_per_bin[0]*k_i_m_t_part1
dm_dt=k_i_m_t*(Cg_i_m_t-Cstar_i_m_t)

# Store values of dm_dt in a matrix, ready for dealing with
polydisperse
# populations where a contribution/loss per size bin
dy_dt_gas_matrix = np.zeros((num_species,num_bins), dtype=float)
dy_dt_gas_matrix[0:num_species,0]=dm_dt

#Initiase an array that passes back a set of derivatives for
Each
# compound in both the gaseous and condensed phases
dy_dt_array = np.zeros((num_species+num_species*num_bins), dtype
=float)
dy_dt_array[num_species:num_species+num_species*num_bins]=dm_dt
[0:num_species]
dy_dt_array[0:num_species]=dy_dt_array[0:num_species]-np.sum(
dy_dt_gas_matrix, axis=1)

return dy_dt_array


# Define the time over which the simulation will take place
t = np.linspace(0, 10000, num=1000)

# Call ODE solver with reference to dy_dt, setting the
# absolute and relative tolerance.
solution = odeint(dy_dt, array, t, rtol=1.0e-6, atol=1.0e-4, tcrit=
    None)
```

**Listing 2.12** Define the RHS of the differential equation and call the ODE solver ODEint in the Scipy module.

Within the function **dydt**, we retrieve the gas phase concentrations held within the cell index range (`0:num_species`). Notice that whilst we only simulate one size bin, we have also used relative cell referencing that enables multiple size bins to be used. Specifically, we extract concentrations in the condensed phase using a temporary array (`temp_array`) as highlighted in code Listing 2.13.

```
temp_array=array[num_species:num_species+num_species*num_bins]
```

**Listing 2.13** Relative cell referencing to extract condensed phase concentrations.

which, for our current example, would be equivalent to the following:

```
temp_array=array[10:20]
```

**Listing 2.14** Relative cell referencing to extract condensed phase concentrations.

When we do cycle through individual size bins in a poly-disperse sectional model, we need to select the appropriate start and end range of our concentrations in `array` that represent each size bin. Following this, the density and size of the droplets are calculated, along with the equilibrium pressures above the droplet and thus rate of change of mass arising through condensational growth.

Running the simulation using the initial conditions and time period defined earlier, we arrive at results presented in Figure 2.6.

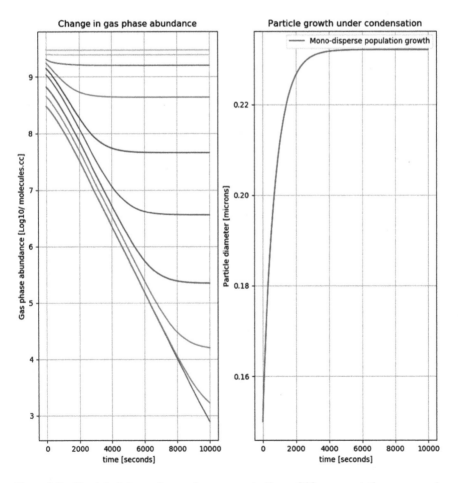

**Figure 2.6** Simulated change in gas phase concentrations of 10 representative compounds (left figure) and subsequent change in particle diameter (right figure) as a function of time).

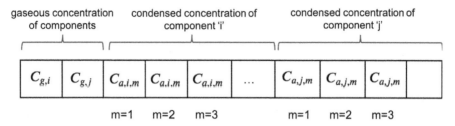

Figure 2.7 Schematic illustration of a numerical array ordered in such a way to hold the concentrations of each specie in the gas phase ($C_{g,i}$, $C_{g,j}$, etc.) and then the concentration of each specie in each size bin, for multiple size bins ($C_{a,i,m=1}$, $C_{a,i,m=2}$ .... $C_{a,j,m=1}$, etc.).

Now we expand this example to incorporate multiple size bins. If we revisit how the information might now be carried as an array, Figure 2.7 illustrates the suggested expansion of Figure 2.5 where the gas phase concentrations $C_{g,i}$, $C_{g,j}$, etc. populate the initial cells before the concentration of every condensate in each size bin is accounted for in specific slices, or blocks, of the proceeding cells.

```
import numpy as np
from scipy.integrate import odeint
# Define a differential equation to be solved
def dy_dt(<<inputs>>):
    ....
    # Now cycle through each size bin
    for size_step in range(num_bins):

        # Update change in gas phase concentrations

        # Update change in condensed phase concentrations

    ....
    # Sum contributions to RHS of ODE

# Define an array of time steps [seconds]
t = np.linspace(<<start time>>, <<end time>>, num=<<number of points
    >>) # time span and steps (start, end, number of entries). Same
    as before, with 100 entries
solution = odeint(dy_dt, array, t, rtol=<<relative tolerance>>, atol
    =<<absolute tolerance>>)
```

Listing 2.15 New structure to incorporate multiple size bins.

The structure of our code now has to change. The template in code Listing 2.15 illustrates use of a loop to calculate the difference in the partial pressure and equilibrium pressure above the droplets in every size bin and thus an update to the RHS of the ODE for every cell in our array.

The complete code that demonstrates the expansion to account for multiple size bins is given in file *chap2_alg5_Droplet_growth_multi_bin.py*. Let us extract relevant extensions in that file. First, in code Listing 2.16 we initialize a sectional log-normal distribution with eight size bins between a lower and upper diameter limit of 10 nm and 1 μm respectively. Each bin is prescribed a number density using the volume ratio log-normal procedures. Indeed, using the same procedures outlined in Chapter 1, we can calculate the centers and particle number density in each bin, assuming a total number density of 100 cm$^{-3}$, a mean diameter of 150 nm and geometric standard deviation of 1.7.

```
# Define a polydisperse size distribution
# Set the smallest size
d1 = 0.01
# Set the largest size
d_Nb = 1.0
# Number of size bins
num_bins = 8

# Volume ratio between bins (Equation 1.17)
V_rat = np.power((d_Nb/d1),3.0/(num_bins-1.0))
# Diameter ratio between bins  (Equation 1.18)
d_rat = V_rat**(1.0/3.0)

# Create an array of diameters
d_i=np.zeros((num_bins), dtype=float)
d_i[0]=d1
for step in range(num_bins):
    if step > 0:
        d_i[step]=d_i[step-1]*d_rat
# Log of Diameter array
log_di = np.log(d_i)
```

Listing 2.16 Initialize an eight bin log-normal sectional distribution.

We calculate the particle radius in each bin using the evolving mass in each bin to update the size as the simulation proceeds. In addition, assuming we have the same inert absorptive core as the previous example, we initialize an array of core concentrations, carried in the same units of molecules cm$^{-3}$, for use in all size dependent calculations of the droplet growth equation. The additional loop added to our ODE function definition is given in code Listing 2.17.

```
# Now cycle through each size bin
for size_step in range(num_bins):

    # Select a slice of y that represents this size bin
    temp_array=array[num_species+size_step*num_species:\
        num_species+num_species*(size_step+1)]
    # Sum the total molecules in the condensed phase, adding core
material
    total_moles=np.sum(temp_array)+core_concentration[size_step]
    # Calculate the mole fractions in the, assumed, liquid phase
for use
    # in calculating the equilibrium pressure above the droplet
    mole_fractions=temp_array/total_moles
    # Calculate the density of the assumed liquid phase
    density_array = np.zeros((num_species+1), dtype=float)
    density_array[0:num_species]=density_org[0:num_species]
    density_array[num_species]=density_core[size_step]
    # Create an array that holds mass concentrations [g/cc], used
for
    # calculation of solution density [kg/m3]
    mass_array = np.zeros((num_species+1), dtype=float)
    mass_array[0:num_species] = (temp_array/NA)*mw_array
    mass_array[num_species] = (core_concentration[size_step]/NA)\
        *core_mw[0]
    total_mass=np.sum(mass_array)
    mass_fractions_array=mass_array/total_mass
    density=1.0/(np.sum(mass_fractions_array/density_array))
    # Now calculate the size [cm] of our particles according
```

```
        # to the condensed phase abundance of material
        # We need to remember that mass is in [g/cc] whilst density
        # is in [kg/m3].
        # Thus we convert mass and number concentrations to kg/m3 and
    /m3
        size_array[size_step]=((3.0*((total_mass*1.0e3)/
            (N_per_bin[size_step]*1.0e6)))/\
            (4.0*np.pi*density))**(1.0/3.0)
        # Calculate the Knudsen number for all condensing molecules
    based
        # on this new size
        # This relies on mean free path for each species [cm] and
        # particle radius [cm]
        Kn=gamma_gas/size_array[size_step]
        # Calculate Non-continuum regime correction (Equation 2.19)
        # Calculate a correction factor according to the continuum
    versus
        # non-continuum regimes
        Inverse_Kn=1.0/Kn
        Correction_part1=(1.33e0+0.71e0*Inverse_Kn)/\
            (1.0e0+Inverse_Kn)
        Correction_part2=(4.0e0*(1.0e0-alpha_d_org))/\
            (3.0e0*alpha_d_org)
        Correction_part3=1.0e0+(Correction_part1+\
            Correction_part2)*Kn
        Correction=1.0/Correction_part3
        # Now calculate a kelvin factor for every semi-volatile
    compound in this
        # size bin (Equation 2.28)
        kelvin_factor=np.exp((4.0E0*mw_array*1.0e-3*sigma)/
            (R_gas*Temp_K*size_array[size_step]*2.0e0*density))
        # Calculate the equilibrium concentration [molecules/cc]
        Pressure_eq=kelvin_factor*mole_fractions*P_sat
        # Calculate the equilibrium concentration equivalent
        Cstar_i_m_t=Pressure_eq*(NA/(R_gas_other*1.0e6*Temp_K))
        # Implement the droplet growth equation (Equation 2.44)
        k_i_m_t_part1 = DStar_org*Correction
        k_i_m_t=4.0e0*np.pi*size_array[size_step]*1.0e2*\
            N_per_bin[size_step]*k_i_m_t_part1
        dm_dt=k_i_m_t*(Cg_i_m_t-Cstar_i_m_t)
        # Now update the contribution to the ODEs being solved
        # Add contributory loss from the gas phase to particle phase
        dy_dt_gas_matrix[0:num_species,size_step]=dm_dt
        # Add a contributory gain to the particle phase from the gas
    phase
        dy_dt_array[num_species+size_step*num_species:\
            num_species+num_species*(size_step+1)]=\
            dm_dt[0:num_species]

    # Subtract the net condensational mass flux from the gas phase
    concentrations
    dy_dt_array[0:num_species]=dy_dt_array[0:num_species]-np.sum(
    dy_dt_gas_matrix, axis=1)
```

**Listing 2.17** Addition of an extra loop to update the RHS for all condensates in each size bin.

In each loop increment, we recalculate the size of particles in each bin according to the partitioned mass. We then store the rate of change of mass of each compound in the gas phase, according to the calculated partitioning to the particular size bin, in a 2D Numpy array (code Listing 2.18).

```
        dy_dt_gas_matrix[0:num_species,size_step]=dm_dt
```

**Listing 2.18** Store the rate of change.

In each loop, we also need to select a slice of the concentration array that represents the concentrations of each compound in a particular size bin. According to how we have structured that array, as illustrated in Figure 2.7, we can use the approach outlined in code Listing 2.19,

```
temp_array=array[num_species+size_step*num_species:\
    num_species+num_species*(size_step+1)]
```

**Listing 2.19** Slice the concentration array for concentrations of each condensate in a size bin defined by the loop integer variable size_step.

where the integer variable `size_step` represents the loop increment. The need for `size_step+1` in the slice boundary definition arises from the first loop increment starting at 0. Following the calculation of rate of change of mass in each size bin, through the droplet the droplet growth equation, we also assign this net flux to the RHS of the output array `dy_dt_array` through the same slice in code Listing 2.20.

```
dy_dt_array[num_species+size_step*num_species:\
    num_species+num_species*(size_step+1)]=\
    dm_dt[0:num_species]
```

**Listing 2.20** Update the RHS in a size bin defined by the loop integer variable size_step.

As with the monodisperse example, we define a time period of simulation and call the ODE solver ODEint. Figure 2.8 plots the change in gas phase concentrations (left plot), the absolute change in droplet diameter (middle plot) and the relative change in diameter for each size bin (right plot). In this figure we can see the compounds with the lowest volatility, defined through $log(C^*)$, decrease with a constant trends, whilst higher volatility bins reach a plateau as the simulation proceeds and the system moves towards a state of equilibrium.

In the preceding text and examples, we have focused on using Python. However, given we introduce existing community models written in both Python and Julia in Chapter 4, it is useful to compare using both languages for solving the same droplet growth equation. With this in mind, in the following example, we demonstrate the polydisperse sectional example using Julia. You can find the complete code in the Julia file `chap2_alg6_Droplet_growth_equation_multi_bin.jl`. Whilst each language has its own syntax and workflow, by comparing the Python and Julia variants of our polydisperse example, you may find some useful similarities. Indeed, to solve the set of droplet growth equations, we can use the same flow presented in Algorithm 2.2. First, we import the relevant libraries to run our simulation:

```
using DifferentialEquations
using DiffEqBase
using DiffEqCallbacks
using DiffEqOperators
using Sundials
using Plots
```

**Listing 2.21** Defining the existing internal modules used in our Julia environment.

where the `DifferentialEquations.jl` package has an extensive range of ODE solvers available. Code Listing 2.22 highlights the similarity between Julia and Python

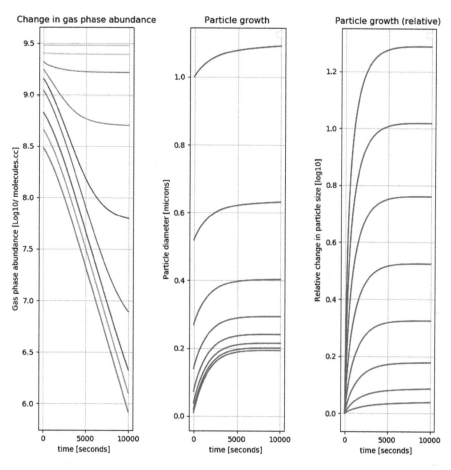

**Figure 2.8** Change in gas phase concentrations (left plot), the absolute change in droplet diameter (middle plot), and the relative change in diameter for each size bin (right plot).

in simply initializing an empty array that will track the concentrations of each compound in the gas phase and subsequent size bins, using the "." broadcasting operator in Julia to allocate the concentrations in this array that have been defined elsewhere.

```
# each gas and the concentration of each gas in each size bin
array = zeros(num_species+num_species*num_bins)
array[1:num_species] .= gas_concentration
```

**Listing 2.22** Initializing an array that will store and track the concentration of each component in the gas and condensed phases.

Likewise, code Listing 2.23 displays the syntax used to create a sectional log-normal distribution with a total number density of 100 cm$^{-3}$.

```
# Define a polydisperse size distribution
# Set the smallest size
```

```
d1 = 0.01
# Set the largest size
d_Nb = 1.0
# Number of size bins
num_bins = 8

# Volume ratio between bins (Equation 1.17)
V_rat =(d_Nb/d1).^(3.0/(num_bins-1.0))
# Diameter ratio between bins  (Equation 1.18)
d_rat = V_rat.^(1.0/3.0)

# Create an array of diameters
d_i=zeros(num_bins)
d_i[1]=d1
for step in range(1,num_bins,step=1)
    if step > 1
        d_i[step]=d_i[step-1]*d_rat
    end
end
# Log of Diameter array
log_di = log.(d_i)

# Define parameters of log-normal distribution
# Geometric standard deviation
sigmag1 = log.(1.7)
# Mean particle diameter [150nm]
mean1 = log.(0.15)
# Calculate the probability density distribution
distribution_1 = (exp.((-1.0.*(log_di .- mean1).^2) ./ (2 * sigmag1^2)) ./ (
    sigmag1 * sqrt.(2 * pi)))
# Seperate out the probability density function
d_width = d_i.*(2^(1.0/3.0))*((V_rat.^(1.0/3.0)-1.0)/((1+V_rat).^(1.0/3.0)))
# Total number of particles [per cm-3]
N_total = 100.0
# Discrete number distribution
```

**Listing 2.23** Creating a sectional log-normal distribution.

As we move through our script, we can now define the derivative function, or the RHS function. In our case, this function accepts a concentration array u, a parameter store (in this case dictionary) p, and current simulation time t, where variables are chosen according to the standard Julia tutorial conventions for DifferentialEquations.jl.

```
# Define the RHS function (that includes droplet growth equation)
function dy_dt!(u,p,t)

    # Retrieve abundance
    array=u
    # Access parameters through parameter dictionary
    num_species,N_per_bin,num_bins=[p[i] for i in ["num_species","N_per_bin","
     num_bins"]]
    core_concentration,density_org,density_core,mw_array,core_mw=[p[i] for i in
     ["core_concentration","density_org","density_core","mw_array","core_mw"]]
    alpha_d_org,gamma_gas,DStar_org=[p[i] for i in ["alpha_d_org","gamma_gas","
     DStar_org"]]
    sigma,R_gas,Temp_K,P_sat,NA,R_gas_other=[p[i] for i in ["sigma","R_gas","
     Temp_K","P_sat","NA","R_gas_other"]]
    # Retrieve gas phase abundance
    Cg_i_m_t = array[1:num_species]
```

```
# We are working with 8 size bins, each of which has an involatile core
size_array = zeros(num_bins)
#Initialise empty dydt arrays
dy_dt_array = zeros(num_species+num_species*num_bins)
dy_dt_gas_matrix = zeros(num_species,num_bins)

# Now cycle through each size bin
for size_step = 1:num_bins

    # Select a slice of y that represents this size bin
    temp_array=array[(num_species+(size_step-1)*num_species)+1:num_species+
num_species*(size_step)]
    # Sum the total molecules in the condensed phase, adding core material
    total_moles=sum(temp_array)+core_concentration[size_step]
    # Calculate the mole fractions in the, assumed, liquid phase for use
    # in calculating the equilibrium pressure above the droplet
    mole_fractions=temp_array./total_moles
    # Calculate the density of the assumed liquid phase
    density_array = zeros(num_species+1)
    density_array[1:num_species].=density_org[1:num_species]
    density_array[num_species+1]=density_core[size_step]
    # Create an array that holds mass concentrations [g/cc], used for
    # calculation of solution density [kg/m3]
    mass_array = zeros(num_species+1)
    mass_array[1:num_species].= (temp_array./NA).*mw_array
    mass_array[num_species+1]= (core_concentration[size_step]/NA)*core_mw[
size_step]
    total_mass=sum(mass_array)
    mass_fractions_array=mass_array./total_mass
    density=1.0./(sum(mass_fractions_array./density_array))
    # Now calculate the size [cm] of our particles according
    # to the condensed phase abundance of material
    # We need to remember that mass is in [g/cc] whilst density
    # is in [kg/m3].
    # Thus we convert mass and number concentrations to kg/m3 and /m3
    size_array[size_step]=((3.0*((total_mass*1.0e3)/(N_per_bin[size_step]*1
.0e6)))/(4.0*pi*density))^(1.0/3.0)
    # Calculate the Knudsen number for all condensing molecules based
    # on this new size
    # This relies on mean free path for each species [cm] and
    # particle radius [cm]
    Kn=gamma_gas./size_array[size_step]
    # Calculate Non-continuum regime correction (Equation 2.19)
    # Calculate a correction factor according to the continuum versus
    # non-continuum regimes
    Inverse_Kn=1.0./Kn
    Correction_part1=(1.33e0.+0.71e0.*Inverse_Kn)./(1.0e0.+Inverse_Kn)
    Correction_part2=(4.0e0.*(1.0e0.-alpha_d_org))./(3.0e0.*alpha_d_org)
    Correction_part3=1.0e0.+(Correction_part1.+Correction_part2).*Kn
    Correction=1.0./Correction_part3
    # Now calculate a kelvin factor for every semi-volatile compound in this
    # size bin (Equation 2.28)
    kelvin_factor=exp.((4.0E0.*mw_array.*1.0e-3.*sigma)/(R_gas.*Temp_K.*
size_array[size_step].*2.0e0.*density))
    # Calculate the equilibrium concentration [molecules/cc]
    Pressure_eq=kelvin_factor.*mole_fractions.*P_sat
    # Calculate the equilibrium concentration equivalent
    Cstar_i_m_t=Pressure_eq.*(NA./(R_gas_other.*1.0e6.*Temp_K))
    # Implement the droplet growth equation (Equation 2.44)
    k_i_m_t_part1 = DStar_org.*Correction
```

```
        k_i_m_t=4.0e0.*pi.*size_array[size_step].*1.0e2.*N_per_bin[size_step].*
    k_i_m_t_part1
        dm_dt=k_i_m_t.*(Cg_i_m_t.-Cstar_i_m_t)
        # Now update the contribution to the ODEs being solved
        # Add contributory loss from the gas phase to particle phase
        dy_dt_gas_matrix[1:num_species,size_step].=dm_dt[1:num_species]
        # Add a contributory gain to the particle phase from the gas phase
        dy_dt_array[(num_species+(size_step-1)*num_species)+1:num_species+
    num_species*(size_step)].=dm_dt[1:num_species]

    end

    # Subtract the net condensational mass flux from the gas phase
      concentrations
    dy_dt_array[1:num_species]=dy_dt_array[1:num_species].-sum(dy_dt_gas_matrix,
        dims = 2)[:,1]

    du=dy_dt_array

    return du
```

Listing 2.24 Defining the derivative function `dy_dt`.

We define the parameters passed to our function through the parameter dictionary in code Listing 2.25, where the period of simulation is also defined in the array `tspan`, and we specify our algorithm to be the **CVODE_BDF** method from the Sundials library. We do not review the entirety of the similarities between the files *chap2_alg5_Droplet_growth_equation_multi_bin.py* and `chap2_alg6_Droplet_growth_equation_multi_bin.jl` here. However, in the exercises associated with this Chapter in the Appendix, you are asked to modify our polydisperse example to account for activity coefficients ($\gamma_i$) that varies with mole fraction of solution which helps to better understand the proposed structure employed here.

```
end
# Create a dictionary that contains properties of condensing gases and
# the liquid mixture
param_dict=Dict("num_species"=>num_species,"N_per_bin"=>N_per_bin,"num_bins"=>
    num_bins,#"dydt"=>dydt,
                "core_concentration"=>core_concentration,"density_org"=>
    density_org,"density_core"=>density_core,
                "mw_array"=>mw_array,"core_mw"=>core_mw,"alpha_d_org"=>
    alpha_d_org,
                "gamma_gas"=>gamma_gas,"DStar_org"=>DStar_org,"sigma"=>sigma
    ,
                "R_gas"=>R_gas,"Temp_K"=>Temp_K,"P_sat"=>P_sat,
                "NA"=>NA,"R_gas_other"=>R_gas_other)
```

Listing 2.25 Initializing the parameter dictionary and conditions of our simulation.

In order to run this script, we can issue the command given in code listing 2.26 in the Julia console.

```
julia> include("chap2_alg6_Droplet_growth_equation_multi_bin.jl")
```

Listing 2.26 Running the multi-bin Julia example.

Figure 2.9 plots the same information as in Figure 2.8, but this time using the Julia simulation and internal plotting package.

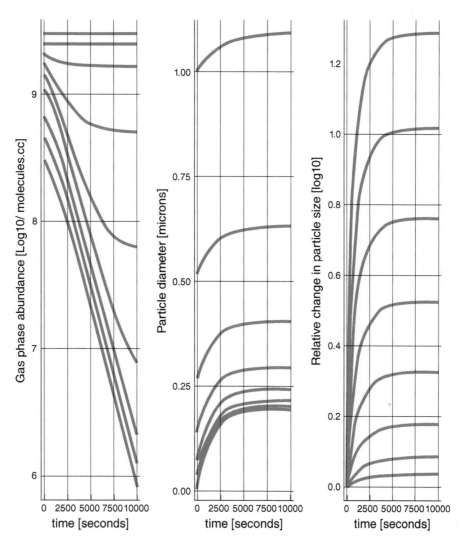

**Figure 2.9** Output from the Julia version of solving the multi-bin droplet growth equation example.

## 2.5 Cloud Condensation Nuclei Activation

### 2.5.1 Köhler Theory

Köhler Theory [23] provides a framework to predict the equilibrium water content at sub- and supersaturated humid conditions as a function of individual particle composition and size. Derived under the assumption that the amount of solute remains constant, the theory combines the competing properties of solubility and the Kelvin effect on the equilibrium vapor pressure above a droplet surface via Equation (2.50).

$$S_w = K_{surf,w} a_w \tag{2.50}$$

where $S_w$ is the saturation ratio of water vapor, $K_{surf,w}$ is the Kelvin factor for water (see Section 2.3), and $a_w$ is the water activity of the solution. The latter is often called the Raoult term and represents the solute effect on the equilibrium water content, manifest through the partial pressure of water vapor at the droplet surface in equilibrium with the aqueous solution. In Chapter 3 we discuss methods for calculating the activity of a solution. Substituting the expression for the Kelvin equation, (2.27), into Equation (2.50) leaves us with Equation (2.51):

$$S_w = exp\left(\frac{4v\bar{\sigma}_{sol}}{RTd_p}\right) a_w \tag{2.51}$$

For an ideal solution, the water activity is simply the mole fraction of water:

$$a_w = x_w = \frac{n_w}{n_{w,sol} + n_{sol}} \tag{2.52}$$

where $n_{w,sol}$ is the number of moles of water and $n_{sol}$ is the number of moles of solute. For a mix of solutes, $n_{sol}$ can be calculated from the concentration of each solute and its molecular weight via Equation (2.53):

$$n_{sol} = \sum_{i=1}^{N_{solutes}} n_i = \sum_{i=1}^{N_{solutes}} \frac{M_{tot,i}}{M_i} \tag{2.53}$$

where $N_{solutes}$ is the total number of solutes, $M_{tot,i}$ is the concentration (mass) of solute $i$, and $M_i$ is the molecular weight. In nonideal solutions, the water activity can be obtained through a number of expressions. As Wex et al. [24] specify, one approach is to use Equation (2.54):

$$a_w = exp\left[-\left(\frac{\phi_{sol}\vartheta n_{sol}}{n_{w,sol}}\right)\right] \tag{2.54}$$

where the osmotic coefficient $\phi_{sol}$ accounts for nonideal behavior and $\vartheta$ represents the number of ions per solute molecule. Whilst $\phi$ varies with solute concentration [24], one approximation is to use a series expansion leading to Equation (2.55):

$$a_w \approx \frac{n_{w,sol}}{\vartheta n_{sol} + n_{w,sol}} \tag{2.55}$$

Similarly, for nonideal solutions one expresses the water activity using

$$a_w = \gamma_w x_w \tag{2.56}$$

where $\gamma_w$ is the mole fraction-based activity coefficient of water in the solution (see Chapter 3). The equilibrium curve produced through application of Equation (2.50) has a maximum at a saturation ratio of water vapor, $S_w$, above 1.0. Before the maximum point, we predict the droplet grows and shrinks to maintain equilibrium with the gas phase. At the maximum point we say the droplet has activated into a cloud droplet and the size is the critical droplet diameter. We now implement Equations (2.50)–(2.56) to produce Figure 2.10 which plots the environment curve for an assumed initial aerosol particle composed of ammonium sulfate $((NH_4)_2 (SO_4))$ with an ideal aqueous solution and surface tension of 72 mN m$^{-1}$. Figure 2.11 also shows the variation in maximum supersaturation of water vapor as a function of initial particle size. This is calculated through a series of calls to a function that iteratively calculates the maximum point

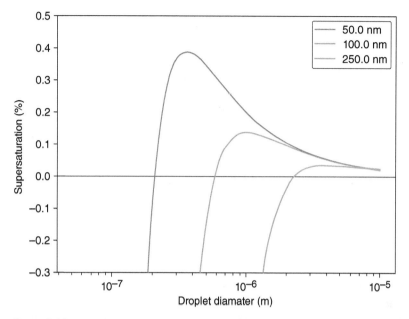

**Figure 2.10** Equilibrium curves calculated using Köhler theory for 50,100 and 250 nm diameter $(NH_4)_2 (SO_4)$ particle, assuming an ideal aqueous solution.

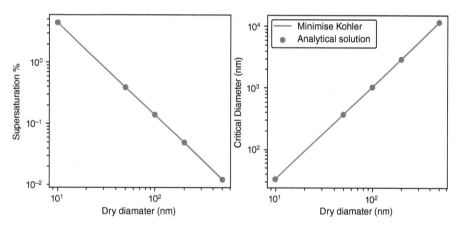

**Figure 2.11** Critical saturation ratio $lnS_c$ as a function of dry particle diameter $D_d$ for a $(NH_4)_2 (SO_4)$ particle.

for an given initial particle size. In the same figure, we also compare these predictions with an analytical expression that relates $S_w$ to initial dry particle diameter $d_p$ taken from Seinfeld and Pandis [2]. This is based on an expansion of Equation (2.50) to arrive at another expression for the Köhler curve through a series of assumptions targeted at dilute and non-volatile solutes. Specifically, if we assume that $n_{sol}$, remains the same, the mole fraction of water $x_w$ can be rewritten as

$$\frac{1}{x_w} = 1 + \frac{n_{sol}}{n_{w,sol}} = 1 + \frac{n_{sol}\bar{v}_w}{\left(\frac{\pi}{6}\right)d_p^3 - n_{sol}\bar{v}_s} \tag{2.57}$$

where $\bar{v}_w$ is the molar volume of water and $\bar{v}_s$ is the molar volume of the solute. If the solution is dilute and the volume of the solute is negligible compared with the volume of the droplet:

$$n_{sol}\bar{v}_s \ll \frac{\pi}{6}d_p^3 \tag{2.58}$$

Then a new form of the Köhler equation emerges as we substitute Equation (2.57) into Equation (2.51):

$$\ln\left(S_w\right) = \frac{4\bar{v}_w\sigma_w}{RTd_p} + \ln\gamma_w - \ln\left(1 + \frac{6n_{sol}\bar{v}_w}{\pi d_p^3}\right) \tag{2.59}$$

In dilute droplets, if we assumed the activity coefficient of water to approach unity, $\ln\gamma_w \approx 0$, whilst approximating $\ln(1+x) \cong x$ as $\frac{6n_{sol}\bar{v}_w}{\pi d_p^3} \to 0$, then we also arrive at Equation (2.62):

$$\ln\left(S_w\right) = \frac{4M_w\sigma_w}{RT\rho_w d_p} - \frac{6n_s M_w}{\pi\rho_w d_p^3} \tag{2.60}$$

where we have also assumed the molar volume of water $\bar{v}_w$ can be approximated, in a dilute droplet, by $\frac{M_w}{\rho_w}$ using the density of pure water $\rho_w$. Using two new composition dependent parameters, $A$ and $B$, we can write Equation (2.62) as

$$\ln\left(S_w\right) = \frac{A}{d_p} - \frac{B}{d_p^3} \tag{2.61}$$

Setting the derivative of Equation (2.62) to 0 allows us to formulate expressions that calculate the critical droplet diameter:

$$d_{pc} = \left(\frac{3B}{A}\right)^{1/2} \tag{2.62}$$

and the critical saturation ratio:

$$\ln S_c = \left(\frac{4A^3}{27B}\right)^{1/2} \tag{2.63}$$

In the following code examples, we review the code used to arrive at Figure 2.10. The additive mixing rule for solution density described in Chapter 4 is used and we assume a constant surface tension of 72 mN m$^{-1}$ and a dry density of 1770 kg m$^{-3}$ for $(NH_4)_2(SO_4)$.

We then calculate and plot the critical saturation ratios given by the original Köhler equation formulation (2.50) as a function of initial particle size to generate Figure 2.11. We compare these predictions with the implementation based on Equation (2.63).

In this instance, it might be useful for us to practice writing a collection of functions within a separate file and importing each function depending on which aspect or calculation within Köhler theory we wish to implement. In our example, we have created the file *Kohler_theory_modules.py* that contains separate functions to implement Equations (2.51) and a combination of (2.62) and (2.63).

In code Listing 2.27 we define a function **Kohler_curve** within the file *Kohler_theory_modules.py* that implements Equation (2.50) and returns the equilibrium saturation ratio of water vapor when provided the composition and size of an assumed liquid droplet. In this function we assume it is appropriate to use the additive, mass-weighted mixing rule for density and rely on the user to supply relevant values for the number of dissociated ions and surface tension of the droplet which is assumed to remain constant.

```python
import numpy as np

def Kohler_curve(Dp,Temp_K,moles_solute,diss_num,molar_vol_solute,
    molar_vol_water,density_solute,density_water,
    molar_weight_water,surf_tens,R_gas):

    # Convert Droplet size to water content and mole fraction
    # First calculate the number of moles of water
    volume_total = (4.0/3.0)*np.pi*np.power((Dp*0.5),3.0)
    volume_water = volume_total - (moles_solute*molar_vol_solute)
    moles_water = volume_water/molar_vol_water
    # Then calculate a mole fraction, and density, assuming additive
    #   mixing rule
    X_w = moles_water / (moles_water+moles_solute*diss_num)
    # Mass fractions. Create a numpy array of mass loadings
    total_mass = moles_solute*molar_vol_solute+moles_water*
    molar_vol_water
    mass_fractions = np.zeros((2),dtype=float)
    mass_fractions[0] = (moles_solute*molar_vol_solute)/total_mass
    mass_fractions[1] = (moles_water*molar_vol_water)/total_mass
    # Now create a numpy array of densities for the solute and water
    density_pure = np.zeros((2),dtype=float)
    density_pure[0] = density_solute
    density_pure[1] = density_water
    # Now calculate the density of the solution
    density_solution = 1.0/(np.sum(mass_fractions/density_pure))
    # Calculate the Kelvin effect
    Kelvin = np.exp((4.0*(molar_weight_water*1.0e-3)*surf_tens)/
            (R_gas*Temp_K*density_water*Dp))
    # Calculate the equilibrium saturation ratio
    Sw = Kelvin*X_w

    return Sw
```

Listing 2.27 Defining a function within a Python file that will be imported by another file.

In a separate file (*chap2_alg7_Kohler_curves.py*), after importing our new module and function **Kohler_curve**, we loop through each of our three initial particle diameters, gradually increasing the droplet size by 1 nm until we reach 10 μ m. Our equilibrium curves are then produced in the resultant figure. Code Listing 2.28 highlights the syntax used to import our own collection of Köhler theory functions, whilst 2.29 highlights this series of loops used to construct the environment curve.

```python
import numpy as np
import matplotlib.pyplot as plt
from scipy.optimize import minimize_scalar
import Kohler_theory_modules
```

Listing 2.28 Define a "master" script that imports our collection of Köhler theory functions.

```
dry_size_list = [50.0,100.0,250.0] #nm
# Initialise a figure canvas to plot on
plt.figure()
for size in dry_size_list:

    mass_particle = (4.0/3.0)*np.pi*\
    np.power((0.5*size*1.0e-9),3.0)*density_solute #kg
    moles_solute = (mass_particle*1.0e3)/molar_weight_solute

    #pdb.set_trace()

    # Initialise a list of saturation ratios, Sw
    Sw_list = []
    droplet_size_list =[]
    # Now iterative through droplet sizes. To do this we add 1nm
 to the
    # initial dry particle size until we reach 10'000nm.
    droplet_size = size+1.0

    while droplet_size <= 10000.0:

        Dp = droplet_size*1.0e-9
        droplet_size_list.append(Dp)

        # Call the Kohler_curve function
        Sw=Kohler_theory_modules.Kohler_curve(\
            Dp,Temp_K,moles_solute,\
            diss_num,molar_vol_solute,\
            molar_vol_water,density_solute,\
            density_water,molar_weight_water,\
            surf_tens,R_gas)
        Sw_list.append(Sw)

        droplet_size=droplet_size+1.0

    # Plot the results on the figure
    # Note we are plotting supersaturation levels
    plt.plot(np.array(droplet_size_list),
    (np.array(Sw_list)-1.0)*100.0,label='%s nm' % size)
# Constrain the y axis to show supersaturation
plt.xscale('log')
plt.ylim(-0.3, 0.5)
ax = plt.gca()
ax.axhline(0, color='black',lw=0.5)
ax.set_ylabel('Supersaturation (%)')
ax.set_xlabel('Droplet diamater (m)')
plt.legend()
plt.show()
```

Listing 2.29 Series of loops to produce equilibrium curves for three different sized particles.

In the case of automatically estimating the maximum point of the environment curve, thus critical supersaturation, we can repeat the same workflow of function **Kohler_curve** but return a slightly modified value for use in a minimization algorithm. We thus repeat the same calculations but include a new function **Kohler_curve_min** that returns the negative value of $S_w$. In the case of constant aerosol properties and additive mixing rules, we could derive an analytical expression for the Jacobian of Equation (2.50) and thus use gradient-based optimization methods. Rather, to retain general use where such an expression might be difficult to derive, we opt to use the constrained version of Brents algorithm available in the Scipy function `minimize_scalar`. Designed as a method for the minimization of scalar function of one variable, in this example we vary $d_p$ and calculate the associated water content, size, and equilibrium value of $S_w$. In code Listing 2.30, we use the Scipy function **minimize_scalar** to construct arrays of the critical saturation ratio $S_{crit,w}$ and diameter $d_{p,crit}$ whilst also comparing with the estimates provided through Equations (2.62) and (2.63).

```python
dry_size_list=[10,50,100,200,500]
iterative_Sc=[]
analytical_Sc=[]
iterative_Dc=[]
analytical_Dc=[]
upper_limit= 50000 #nm

for size in dry_size_list:

    Dp = size*1.0e-9 # Convert from nm to m
    mass_particle = (4.0/3.0)*np.pi*\
        np.power((0.5*Dp),3.0)*density_solute #kg
    moles_solute = (mass_particle*1.0e3)/\
        molar_weight_solute #moles

    # Calculate the critical saturation ratio by Brents method
in minimize_scalar
    result = minimize_scalar(\
        Kohler_theory_modules.Kohler_curve_min,\
        bounds=(size, upper_limit), \
        args=(Temp_K,moles_solute,stoich_coeff,\
        molar_vol_solute,molar_vol_water,\
        density_solute,density_water,\
        molar_weight_water,surf_tens,R_gas),\
        method='bounded')
    iterative_Sc.append(((result.fun*-1.0)-1.0)*100.0)
    iterative_Dc.append(result.x)

    # Calculate the critical saturation ratio using the
analytical 2 parameter solution
    Sc_it, Dpc_it = Kohler_theory_modules.Crit_2param(Dp,\
        Temp_K,moles_solute,surf_tens,diss_num,\
        molar_vol_water,density_water,\
        molar_weight_water,R_gas)
    analytical_Sc.append((Sc_it-1.0)*100.0)
    analytical_Dc.append(Dpc_it*1.0e9)
```

**Listing 2.30** Constructing a list of critical saturation ratios and diameters through an iterative solver or analytical approximation.

In file *chap2_alg7_Kohler_curves.py*, you will find we use conditional checks on the value of a variable that defines whether we want to build up an array of saturation ratios for linearly separated droplet sizes or compare the results from using a minimization algorithm to find the critical droplet diameter and saturation ratio through (2.51) with the analytical approximation [Equations (2.62) and (2.63)].

### 2.5.2 Hygroscopic Growth Factors and Kappa Köhler Theory

In subsaturated humid conditions ($S_w < 1$), the measured change in particle size under "dry" and humid conditions is often used as a measure for hygroscopicity. Combined with variants of Köhler theory that may or may not include assumptions relevant to the composition and conditions of the droplet, comparing empirical and theoretical growth factors provides valuable insights into appropriate design of reduced complexity frameworks used in regional and global climate models [25].

The hygroscopic growth factor (GF) can simply be obtained from the ratio of a measured wet to "dry" diameter:

$$GF = \frac{d_p}{d_d} \tag{2.64}$$

where $d_p$ is the diameter of the droplet at a specific relative humidity and $d_d$ the diameter of the particle under dry conditions. Petters and Kreidenweis [25] proposed a framework to model hygroscopic growth of aerosol particles using one parameter $\kappa_K$, that is defined through its effect on the water activity of the solution:

$$\frac{1}{a_w} = 1 + \kappa_K \frac{V_s}{V_w} \tag{2.65}$$

where $V_s$ and $V_w$ are the volumes of the dry particle (both soluble and insoluble) and water, respectively. Using this expression within traditional Köhler theory leads to the formulation of Kappa Köhler theory. Using the Zdanovskii, Stokes, and Robinson (ZSR) mixing assumption [26] that the total water content is the sum of water content from individual components:

$$V_w = \frac{a_w}{1 - a_w} \sum_i \kappa_{K,i} V_i \tag{2.66}$$

where $\kappa_{K,i}$ is the hygroscopicity parameter for the individual solute $i$ and the total volume of the system is

$$V_T = \sum_i V_i + V_w = V_s + V_w \tag{2.67}$$

where $V_i$ is the dry volume of component $i$. If we use the volume equivalent diameters $d_d^3 = 6\frac{V_s}{\pi}$ and $d^3 = 6\frac{V_T}{\pi}$ in (2.65), we arrive at a general expression for Kappa Köhler theory:

$$S_w = \frac{d_p^3 - d_d^3}{d_p^3 - d_d^3(1 - \kappa_K)} exp\left(\frac{4v\bar{\sigma}_w}{RTd_p}\right) \tag{2.68}$$

For mixed solutes, the total value for $\kappa_K$ is calculated from a simple additive mixing rule, given by Equation (2.69).

$$\kappa = \sum_i \xi_i \kappa_{K,i} \tag{2.69}$$

where $\varepsilon_i$ is the dry component volume fraction:

$$\varepsilon_i = \frac{V_i}{V_s} \tag{2.70}$$

Equation (2.68) can also be rewritten in terms of the hygroscopic growth factor (2.64) as

$$S_w = \frac{GF^3 - 1}{GF^3 - (1 - \kappa_K)} exp\left(\frac{4v\bar{\sigma}_w}{RTD_p}\right) \tag{2.71}$$

Whilst Equation (2.65) assumes water uptake to be continuous over the full range of humidities, Petters and Kriedenweiss [25] introduced a term to account for the dissolved fraction of the solute $\omega_i$ which is unity if completely dissolved and for all other cases is calculated using Equation (2.74).

$$\omega_i = C_{sol,i} \frac{V_w}{V_i} \tag{2.72}$$

where $C_i$ is the solubility of the solute in water, expressed as the volume of compound per unit volume of water. This leads to a new expression for the total volume of water:

$$V_w = \frac{a_w}{1 - aw} \sum_i \kappa_{K,i} V_i H(x_i) \tag{2.73}$$

where parameter $H(x_i)$ is defined as

$$H(\omega_i) = \begin{cases} \omega_i \, if \, \omega_i < 1 \\ 1 \, if \, \omega_i \geq 1 \end{cases} \tag{2.74}$$

For use in the standard $\kappa$-Köhler theory expression (2.68), $\kappa_K$ is calculated through mixing rule (2.75).

$$\kappa_K = \sum_i \varepsilon_i \kappa_{K,i} H(\omega_i) \tag{2.75}$$

where the dissolved volume fraction of the solute is calculated using (2.76) (assuming no interactions with a variable mixture composition [see Chapter 3], no dissolution kinetics, and morphology enhanced solubility).

$$\omega_i = (GF^3 - 1) \frac{sol, i}{\varepsilon_i} \tag{2.76}$$

Thus, when $\omega_i$ is unity for all substances, we arrive back at Equation (2.69).

To generate Figure 2.12 we can implement a new function that couples Equations (2.71)–(2.76) within our existing file *Kohler_theory_modules.py* as demonstrated in code Listing 2.31:

```python
def Kappa_kohler(Dp,D_dry,Temp_K,solubility_C,Kappa_array,
    surf_tens,vol_frac,density_water,molar_weight_water,R_gas):

    # Calculate the growth factor
    g=Dp/D_dry
    # Calculate the dissolved volume fraction
    xi=((g**3.0)-1.0)*(solubility_C/vol_frac)
    H_xi = np.zeros(len(vol_frac),dtype=float)
    step=0
    for entry in xi:
        if entry < 1.0:
            H_xi[step]=xi[step]
        else:
            H_xi[step]=1.0
        step+=1

    # Calculate the Kappa parameter
    Kappa = np.sum(vol_frac*Kappa_array*H_xi)

    # Calculate the Kelvin effect
    Kelvin = np.exp((4.0*(molar_weight_water*1.0e-3)*surf_tens)/
        (R_gas*Temp_K*density_water*Dp))

    # Calculate the equilibrium saturation ratio
    Sw = Kelvin*((Dp**3.0-D_dry**3.0)/(Dp**3.0-(D_dry**3.0)*(1.0-
    Kappa)))
    return Sw
```

**Listing 2.31** Implementing a solubility based $\kappa$-Köhler theory for a multicomponent droplet.

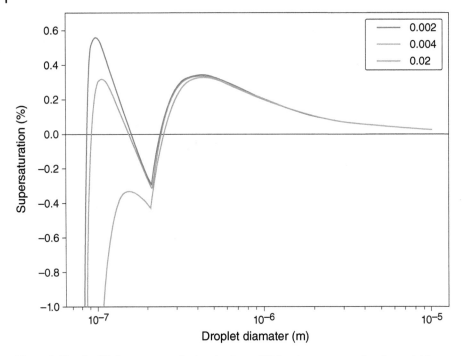

**Figure 2.12** Equilibrium curves calculated using $\kappa$-Köhler theory accounting for variable solubility.

In this function we pass in arrays `solubility_C`, `Kappa_array` and `vol_frac` that can hold information on the solubility, hygroscopicity parameter $\kappa_K$, and volume fraction of each component from a binary to multicomponent droplet. Please note, of course, that the location of each component within each array has to be consistent. For example, in replicating the binary NaCl-Succinic acid example provided by Petter and Kreidenweiss [25], we choose to use the first cell for NaCl and the second for succinic acid. Thus, within our separate script *chap2_alg7_Kohler_curves.py*, and illustrated in code Listing 2.32, we can now call this new function within a loop designed to map increases in our droplet diameter to changes in $S_w$:

```
dry_diameter = 80.0
volume_frac_list = [0.002,0.004,0.02] #nm
solubility_C=[1.6e-1,5.6e-2]
Kappa_array=[1.28,0.23]
# Initialise a figure canvas to plot on
plt.figure()

for vol_frac in volume_frac_list:

    # Initialise a list of saturation ratios, Sw
    Sw_list = []
    droplet_size_list =[]
    # Now iterative through droplet sizes. To do this we add 1nm
to the
    # initial dry particle size until we reach 10'000nm.
    droplet_size = dry_diameter+1.0

    vol_frac_array=np.zeros((2),dtype=float)
```

```
        vol_frac_array[0]=vol_frac
        vol_frac_array[1]=1.0-vol_frac

        while droplet_size <= 10000.0:

            Dp = droplet_size*1.0e-9
            droplet_size_list.append(Dp)

            # Call the Kohler_curve function
            Sw=Kohler_theory_modules.Kappa_kohler(\
                Dp,dry_diameter*1.0e-9,Temp_K,\
                solubility_C,Kappa_array,surf_tens,\
                vol_frac_array,
                density_water,molar_weight_water,R_gas)
            Sw_list.append(Sw)

            droplet_size=droplet_size+1.0

        # Plot the results on the figure
        # Note we are plotting supersaturation levels
        plt.plot(np.array(droplet_size_list),
            (np.array(Sw_list)-1.0)*100.0,label=r'%s' % vol_frac)
    # Constrain the y axis to show supersaturation
    plt.xscale('log')
    plt.ylim(-1.0, 0.7)
    ax = plt.gca()
    ax.axhline(0, color='black',lw=0.5)
    ax.set_ylabel('Supersaturation (%)')
    ax.set_xlabel('Droplet diamater (m)')
    plt.legend()
    plt.show()
```

**Listing 2.32** Calling our new function that returns the value for $S_w$ based on $\kappa$-Köhler theory for a multicomponent droplet.

# Bibliography

[1] Manish Shrivastava, Christopher D. Cappa, Jiwen Fan, Allen H. Goldstein, Alex B. Guenther, Jose L. Jimenez et al. Recent advances in understanding secondary organic aerosol: implications for global climate forcing. *Reviews of Geophysics*, 55:509–559, 6, 2017.

[2] Manabu Shiraiwa, Andreas Zuend, Allan K. Bertram, and John H. Seinfeld. Gas-particle partitioning of atmospheric aerosols: Interplay of physical state, non-ideal mixing and morphology. *Physical Chemistry Chemical Physics*, 15:11441–11453, 7, 2013.

[3] Guiying Rao and Eric P. Vejerano. Partitioning of volatile organic compounds to aerosols: A review, 12, 2018.

[4] M. Hallquist, J. C. Wenger, U. Baltensperger, Y. Rudich, D. Simpson, M. Claeys et al. The formation, properties and impact of secondary organic aerosol: Current and emerging issues. *Atmospheric Chemistry and Physics*, 9:5155–5236, 2009.

[5] Simon O'Meara, David O. Topping, and Gordon McFiggans. The rate of equilibration of viscous aerosol particles. *Atmospheric Chemistry and Physics*, 16:5299–5313, 4, 2016.

[6] Ulrich K. Krieger, Claudia Marcolli, and Jonathan P. Reid. Exploring the complexity of aerosol particle properties and processes using single particle techniques. *Chemical Society Reviews*, 41:6631–6662, 9, 2012.

**[7]** Yuqing Qiu and Valeria Molinero. Morphology of liquid-liquid phase separated aerosols. *Journal of the American Chemical Society*, 137:10642–10651, 7, 2015.

**[8]** Pauli Virtanen, Ralf Gommers, Travis E. Oliphant, Matt Haberland, Tyler Reddy, David Cournapeau et al. SciPy 1.0: fundamental Algorithms for Scientific Computing in Python. *Nature Methods*, 17(3):261–272, March, 2020.

**[9]** Angelina Leonardi, Heather M. Ricker, Ariel G. Gale, Benjamin T. Ball, Tuguldur T. Odbadrakh, George C. Shields et al. Particle formation and surface processes on atmospheric aerosols: A review of applied quantum chemical calculations. *International Journal of Quantum Chemistry*, 120:e26350, 10, 2020.

**[10]** Prashant Kumar, Athanasios Nenes, and Irina N. Sokolik. Importance of adsorption for CCN activity and hygroscopic properties of mineral dust aerosol. *Geophysical Research Letters*, 36, 12, 2009.

**[11]** Mingjin Tang, Daniel J. Cziczo, and Vicki H. Grassian. Interactions of water with mineral dust aerosol: Water adsorption, hygroscopicity, cloud condensation, and ice nucleation, 4, 2016.

**[12]** C. D. Hatch, P. R. Tumminello, M. A. Cassingham, A. L. Greenaway, R. Meredith, and M. J. Christie. Technical note: Frenkel, Halsey and Hill analysis of water on clay minerals: toward closure between cloud condensation nuclei activity and water adsorption. *Atmospheric Chemistry and Physics*, 19(21):13581–13589, 2019.

**[13]** Sami Romakkaniemi, Kaarle Hämeri, Minna Väkevä, and Ari Laaksonen. Adsorption of water on 8-15 nm NaCl and $(NH_4)_2SO_4$ aerosols measured using an ultrafine tandem differential mobility analyzer. *Journal of Physical Chemistry A*, 105:8183–8188, 9, 2001.

**[14]** J. H. Seinfeld and S. N. Pandis. *Atmospheric Chemistry and Physics: From Air Pollution to Climate Change.* J. Wiley & Sons, New York, USA, 1998.

**[15]** Ari Laaksonen, Jussi Malila, Athanasios Nenes, Hui Ming Hung, and Jen Ping Chen. Surface fractal dimension, water adsorption efficiency, and cloud nucleation activity of insoluble aerosol. *Scientific Reports*, 6:1–5, 5, 2016.

**[16]** N. M. Donahue, A. L. Robinson, C. O. Stanier, and S. N. Pandis. Coupled partitioning, dilution, and chemical aging of semivolatile organics. *Environmental Science and Technology*, 40:2635–2643, 5, 2006.

**[17]** M. Barley, D. O. Topping, M. E. Jenkin, and G. McFiggans. Sensitivities of the absorptive partitioning model of secondary organic aerosol formation to the inclusion of water. *Atmospheric Chemistry and Physics*, 9:2919–2932, 9, 2009.

**[18]** James F. Pankow. An absorption model of the gas/aerosol partitioning involved in the formation of secondary organic aerosol. *Atmospheric Environment*, 28:189–193, 1, 1994.

**[19]** M. Z. Jacobson. *Fundamentals of Atmospheric Modeling, Second Edition.* Cambridge University Press, New York, 2005.

**[20]** N.A. Fuchs and A.G. Sutugin. High-dispersed aerosols, 1, 1971.

**[21]** Rahul A. Zaveri, Richard C. Easter, Jerome D. Fast, and Leonard K. Peters. Model for simulating aerosol interactions and chemistry (mosaic). *Journal of Geophysical Research*, 113:D13204, 7, 2008.

**[22]** Pauli Virtanen, Ralf Gommers, Travis E. Oliphant, Matt Haberland, Tyler Reddy, David Cournapeau et al. SciPy 1.0: Fundamental algorithms for scientific computing in Python. *Nature Methods*, 17:261–272, 2020.

[23] Hilding Köhler. The nucleus in and the growth of hygroscopic droplets. *Transactions of the Faraday Society*, 32:1152–1161, 1, 1936.

[24] Heike Wex, Frank Stratmann, David Topping, and Gordon McFiggans. The kelvin versus the raoult term in the Köhler equation. *Journal of the Atmospheric Sciences*, 65:4004–4016, 12, 2008.

[25] M. D. Petters and S. M. Kreidenweis. A single parameter representation of hygroscopic growth and cloud condensation nucleus activity - part 2: Including solubility. *Atmospheric Chemistry and Physics*, 8:6273–6279, 2008.

[26] R. H. Stokes and R. A. Robinson. Interactions in aqueous nonelectrolyte solutions. i. solute-solvent equilibria. *The Journal of Physical Chemistry*, 70:2126–2131, 7, 1966.

# 3

# Thermodynamics, Nonideal Mixing, and Phase Separation

## 3.1 Thermodynamics and Nonideal Mixing

Applications of thermodynamic theory are a key part of aerosol science. This is perhaps most obvious when concerned with the theory and modeling of equilibrium processes, such as equilibrium gas–particle partitioning discussed in Chapter 2. In this chapter, we will have a closer look at the basic equations and principles governing the thermodynamics of mixing between different aerosol species from the perspective of single particle systems. While focusing on the chemical thermodynamics and related models applicable to liquid or amorphous viscous (liquid) phases, we will discover that such approaches can be integrated as a specific condensed-phase module in a multiprocess aerosol box model.

The use of thermodynamic theory will enable us to model the nonideal mixing among chemical species within aerosol particles (or bulk solutions) and to make model predictions about the equilibrium chemical compositions in different compartments of an aerosol particle or the gas–aerosol system as a whole. In addition to the thermodynamics of mixing, we will also discuss the drivers of condensed-phase diffusion and introduce methods to model diffusion within the interior of single particles.

### 3.1.1 Chemical Thermodynamics

Classical thermodynamics is well known for its success in describing accurately the rules for energy transfer in macroscopic systems, such as heat engines. This theory provides the framework to express and quantify relationships between temperature, pressure, work, and other forms of energy. The adoption of thermodynamics for use in describing mixing effects and phase equilibria in the field of physical chemistry is largely attributed to the fundamental work by Josiah Willard Gibbs during the 19th century [1]. The term chemical thermodynamics typically refers to the use of thermodynamics in the context of chemical reactions, mixing effects, and phase equilibria. Here, we will focus on practical applications of chemical thermodynamics in single and two-phase systems composed of more than one chemical component at given temperature and total pressure.

Before we turn to applications in aerosol science, we have to introduce a few important concepts and definitions. In thermodynamics, a so-called open system is a compartment of interest within which the pressure temperature and the amount of

*Introduction to Aerosol Modelling: From Theory to Code.*
First Edition. Edited by David Topping and Michael Bane.
© 2022 John Wiley & Sons Ltd. Published 2022 by John Wiley & Sons Ltd.

matter may change. For example, consider a glass bottle partially filled with water: the water molecules may freely partition between the liquid water and the gas phase above it. In this sense, the liquid water phase is an open system, since the amount of water it contains may change over time and is therefore not a conserved quantity. The same applies to the gas phase in the bottle, it may even exchange water molecules with the bottle's surroundings. However, if we close the bottle with a lid, suppressing any material exchange with its *environment*, the number of water molecules in the bottle (in absence of chemical reactions) will be constant. Therefore, we could state that the bottle as a whole is a closed (multiphase) system. In that case, it is still true that the number of water molecules in the liquid phase may change over time due to dynamic gas–liquid exchange of molecules, especially when temperature varies. Hence, both the liquid and the gas phases inside the bottle are open subsystems within a closed system. Now, if we solely focus on the liquid phase, we could also decide to declare that phase to be our "system." The choice of what is part of a system and what not is somewhat arbitrary; typically we make the choice in such a way that we can focus on the problem of interest and declare everything else as being part of the environment. Also, note that the term *phase* has specific meaning in thermodynamics. It refers to a compartment that is well-mixed (homogeneous) on the macroscopic level, that is, the density, the concentrations of different molecules, and the temperature within different samples (containing sizeable amounts of molecules) of that compartment are the same. In contrast, if there are for example two liquid compartments of substantially different composition existing over sufficiently long times (not just a fluctuation for a nanosecond), we would have a case of liquid–liquid phase separation (LLPS), a topic to which we will come back to in this chapter. Thinking of aerosol systems, we have always at least two phases present, the gas phase and the condensed, liquid, semi-solid, or solid phase (or phases) constituting the particles.

A key variable of an open system is its so-called Gibbs energy, $G$ (sometimes referred to as Gibbs free energy). It can be defined in terms of the fundamental equation in differential form as [e.g., 1],

$$dG = -SdT + Vdp + \sum_j \mu_j dn_j.$$

(3.1)

Here, $S$ denotes the entropy, $T$ the temperature, $V$ the volume, and $p$ the pressure of the system. Changes in Gibbs energy due to changes in chemical composition are accounted for by the last term on the right hand side, in which $\mu_j$ denotes the chemical potential and $n_j$ the molar amount of component $j$, while the sum covers all system components. We are excluding in our treatment the contributions from other sources of change to the Gibbs energy, such as variations in electric or gravitational fields. The chemical potential is defined as the partial molar Gibbs energy,

$$\mu_j = \left( \frac{\partial G}{\partial n_j} \right)_{T,p,n_{k \neq j}}.$$

(3.2)

While perhaps less common to us than variables like pressure or temperature, which are relatively easily measured, the chemical potential is a key conceptual variable and its usefulness will soon become clear.

### 3.1.1.1 Thermodynamic Equilibrium

A closed system is said to have reached a state of *thermodynamic equilibrium* when there is no longer any observable change in its temperature, pressure, or chemical composition—no matter how long one waits. Formally, this means that $dT = 0, dp = 0$, and $dn_j = 0$ for all $j$. Applied to Equation (3.1), this yields $dG = 0$, meaning that the Gibbs energy of the system has reached a constant value. At constant temperature, pressure, and composition, any textbook on chemical thermodynamics will further point out that the constant Gibbs energy of such a system represents a minimum value for the given conditions—a consequence of the second law of thermodynamics. In other words, any perturbation from this state will result in a higher Gibbs energy for the system. Therefore, finding the stable equilibrium state of a closed system can be treated as finding the (global) minimum in its Gibbs energy. This is a key result of classical thermodynamics.

By integrating either Equation (3.1), or more directly by using the definition of the chemical potential (Equation (3.2)), under conditions of thermodynamic equilibrium, one can show that the total Gibbs energy of a closed system is expressed by

$$G = \sum_j \mu_j n_j. \tag{3.3}$$

This is a generally valid expression for $G$ at equilibrium even if the system contains several phases; therefore, knowing the molar amounts and chemical potentials of the different components involved allows for the computation of the value of $G$. Note that we are usually not interested in determining an absolute value of $G$; rather, we target the comparison of Gibbs energies for different (nonequilibrium) configurations of a system, for example, in the search of its minimum possible value at equilibrium. Since the molar composition of a system is usually known (e.g., as external model input or from a measurement), Equation (3.3) points out the need for determining the chemical potentials of the different components. When two or more phases are in equilibrium, it follows that the chemical potentials of the same component in different phases must be equal. Denoting distinct phases by a superscript Greek letter ($\alpha$, $\beta$, etc.), we can state

$$\mu_j^\alpha = \mu_j^\beta = \cdots = \mu_j^\omega. \tag{3.4}$$

While the chemical potentials are the same, note that the concentrations of any component $j$ in phases $\alpha$ and $\beta$ are usually different. Depending on the amounts of other components present in each phase and their interactions, the concentration differences for $j$ may range from marginal to substantial. However, considering all component concentrations or mole fractions collectively, there must be at least a slight difference in the phase compositions of some components, otherwise we would not have distinct phases. Equation (3.4) outlines a possible approach for determining a multiphase equilibrium state: we need to find distinct phase compositions for which the equality of Equation (3.4) is fulfilled. Furthermore, if such a multiphase state exhibits a lower Gibbs energy than a configuration with fewer phases (e.g., a single liquid phase) it means that it is thermodynamically favorable. To make use of this approach we will need to turn our attention to the problem of computing chemical potentials.

### 3.1.1.2 Chemical Potentials and Activities

We will treat the gas phase, for example, air surrounding aerosol particles, as an ideal gas mixture. This is a typical assumption in atmospheric chemistry, since air at standard sea-level pressure (or lower) is well represented as an ideal gas [2]. An expression for the chemical potential $\mu_j^g$ (units of J·mol$^{-1}$) of a component $j$ in such an ideal gas mixture is as follows:

$$\mu_j^g = \mu_j^{g,\ominus} + RT \ln \left( \frac{p_j}{p_j^{\ominus}} \right). \tag{3.5}$$

Here, $R$ is the universal gas constant (SI units of J mol$^{-1}$ K$^{-1}$), $p_j$ is the partial pressure of $j$, and $p_j^{\ominus}$ is a reference pressure, such as $10^5$ Pa or, in case of vapors, often chosen as the liquid-state pure-component saturation vapor pressure at temperature $T$. The superscript $g$ denotes the gas phase (not a mathematical exponent). $\mu_j^{g,\ominus} = \mu_j^{g,\ominus}(p_j^{\ominus}, T)$ is the so-called standard chemical potential of $j$. Usually, chemical potentials are expressed in the functional form of Equation (3.5), which indicates that we determine the chemical potential relative to a defined standard state. Mathematically, we can see that $\mu_j^g = \mu_j^{g,\ominus}$ for $\ln \left( p_j/p_j^{\ominus} \right) = 0$, which is the case for $p_j = p_j^{\ominus}$. Therefore, the value of the standard chemical potential depends on the chosen reference pressure; that is, if we decide to use a different reference pressure, we have to use a different standard chemical potential value that is consistent with that reference pressure. Note that the value of the chemical potential itself is independent of such arbitrary choices of reference and standard states—a fact that can be exploited for conversions between different concentration scales and associated standard and reference states [e.g., 3].

Similar to the expression for the gas phase, we define the chemical potential for a component in a mixed liquid phase as [4]

$$\mu_j^{\alpha} = \mu_j^{\alpha,\ominus} + RT \ln \left( a_j^{\alpha} \right), \tag{3.6}$$

where $a_j$ is the component's "activity"—a corrected mole fraction. For non-electrolytes such as water and organic compounds, the activity is usually expressed on mole fraction basis, $a_j^{\alpha} = x_j^{\alpha} \gamma_j^{\alpha}$. The activity coefficient, $\gamma_j^{\alpha}$, is a correction factor accounting for nonideal mixing effects in solution; typically this is a complex function of mixture composition and temperature. The reference state is defined as the pure component liquid for which the mole fraction $x_j = 1$ and the activity coefficient is unity ($\gamma_j = 1$). Consistent with this, the standard chemical potential is then given by $\mu_j^{\alpha,\ominus} = \mu_j^{\alpha,\ominus}(p_j^{\ominus}, T, x_j^{\circ})$, with $x_j^{\circ} = 1$ (unit mole fraction). Note that the choice of composition scale, here mole fraction, affects the basis (or function) of the associated activity coefficients, it would differ if we were to choose mass fraction or volume fraction rather than mole fraction. In the following, we will use mole fraction basis for all nonelectrolyte (nonionic) components.

For electrolyte components, which tend to dissociate partially or completely into several ions in aqueous solutions, a common choice is to use the molality scale to express composition [4]. Since ions carry an electric charge, additional complexity arises, much of which can be avoided by focusing strictly on electroneutral phases and neutral anion–cation combinations. For a hypothetical "salt" unit $MX$, consisting of $\nu_+$ cations $M$ and $\nu_-$ anions $X$, we can define the molality-based chemical potential as [5]

$$\mu_{MX}^{\alpha} = \mu_{MX}^{\alpha,\Theta,(m)} + RT \ln\left[\left(a_M^{(m)}\right)^{\nu_+}\left(a_X^{(m)}\right)^{\nu_-}\right]. \tag{3.7}$$

Here, the notation with superscript $(m)$ clarifies that these variables are defined on molality basis. For ions as solutes in an aqueous solution, their molality $m_i$ is defined as moles of ion per unit mass of solvent. Usually, the solvent is pure water, but sometimes the whole mixture of nonelectrolytes serves as solvent in this definition. The activity of an ion $i$ is defined using the molality scale normalized by the unit molality value ($m^{\circ} = 1$ mol kg$^{-1}$), as $a_i^{(m)} = \frac{m_i}{m^{\circ}}\gamma_i^{(m)}$. This means that activities are dimensionless, yet scale-dependent quantities. The reference state is chosen as the limiting case of an infinitely dilute solution of ion $i$ in a reference solvent, usually pure water (chosen here). For that reference state, we define the molality-based activity coefficient via $\gamma_i^{(m)} \to 1$ as $m_i \to 0$ and as the mole fraction of water, $x_w$, approaches 1. For electrolytes, the standard chemical potential is defined as the $\mu_{MX}^{\alpha}$ value of a 1-molal solution of $MX$ in pure water behaving ideally (i.e., assuming that the argument of the ln [..] term in Equation (3.7) is equal to 1.0) [5]. This defines a hypothetical standard state since the actual 1-molal aqueous solution of electrolyte $MX$ will likely not form an ideal solution. Nevertheless, such a standard state is a valid choice against which we may quantify chemical potentials of electrolyte solutes consistently. We note that in practice the superscripts denoting molality or mole fraction basis as well as the normalization of concentrations by the concentration unit (e.g., $m^{\circ}$) are often omitted when common notation is used. This simplifies the notation at the added risk of leading to occasional misunderstandings.

With these expressions for chemical potentials established, we are ready to apply the chemical equilibrium relationship (Equation (3.4)) to specific situations. For example, let us consider an aqueous solution in equilibrium with the gas phase under typical pressure and temperature conditions, say $10^5$ Pa and 20° C. In this case, liquid water and water vapor will be in vapor–liquid equilibrium (VLE) and the same must hold for organic vapors involved. For a VLE of non-electrolytes such as water, we can equate Equation (3.5) with Equation (3.6), that is, $\mu_j^{\alpha} = \mu_j^{g}$ or

$$\mu_j^{\alpha,\Theta} + RT \ln\left(a_j^{\alpha}\right) = \mu_j^{g,\Theta} + RT \ln\left(\frac{p_j}{p_j^{\Theta}}\right). \tag{3.8}$$

With no loss of generality, it is convenient to define the standard chemical potential for component $j$ to be the same in the liquid and gas phases by using as standard pressure the pure-component liquid-state saturation vapor pressure of $j$ at temperature $T$, then $p_j^{\Theta} = p_j^{\circ}(T)$, such that we can simplify Equation (3.8) to yield

$$a_j^{\alpha} = \frac{p_j}{p_j^{\circ}} \quad \text{or} \quad p_j = p_j^{\circ} x_j^{\alpha}\gamma_j^{\alpha}. \tag{3.9}$$

This VLE relationship is known as modified Raoult's law—a key equation in equilibrium gas–particle partitioning. It describes how the saturation vapor pressure (or equilibrium partial pressure) over a liquid mixture depends on the mixture composition in terms of mole fraction $x_j$ as well as effects from non-ideal mixing expressed by the activity coefficient $\gamma_j$. Equation (3.9) highlights that we can express useful VLE conditions in terms of activities and pressure ratios without a need for quantifying the standard chemical potentials. For an ideal liquid solution ($\gamma_j = 1$ for all $j$), Equation (3.9) reduces to Raoult's law (the ideal, classical version). In practice, activity coefficients may differ substantially from unity, such that their consideration becomes important. This is certainly true in the case of systems which exhibit LLPS, further discussed in Section 3.4, since LLPS is impossible in ideal solutions. Modified Raoult's law can also readily be converted to a form using mass concentrations instead of partial pressures [6, 7],

$$C_j^g = C_j^\circ x_j^\alpha \gamma_j^\alpha = C_j^\circ \frac{C_j^\alpha}{M_j \sum_k \frac{C_k^\alpha}{M_k}} \gamma_j^\alpha,$$
(3.10)

where $C_j^g$ denotes the mass concentration in the gas phase (typically expressed in units of $\mu g\,mol^{-1}$ in the context of aerosols), $C_j^\circ$ is the pure-component liquid-state saturation vapor concentration at $T$, $M_j$ is the molar mass, and the sum includes all components in phase $\alpha$.

## 3.2 Activity Coefficient Model

While describing the composition of a phase is straightforward, describing the effects of interactions among molecules and/or ions of different shapes, sizes and polarities is far more complicated. We can usually only employ a semi-empirical model for the computation of activity coefficients, typically including model parameters fitted to experimental data. There are a number of choices for activity coefficient models, from very simple and limited, to fairly complex and computationally demanding. First, we will consider a relatively simple, yet practical activity coefficient model for aqueous organic mixtures, a version of the Binary Activity Thermodynamics (BAT) model developed by Gorkowski et al. [7]. The BAT model is a form of a Duhem–Margules Gibbs excess energy model [8] tailored for the computation of nonideal mixing within binary aqueous organic solutions characterized by a few organic molecular properties only. The considered properties include molar mass, oxygen-to-carbon (O:C), and hydrogen-to-carbon (H:C) ratios, which are often available from ambient atmospheric measurements and or available as (input) data fields in large-scale models. The theoretical basis and design details of that model are described in Ref. [7]. We will first focus on the key equations of this model and then turn our interest to its implementation in code.

Common to any binary Duhem–Margules Gibbs excess energy model, including the BAT model, the natural logarithms of the activity coefficients on mole fraction basis are given by expressions involving the Gibbs excess energy of mixing, $G^E$, normalized by the universal gas constant and temperature ($RT$),

$$\ln(\gamma_w) = \frac{G^E}{RT} - x_o \frac{d\left(\frac{G^E}{RT}\right)}{dx_o},$$
(3.11)

$$\ln(\gamma_o) = \frac{G^E}{RT} + (1 - x_o) \frac{d\left(\frac{G^E}{RT}\right)}{dx_o}.$$
(3.12)

Here, subscripts $w$ and $o$ indicate the water and organic component, respectively. The mole fractions are directly coupled, $x_w = 1 - x_o$, in a binary system and the second term on the right-hand side involves the derivative with respect to $x_o$. Equations

(3.11) and (3.12) make clear that we need an expression for $G^E$ that accounts for the nonideal interactions among water and the organic component $o$ as a function of mixture composition. In the BAT model, the composition is expressed by a scaled volume fraction, $\phi_o$, which captures a part of the molecular size or shape differences (while in case of system-specific fitted models often mole fraction is used). The scaled volume fraction of the organic component is related to the mole fraction and the molecular properties (density $\rho$ and molar mass $M$) of the organic and water components as follows [7]:

$$\phi_o = \frac{x_o}{x_o + (1 - x_o)\frac{\rho_o}{\rho_w}\frac{M_w}{M_o}\left[s_1(1 + \vartheta)^{s_2}\right]},\qquad(3.13)$$

where $\vartheta$ denotes the elemental O:C ratio of the organic compound; $s_1$ and $s_2$ are constant coefficients (determined by a model fit to data from experiments or a reference model). The $G^E$ function is expressed as

$$\frac{G^E}{RT} = \phi_o(1 - \phi_o)\left[c_1 + c_2(1 - 2\phi_o)\right],\qquad(3.14)$$

where $c_1$ and $c_2$ are constants for a specific organic compound, parameterized by functions of the following form:

$$c_i = a_{1,i} \cdot \exp\left[a_{2,i}\,\vartheta\right] + a_{3,i} \cdot \exp\left[a_{4,i}\frac{M_w}{M_o}\right].\qquad(3.15)$$

Here, $\exp[\dots]$ denotes the natural exponential function and the $a_{1,i}, a_{2,i}, \dots$ are constant coefficients with $i = 1$ or $2$ (for $c_1$ or $c_2$). While $\vartheta$ (= O:C) serves as a simple proxy for the polarity of the organic compound, the molar mass ratio $\frac{M_w}{M_o}$ is a proxy for the relative size difference of the two types of molecules. Both properties are considered to be of importance for estimating a deviation from ideal (mole-fraction-based) mixing. One advantage of this form of a parameterized excess Gibbs energy function is that a set of coefficients can be determined such that the BAT model describes a variety of binary water–organic mixtures approximately (and not just one). That is, the model offers a degree of predictive power, allowing one to estimate the nonideal behavior of a variety of organic compounds mixed with water. Gorkowski et al. [7] list several sets of fitted coefficients in their electronic supplement with the purpose of covering three different O:C ratio regimes in more detail, including a blending method for coefficients to provide smooth transitions between regimes. They also offer a conversion function of the parameters to account more specifically for the dominant type of oxygen-bearing functional group relative to that of a hydroxyl group used as reference. For the sake of simplicity, in our examples and code, we will use a single set of coefficients, listed in Table 3.1.

Given the defined function for $\frac{G^E}{RT}$, we can now derive the expression for its derivative with respect to $x_o$ (for use in Equations (3.11) and (3.12)),

$$\frac{d\left(\frac{G^E}{RT}\right)}{dx_o} = \left[(1 - 2\phi_o)\left[c_1 + c_2(1 - 2\phi_o)\right] - 2c_2\phi_o(1 - \phi_o)\right] \cdot \frac{d\phi_o}{dx_o},\qquad(3.16)$$

$$\frac{d\phi_o}{dx_o} = \frac{\rho_o}{\rho_w}\frac{M_w}{M_o}\left[s_1(1 + \vartheta)^{s_2}\right]\left(\frac{\phi_o}{x_o}\right)^2.\qquad(3.17)$$

**Table 3.1** Coefficients for the BAT model expressions (Equations (3.13–3.15)) for $T \approx 298$ K.

| Coefficient | $n$ | | | |
|---|---|---|---|---|
| | **1** | **2** | **3** | **4** |
| $a_{n,1}$ | 5.88511 | −4.73125 | −5.20165 | −30.8230 |
| $a_{n,2}$ | −0.98490 | −6.22721 | 2.32029 | −25.8404 |
| $s_n$ | 4.06991 | −1.23723 | | |

Note: the listed values serve as an example; they are based on an initial BAT model fit by Gorkowski et al. [7], who provide distinct sets of coefficients for different O:C domains as well as mappings for specific (predominant) organic functional groups in their work.

With the characterization of this version of the BAT model by Equations (3.11)–(3.17), we can proceed to an implementation in (Fortran) code.

## 3.3 BAT Model Implementation

For the BAT model implementation and associated example programs, we will use several Fortran files (extension of ".f90"). Many of the procedures, such as functions or subroutines, are placed into modules. For flexibility in terms of the desired numerical precision of floating point numbers, we first introduce a module `Mod_NumPrec_Types` defining compiler-independent numerical precision parameters, namely the "working precision" variable `wp`) and a derived data type (for global use in our program).

```fortran
module Mod_NumPrec_Types

implicit none

!define a working precision (wp) level to be used with floating
    point (real) variables, e.g. 1.0 should be stated as 1.0_wp.
!number_of_digits = desired minimum level of precision in terms of
    number of floating point decimal digits.
integer,parameter,private :: number_of_digits = 12
integer,parameter,public  :: wp = selected_real_kind(
    number_of_digits)

!definition of public derived types:
public
!type used for setting options of BAT_light wrapper functions
type :: foptions
    logical   :: determine_local_max
    integer   :: component_no
    real(wp) :: M_org
    real(wp) :: OtoC
    real(wp) :: density_org
    real(wp) :: refval
end type foptions

end module Mod_NumPrec_Types
```

**Listing 3.1** Module for definitions of floating point number (`real` type) precision and a derived type used later in this program (excerpt from file chap3_alg1_Mod_NumPrec_Types.f90).

Note that Fortran is a case-insensitive language, that is, use of upper- or lower case letters is not distinguished by the compiler (e.g., wp, WP, and wP would all refer to the same variable). In some cases, we will make use of mixed-case procedure or variable names solely for the purpose of better readability.

Next, we will introduce a module `Mod_BAT_model` placed into a different file. This module will, among others, contain subroutine `BAT_light`, which implements our simplified, that is "light," version of the BAT model. Hereafter, by referring to the BAT model, we mean this BAT light version. We use the set real type kind (kind = wp) from module `Mod_NumPrec_Types` in this `Mod_BAT_model`, which also means that variable wp is available in all contained procedures (see line 3 of Listing 3.2). The code snippet 3.2 outlines the general structure of this module; subsequently, we will add contents and additional subroutines/functions to this module.

```fortran
module Mod_BAT_model

use Mod_NumPrec_Types, only : wp

implicit none
private      !by default, make variables and procedures private

!public procedures (accessible by using this module):
public :: BAT_light, density_est !, ...

! . . . . . . . . . . . . . . . . . . . . . . . . . . . . . . . . . . . . . . . . . . . . . . .
contains

pure subroutine BAT_light(x_org, M_org, OtoC, rho_org, ln_actcoeff, &
    activity)
! ... (content not shown)
end subroutine BAT_light
! . . . . . . . . . . . . . . . . . . . . . . . . . . . . . . . . . . . . . . . . . . . . . . .

pure elemental function density_est(M, OtoC, HtoC) result(rho)
! ... (content not shown)
end function density_est
! . . . . . . . . . . . . . . . . . . . . . . . . . . . . . . . . . . . . . . . . . . . . . . .

! ... (additional procedures)

end module Mod_BAT_model
```

Listing 3.2 Code snippet outlining the structure of module `Mod_BAT_model`.

At this point, we can implement Equations (3.11)–(3.17) of the BAT model as a pure subroutine `BAT_light` (see Listing 3.3). Note, while the "pure" attribute is not necessary for valid Fortran code, it indicates to the compiler (and debugger) that this subroutine has no side effects on the state of the program other than through the exchange of information via the explicit interface of the subroutine. This may enable additional optimizations by a compiler, as well as the safe use of this subroutine within parallelized code sections. We will not further focus on such aspects in this chapter, but we note that it is good practice to provide such attributes when feasible.

Looking at the provided code file (chap3_alg2_Mod_BAT_model.f90), you will notice a comment section just prior to the start of the subroutine including information about the procedures' purpose, additional notes, and an authorship statement. Such informative comments are usually added in our code files to summarize the purpose of a subroutine or function, yet for brevity they will be omitted from the shown listings in this chapter.

```fortran
pure subroutine BAT_light(x_org, M_org, OtoC, rho_org, ln_actcoeff,
    activity)

implicit none

!interface arguments:
!x_org      [-], mole fraction of the organic component (input)
!M_org      [g/mol], molar mass of the organic (input)
!OtoC       [-], elemental oxygen-to-carbon ratio of organic (input)
!rho_org    [g/cm^3], liquid-state density of organic (input)
!ln_actcoeff[-], ln[activity coeff.] (output); array elements: 1
    water, 2 org
!activity   [-], mole-fraction-based activity (output)
real(wp),intent(in) :: x_org
real(wp),intent(in) :: M_org
real(wp),intent(in) :: OtoC
real(wp),intent(in) :: rho_org
real(wp),dimension(2),intent(out) :: ln_actcoeff
real(wp),dimension(2),intent(out) :: activity

!local constants:
![g/mol], molar mass of water
real(wp),parameter :: M_w = 18.015_wp
![g/cm^3], liquid-state density of water (here assumed
! temperature-independent)
real(wp),parameter :: rho_w = 0.997_wp
!fitted BAT model parameters [-] (as listed in Table 3.1):
real(wp),parameter :: s1 = 4.06991_wp
real(wp),parameter :: s2 = -1.23723_wp
!2-D array of a_n,i coefficients (column-major order)
real(wp),dimension(4,2),parameter :: apar = &
 & reshape([ 5.88511_wp, -4.73125_wp, -5.20165_wp, -30.8230_wp, &
 &          -0.98490_wp, -6.22721_wp, 2.32029_wp, -25.8404_wp], &
 &          shape = [4,2])
!local variables:
real(wp) :: dGEbyRT_dxorg, GEbyRT, M_ratio, one_minus_2phi, &
            & phi_param, phi_org, phi_by_x
real(wp),dimension(2) :: cpar   !array for c1, c2 parameters defined
    by Eq. (3.15)
!.....................................................

!step 1) calculate recurring equation terms and model coefficients
M_ratio = M_w/M_org
phi_param = (rho_org/rho_w) * M_ratio * s1*(1.0_wp + OtoC)**s2
!Eq. (3.13):
phi_org = x_org / ( x_org + (1.0_wp - x_org)*phi_param )
!Eq. (3.15):
cpar(:) = apar(1,:)*exp(apar(2,:)*OtoC) + &
          & apar(3,:)*exp(apar(4,:)*M_ratio)

!step 2) calculate normalized Gibbs excess energy term and its
    derivative
one_minus_2phi = 1.0_wp - 2.0_wp*phi_org
GEbyRT = phi_org*(1.0_wp - phi_org)*( cpar(1) + cpar(2)*
    one_minus_2phi )

! express (phi_org/x_org) in form avoiding division by zero:
phi_by_x = 1.0_wp/( x_org + (1.0_wp - x_org)*phi_param )
dGEbyRT_dxorg = ( one_minus_2phi*(cpar(1) + cpar(2)*one_minus_2phi)
    &
    & -2.0_wp*cpar(2)*phi_org*(1.0_wp - phi_org) )* phi_param*
    phi_by_x**2
```

```
!step 3) calculate natural log of activity coefficients and
   activities
!Eqs. (3.11, 3.12):
ln_actcoeff(1) = GEbyRT - x_org*dGEbyRT_dxorg
ln_actcoeff(2) = GEbyRT + (1.0_wp - x_org)*dGEbyRT_dxorg
activity(1) = (1.0_wp - x_org)*exp(ln_actcoeff(1))   !water
activity(2) = x_org*exp(ln_actcoeff(2))              !organic

end subroutine BAT_light
```

**Listing 3.3** Fortran implementation of the BAT model equations as a pure subroutine (excerpt from file chap3_alg2_Mod_BAT_model.f90).

As shown in lines 12–36 of Listing 3.3, all variables of a Fortran scoping unit (here a subroutine) have to be declared (explicitly) at the top of that unit. We have split it into the declaration of interface arguments, which also carry attributes such as intent(in), and the declaration of local variables and parameters. In Fortran, the default lower bound of an array is 1 (not 0 like in C or Python); if desired, the bounds could also be explicitly specified to other values. This means that the array ln_actcoeff (line 17) has a lower bound of 1 and upper bound of 2 (inclusive). Here, we use the first element to hold the property of water and the second one for the organic component. The input arguments characterize the binary solution composition (x_org) and physical properties of the organic compound, while we will output the natural logarithm of the activity coefficients for water and the organic compound, as well as their activities using the arrays ln_actcoeff and activity. All real (floating point) scalar and array variables are declared as real(wp), which is short notation for real(kind = wp), and literal constants like the molar mass of water (M_w = 18.015_wp) have an appended "_wp." The latter is important to set the precision level of the floating point numbers compatible with that of the parameter or variable to which it is assigned. For example, if wp indicates a single-precision kind, wp = 4 (with gfortran or ifort), the value will be represented by a single-precision real number (in the binary representation of the machine), while wp = 8 would indicate a so-called double-precision floating point value. The beauty of using module Mod_NumPrec_Types for setting wp is that we can set there the real kind used in (most of) our program units, making switching between single- and double-precision versions of our program straightforward.

In this subroutine, the BAT model expressions are implemented in three steps. Step 1, recurring terms and coefficients are computed, such as the molar mass ratio of water to organic component, as well as the scaled volume fraction $\phi_o$ from Equation (3.13). Step 2, the expressions for the normalized Gibbs excess energy and its derivative are evaluated. Equation (3.17) contains the fraction $\frac{\phi_o}{x_o}$, which would lead to a division-by-zero error for an input $x_o = 0$ if we had simply divided variables phi_org by x_org. Instead, there is a better option in this case. Using the definition of $\phi_o$, we can express $\frac{\phi_o}{x_o}$ algebraically in a different way (Equation (3.18)), which avoids any potential division by zero issues, as shown on line 53 of Listing 3.3.

$$\frac{\phi_o}{x_o} = \frac{1}{x_o + (1 - x_o)\,\phi_{\text{param}}}, \tag{3.18}$$

with $\phi_{\text{param}}$ being a constant for a certain organic compound defined by

$$\phi_{\text{param}} = \frac{\rho_o}{\rho_w}\frac{M_w}{M_o}s_1(1 + \vartheta)^{s_2}. \tag{3.19}$$

In Step 3 of the subroutine code, we combine the terms and use Equations (3.11) and (3.12) to compute the expressions for the natural logarithms of the activity coefficients of water and the organic compound, as well as their activities. Both of those arrays are then returned to the calling procedure via the subroutine interface.

### 3.3.1 Example 1: Calculation of Binary Mixture Activities Using the BAT Model

With the `BAT_light` subroutine ready for use, we can now call this subroutine from another program unit. In this example, we will first create a subroutine `BAT_example_1` to compute the activities of water and pinic acid (or rather an organic compound with the estimated physical properties of pinic acid). Second, we will add the subroutine to an executable program, and third, we will create plots of the BAT output from within our Fortran program using the Dislin library. The complete subroutine code for Example 1 is provided in file chap3_alg3_BAT_Example_1.f90. The variable declarations and the first part of the subroutine are shown in Listing 3.4.

```fortran
subroutine BAT_example_1()

use Mod_NumPrec_Types, only : wp
use Mod_BAT_model, only : BAT_light, density_est

implicit none
!local variables:
integer :: i, np, unt
real(wp) :: density_org, HtoC, M_org, OtoC, xinc
real(wp),dimension(:),allocatable :: x_org_vec
real(wp),dimension(:,:),allocatable :: activities, ln_actcoeff
!.........................................

!## Example 1 ##
!(1.1): set input parameters for the organic component, pinic
!       acid, C9H14O4, SMILES: CC1(C)C(C(O)=O)CC1CC(O)=O, in
!       the units required by the BAT_light model
M_org       = 186.207_wp                       ![g/mol], molar mass
OtoC        = 4.0_wp/9.0_wp                     ![-] O:C ratio
HtoC        = 14.0_wp/9.0_wp
density_org = density_est(M_org, OtoC, HtoC)    ![g/cm^3]
```

Listing 3.4 Example 1 subroutine (part 1); variable declarations and setting of organic compound parameters (excerpt from file chap3_alg3_BAT_Example_1.f90).

In the code section labeled as Step (1.1), we set the physical property values needed by the BAT model in the same units as indicated in the interface of subroutine `BAT_light`. Since the BAT model has been designed for enabling activity coefficient calculations when organic compounds are characterized by a few physical properties only, we opt to estimate the liquid-state density (here of pinic acid) by using a relatively simple, limited-information method based on that developed by Girolami [9]. Therefore, the function `density_est` implemented in module `Mod_BAT_model` is used on line 21 of Listing 3.4. This function requires only inputs of a compound's molar mass, its O:C ratio and, if available, its H:C ratio (see the implementation details in file chap3_alg2_Mod_BAT_model.f90).

After the parameters are set, we turn to calculating the activity coefficients and activities for a set of equally spaced points covering the full range of binary solution compositions from pure water to pure organic. Listing 3.5 shows Steps (1.2)–(1.4) in that subroutine. In Step (1.2), we set the number of points `np` and use it to allocate the dynamic arrays for storing the organic mole fraction values and BAT outputs. Line 7 in Listing 3.5 shows the use of an implied loop to populate the array `x_org_vec` with equally spaced points between 0.0 and 1.0. Note that for allocatable arrays of local scope, like `x_org_vec(:)`, de-allocation is implicit at the end of the subroutine.

```fortran
!(1.2): allocate and initialize input and output arrays to cover
!       composition range;
np = 101
allocate( x_org_vec(np), ln_actcoeff(np,2), activities(np,2) )
xinc = 1.0_wp/real(np-1, kind=wp)              !the x_org increment
!population of array values via implied loop:
x_org_vec = [(i*xinc, i = 0,np-1)]

!(1.3): use BAT_light to compute activities, looping over x_org:
do i = 1,np
    call BAT_light(x_org_vec(i), M_org, OtoC, density_org, &
    ln_actcoeff(i,:), activities(i,:))
enddo

!(1.4): write the data to a text file (comma-separated values):
open(newunit = unt, file = './example1_BAT_output.csv', status = '&
    unknown')
write(unt,'(A)') 'x_org,  act_coeff_water,  act_coeff_org,  a_w,  &
    a_org,'
do i = 1,np
    write(unt,'(4(ES12.5,","),ES12.5)') x_org_vec(i), exp( &
    ln_actcoeff(i,:)), activities(i,:)
enddo
close(unt)

write(*,'(A)') 'Output saved to file example1_BAT_output.csv'
```

Listing 3.5 Example 1 subroutine (part 2); code for computation of water and pinic acid activities over a range of solution compositions (excerpt from file chap3_alg3_BAT_Example_1.f90).

The `BAT_light` subroutine is called in Step (1.3) within a loop, iterating in the calls over `x_org_vec(i)`, with the corresponding `BAT_light` outputs saved in the two-dimensional arrays `ln_actcoeff(:,:)` and `activities(:,:)`. The first dimension of `activities(i,:)` denotes the data point index (here i), while the second dimension contains the two entries for water (1) and organic (2) aligned with the input composition of `x_org_vec(i)`. A side note: since `BAT_light` is a pure subroutine, in principle, the do-loop could easily be parallelized using a Fortran "`do concurrent (i = 1:np)`" loop in conjunction with the openMP parallelization compiler flag.

Step (1.4) in Listing 3.5 shows how one can create a simple comma-separated-values (.csv) text file in Fortran and how to write selected columns of BAT output data as a formatted table to that file (example1_BAT_output.csv). Here, the file will be stored in the folder from which the program (or project in case of Visual Studio) was executed. The generated file can be opened with your text editor of choice (e.g., Notepad++ on a Windows machine). Lines 22–23 provide feedback to a (default) console output window informing a user that the file has been written. We can now call the subroutine `BAT_example_1` from a main program unit, which will allow us to compile our code into an executable program. For a first test, the code shown in Listing 3.6 provides such a program.

```
program Prog_BAT_Example_1

implicit none
external :: BAT_example_1
!...........................................
call BAT_example_1()
write(*,*) 'Done; press Enter to end the program'
read(*,*)    !pause to wait for user action.

end program Prog_BAT_Example_1
```

Listing 3.6 A main program for running example 1 (File chap3_alg4a_Prog_BAT_Example1.f90).

At this point, let us compile, link, and run the Example 1 program we have written so far. On a Linux terminal, change the directory to the local folder containing the provided .f90 files. Then, using a recent version of the gfortran compiler, we can compile the modules and subroutines and build an executable file Prog_BAT_Example1.out by entering the command line shown in Listing 3.7.

```
gfortran -o Prog_BAT_Example1.out -O0 -Wall
    chap3_alg1_Mod_NumPrec_types.f90 chap3_alg8_Mod_MINPACK.f90
    chap3_alg2_Mod_BAT_model.f90 chap3_alg3_BAT_Example_1.f90
    chap3_alg4a_Prog_BAT_Example1.f90
```

Listing 3.7 Gfortran compiler command line for Example 1.

This is an example of a command line that could be used for debugging the code, since the "optimization level zero" (-O0) and the "warn all" (-Wall) option flags will provide details on warnings/errors. In the present case, the few warnings stemming from module Mod_MINPACK can be ignored. For an optimized "release" build version of the program, use the second command-line example shown in file chap3_alg4b_Example1_command_line.txt. When working with a MS Windows PC, either use the provided Visual Studio solution (.sln file contained in the zip-archive chap3_alg0_BAT_Example_Programs_VS_proj.zip file) and set example_no = 1 in file chap3_alg4_Prog_BAT_Examples.f90 located in subdirectory BAT_source_code. Alternatively, one could use the gfortran compiler from within a Linux terminal environment via the "Windows Subsystem for Linux" (WSL); see documentation under Ref. [10].

Entering the command line from Listing 3.7 results in the executable file Prog_BAT_Example1.out, which we can run in the Linux terminal by entering command ./Prog_BAT_Example1.out. After program execution has finished, we will find the generated output file example1_BAT_output.csv in our local folder. At this point we could further process the output and use a plotting tool to visualize the data from our output file. In the following, as a convenient option, we show how to generate plots directly from our Fortran program.

### 3.3.1.1 Plotting from Fortran with Dislin

Having the ability to quickly plot data from a selection of arrays is convenient. Fortran does not offer any plotting tools intrinsic to the language; however, there are third-party libraries that can be linked to a Fortran program to create plots directly from our code.

Here we will introduce the use of the versatile Dislin software for data plotting [11, 12]. Dislin is available for most operating systems and many programming languages are supported (including Fortran, C/C++, Python, Julia, etc.).

The use of the Dislin library and associated program functionalities requires the installation of Dislin on your system; information on distributions, downloads, and installation are available from the Dislin website [11]. For the following examples, we will assume that Dislin has been installed on your system. For Windows users, the provided Visual Studio solution refers to Dislin being installed under its default location: "C:\dislin\". For Linux users, we expect the gfortran version to be installed under "/usr/local/dislin/gf" (e.g., in Ubuntu).

A first, very simple example of generating a plot with Dislin is included in file chap3_alg3_BAT_Example_1.f90 under Step (1.5). In order for the first compilation (from Section 3.3.1) to work with or without Dislin being installed, we have commented-out the code sections requiring Dislin. Therefore, first, uncomment the block of code in the Step (1.5) section of the subroutine BAT_example_1. This code section should now look like as shown in Listing 3.8.

```
!(1.5, optional): generate a quick, simple plot of the activities
!        vs x_org using the DISLIN graphics library
!        (https://www.dislin.de/index.html);
!        this requires an installed and linked DISLIN library.
block
    integer,parameter :: dp = kind(1.0D0)
    real(dp),dimension(:),allocatable :: xval, yval
    external :: metafl, qplot
    !...............
    !apply floating point number conversion to double
    !precision kind used here with Dislin;
    xval = real(x_org_vec, kind=dp)
    yval = real(activities(:,1), kind=dp)         !water activity
    !simple x--y scatter plot:
    call metafl('xwin')                           !'xwin' or 'pdf'
    call qplot(xval, yval, np)
    yval = real(activities(:,2), kind=dp)         !organic activity
    call qplot(xval, yval, np)
end block
```

**Listing 3.8** Example 1 subroutine (part 3); code for quick plot with Dislin (excerpt from file chap3_alg3_BAT_Example_1.f90).

We have used a Fortran `block` ... `end block` construct within the subroutine. This is convenient in this case because we can declare a few extra variables (e.g., dp) and external subroutines that are only defined within this block statement, making it a self-contained program unit within our subroutine, while having access to the variables from outside the block statement in the subroutine, such as x_org_vec. This block of code makes use of Dislin's "quick plots" so that we can generate a x–y scatter plot of the water activity data with the call of qplot on line 14 of Listing 3.8. On line 16, qplot is called again, but now to make a second plot to display the organic activity versus mole fraction of organic curve. On line 13, we have set the output to go to a separate (xwin) window; this may cause an error ("X-Window display cannot be opened!") when you run the program from a WSL command prompt. In that case, replace "xwin" by "pdf" on line 13 to generate a ".pdf" plot file instead of output to the screen.

We can compile the program again; see the command line in Listing 3.9. Near the end of this command line, there are additional flags to inform the compiler/linker about the location of the Dislin 64-bit library (change that path if it differs on your system).

```
gfortran -o Prog_BAT_Example1.out -O3
    chap3_alg1_Mod_NumPrec_types.f90 chap3_alg8_Mod_MINPACK.f90
    chap3_alg5_Mod_Dislin_plots.f90 chap3_alg2_Mod_BAT_model.f90
    chap3_alg3_BAT_Example_1.f90 chap3_alg4a_Prog_BAT_Example1.f90
    -ldislin_d -I/usr/local/dislin/gf/real64
```

Listing 3.9 Command line for Example 1 with quick plots, including linking to the Dislin 64-bit library (for gfortran).

Running the built program (./Prog_BAT_Example1.out) generates the example1_BAT_output.csv file, as in the prior run, and it also displays the Dislin "quick plot" for water activity (in the "xwin" display mode). The second plot is shown after the first is closed (or after a right-click with the mouse on the first plot window). The first plot is shown in Figure 3.1.

As illustrated by Figure 3.1, the application of Dislin's quick plot (qplot) subroutine leads to a simple plot of a single curve with no axis labels shown. While a few more commands could be used to add a title and axis labels alongside the qplot subroutine call, clearly there are limitations. The main use of the Dislin's quick plots (there are other types for contours, surfaces, etc.) is in the simplicity of generating a plot for the purpose of a preliminary data visualization with the ease of just a few lines of extra code. If we wish to make more appealing figures, we could use the full tool set of Dislin, which offers control of nearly all aspects of a plot.

For the examples of this chapter, we would like to be able to plot several curves, each with defined attributes, in a single figure. For this purpose, we have written a custom module Mod_Dislin_plots containing two subroutines (add_plot_xydata and dislin_plot) which take care of calling various Dislin subroutines and aide

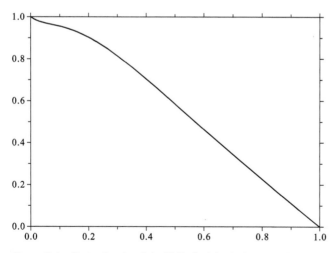

Figure 3.1    Example of a plain Dislin "quick plot" output from Step (1.5) of subroutine BAT_Example_1. The curve shows our BAT model prediction of water activity versus mole fraction of organic (pinic acid).

us in generating nicer $x$–$y$ curve/symbol plots. We will not further discuss the details of that module provided in file chap3_alg5_Mod_Dislin_plots.f90. You can explore its implementation details and features based on that code. Making use of this module, generating a figure with multiple curves will only require a few lines of code in our BAT example program. Let us look at an example: in subroutine BAT_example_1, we will comment-out section (1.5) and instead uncomment the code block shown under step (1.6). That part should now look like the code in Listing 3.10.

```
!(1.6, optional): Generate a nicer x--y curve plot, showing
!   several curves in the same plot (using Dislin but without use
!   of quick plots).
!   Here we use a set of module procedures from Mod_Dislin_plots
!   to make such plots with relatively few statements, while
!   being able to set various curve and plot properties.
block
    use Mod_Dislin_plots, only : add_plot_xydata, dislin_plot
    character(len=75) :: xlabel, ylabel
    integer,dimension(3),parameter :: rgb_blue =  [40, 40, 255], &
                                &   rgb_green = [0, 178, 0]
    !.................................
    
    !first, add ideal mixing activity curve data for comparison:
    !all arguments, including optional ones, and their meaning
    !are outlined near the top of subroutine 'add_plot_xydata'
    !in module Mod_Dislin_plots.
    call add_plot_xydata(xv=x_org_vec, yv=(1.0_wp - x_org_vec), &
        & ltext='ideal, $a_w = x_w$', pen_wid=2.0_wp, &
        & lstyle='dotted', plot_symb='curve')
    call add_plot_xydata(xv=x_org_vec, yv=x_org_vec, &
        & ltext='ideal, $a_{\rm org} = x_{\rm org}$', &
        & pen_wid=2.0_wp, lstyle='dashed', plot_symb='curve')
    
    !second, add non-ideal water and organic activity curves:
    call add_plot_xydata(xv=x_org_vec, yv=activities(:,1), &
        & ltext='water activity, $a_w$', pen_wid=8.0_wp, &
        & rgb_col=rgb_blue, lstyle='solid', plot_symb='curve', &
        & symb_id=15)
    call add_plot_xydata(xv=x_org_vec, yv=activities(:,2), &
        & ltext='organic activity, $a_{\rm org}$', &
        & pen_wid=8.0_wp, rgb_col=rgb_green, lstyle='solid', &
        & plot_symb='curve')
    
    !set overall plot properties and generate Dislin plot:
    xlabel = 'mole fraction of organic, $x_{\rm org}$'
    ylabel = 'activity'
    call dislin_plot(xlabel, ylabel, yaxis_mod=0.67_wp, &
        & legend_position=3, metafile='pdf', &
        & out_file_name='example1_activity_curves')
end block
```

Listing 3.10 Example 1 subroutine (part 4); code for multicurve plot using Mod_Dislin_plots (excerpt from file chap3_alg3_BAT_Example_1.f90).

In the call to subroutine add_plot_xydata, we transfer the curve properties via name association, since, except for the first three arguments, all other arguments are declared as optional and if missing, default values will be assigned. One more time, we can compile and link using the command line shown in Listing 3.11, which now also includes the source file chap3_alg5_Mod_Dislin_plots.f90.

```
gfortran -o Prog_BAT_Example1.out -O3
    chap3_alg1_Mod_NumPrec_types.f90 chap3_alg8_Mod_MINPACK.f90
    chap3_alg5_Mod_Dislin_plots.f90 chap3_alg2_Mod_BAT_model.f90
    chap3_alg3_BAT_Example_1.f90 chap3_alg4a_Prog_BAT_Example1.f90
    -ldislin_d -I/usr/local/dislin/gf/real64
```

**Listing 3.11** Command line for Example 1 with code section (1.6), including module Mod_Dislin_plots (gfortran invoked from a Linux terminal).

After program execution, we will find the generated plot in file example1_activity_curves.pdf; see a slightly modified version of that plot in Figure 3.2. This figure shows four curves for both the ideal, mole-fraction-based activities and the nonideal BAT-calculated activities. It is an improvement over the quick plots, yet only required one subroutine call per curve plus one for the final generation of the plot.

The direct comparison of ideal and BAT-calculated activity curves in Figure 3.2 indicates more clearly that such a binary system of water + pinic acid exhibits moderately nonideal behavior, especially at mole fractions of pinic acid of less than 0.3. How would the activity curves deviate from the ideal mixing case if the oxygen-to-carbon ratio of the organic were higher or lower than that of pinic acid (which is $4/9 \approx 0.444$)?

Given that the O:C ratio is one of the few inputs of the BAT model, we can explore such questions for a range of hypothetical organic molecules. For example, in subroutine BAT_example_1, we could change the O:C ratio (on line 19 of Listing 3.4) to $5/9$ or to $3/9$ in order to explore the influence of this modification on the BAT predictions, while keeping the molar mass and the H:C ratio at the set values for pinic acid. Doing so, we will also recalculate the corresponding density estimation based on the modified O:C (line 21 of Listing 3.4). Thus, our modified inputs will differ both in the organic component's O:C and density values from those of pinic acid. Figure 3.3 shows the outcome of this calculation experiment in terms of component activity curves and their level of deviation from the ideal mixture case. Using our Example 1 program, you are invited to further explore the impacts of changes to input variables on the resulting degree of mixture nonideality.

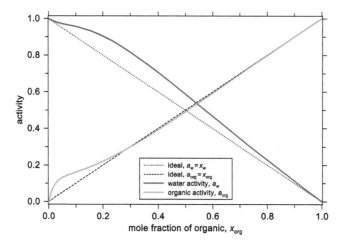

**Figure 3.2** Multicurve Dislin plot from code section (1.6) of subroutine BAT_Example_1 featuring the use of different curve attributes. The BAT (light) model predictions are for a binary aqueous organic solution with pinic acid as the organic component.

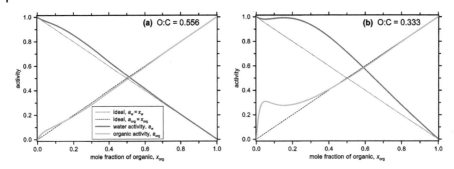

**Figure 3.3** BAT model predictions for the water and organic activities as in Figure 3.2, but using modified organic component property inputs. (a) A higher O:C ratio of 5/9, $\rho_o \approx 1.252$ g cm$^{-3}$; (b) a lower O:C ratio of 3/9, $\rho_o \approx 1.164$ g cm$^{-3}$.

Comparison of panels (a) and (b) of Figure 3.3 suggests that a decrease in O:C ratio, which is here accompanied by a decrease in estimated density, leads to a larger deviation from ideal mixing. In fact, Figure 3.3(b) shows that with an O:C ratio of 0.333, the activity curves exhibit a local maximum and a local minimum in the composition range of $0.02 < x_{\text{org}} < 0.2$. Such local activity curve extrema are a clear signature of the existence of a miscibility gap. In other words, that system will undergo LLPS in a certain composition range. In that phase separated range, the shown activity curves only represent the activities of a metastable or unstable single-phase solution, not those pertaining to the stable equilibrium state. In the following section, we will explore the detection and computation of the extent of LLPS.

## 3.4 Phase Separation

LLPS, also known as liquid–liquid equilibrium (LLE), describes the state of a liquid system that shows two (or more) distinct liquid phases under equilibrium conditions. A common example is that of attempting to mix oil and water in a container, say when mixing a salad dressing containing olive oil and vinegar (which is water-rich). Despite vigorous stirring, after you stop, an oil-rich and a water-rich phase will emerge, showing that those two liquids are virtually immiscible. In other cases, two or more substances may be miscible in certain proportions, but phase separate in a part of the possible mixture composition space. This can occur frequently in atmospheric aerosol particles containing both dissolved inorganic salts and acids and organic compounds of various polarities. Typically, such particles will exhibit LLPS into an aqueous phase rich in ions, water, and highly oxidized (polar) organics and a lower polarity, organic-rich phase [e.g., 13–15]. When different organic compounds of sufficiently distinct polarities are present in a particle, even three liquid phases may coexist [16].

One consequence of LLPS is that it may substantially impact the gas–particle partitioning of semivolatile organics and water compared to assuming the particles to consist of a single liquid phase acting as the absorptive medium for VLE partitioning [e.g., 6, 17]. Phase separation may also impact an aerosol particle's hygroscopic growth and cloud droplet activation properties [e.g., 18–20].

You might wonder why the local extrema shown by the activity curves in Figure 3.3(b) indicate that an LLPS will result when water and the organic are mixed in certain proportions. Let us look first at another example and then discuss the theoretical basis for the link between activity curve shape and LLPS.

### 3.4.1 Example 2: Detection and Computation of LLPS in a Binary System

Using the BAT model, we will focus on Example 2, working with a binary water + 1-pentanol system. The details of our BAT Example 2 implementation are shown in subroutine `BAT_example_2` of file chap3_alg6_BAT_Example_2.f90. Code sections/steps (2.1)–(2.3) (see comments in chap3_alg6_BAT_Example_2.f90) are analogous to the code of BAT example 1, except for changes to the organic compound properties to represent 1-pentanol, as shown in Listing 3.12.

```
!## Example 2 -- consideration of LLPS ##
!(2.1): set input parameters for the organic component,
!        1-pentanol, C5H12O, SMILES: CCCCCO
M_org       = 88.15_wp                        ![g/mol], molar mass
OtoC        = 1.0_wp/5.0_wp                    ![-], O:C ratio
HtoC        = 12.0_wp/5.0_wp
density_org = density_est(M_org, OtoC, HtoC)  ![g/cm^3]

!(2.2): allocate and initialize input and output arrays to cover
!       composition range with a few points, as in example 1.
np = 1001                                      !no. of points
xinc = 1.0_wp/real(np-1, kind=wp)             !the x_org increment
allocate( x_org_vec(np), ln_actcoeff(np,2), activities(np,2) )
x_org_vec = [(i*xinc, i = 0,np-1)]

!(2.3): call the BAT model to compute activities, looping over x_org
!       points:
do i = 1,np
    call BAT_light(x_org_vec(i), M_org, OtoC, density_org, &
    & ln_actcoeff(i,:), activities(i,:))
enddo
```

Listing 3.12 Example 2 subroutine (part 1); computation of BAT predictions for a binary water + 1-pentanol system (excerpt from file chap3_alg6_BAT_Example_2.f90).

Next, we plot the activity curves for this binary system (Step (2.4)) using the subroutines from `Mod_Dislin_plots`. For generating a smooth curve in Figure 3.4, we have temporarily set the number of BAT evaluation points (for $x_{org}$ within interval [0.0, 1.0]) to np = 1001.

Figure 3.4 shows that the predicted activities deviate considerably from ideal mixing behavior. Both the water and 1-pentanol activity curves show values exceeding 1.0, each in a certain range of $x_{org}$. Since a pure component would have an activity of precisely 1.0, exceeding that value means that the system would have a lower Gibbs energy if the corresponding component were to form a pure phase in that mixture composition range, leaving a second, mixed phase of a different composition. Therefore, it is obvious that this system will show a miscibility gap; however, it is not readily clear what the limits of the miscibility gap in terms of an $x_{org}$ interval are. Furthermore, while predicting solution activities above 1.0 is a clear sign that LLPS would occur in the real system

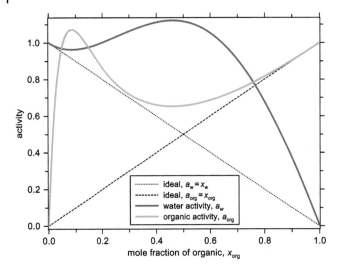

**Figure 3.4** Activity curves predicted by BAT (light) for the binary water + 1-pentanol system. Note that the local maximum and minimum on each curve indicate that LLPS should occur in this system.

in some composition or water activity range (assuming that the model is sufficiently accurate), it is not a necessary condition for LLPS. For example, see Figure 3.3(b) for a case where activities do not exceed 1.0, yet we claim that LLPS will occur (a claim we could put to the test later numerically).

The compilation command line to reproduce the calculations and plots from example 2 is shown in Listing 3.13 (for the gfortran compiler). In file chap3_alg4_Prog_BAT_Examples.f90, set `example_no = 2` on line 20 in order to select example 2 during the program execution. Then run the program using command `./Prog_BAT.out`. The ifort compiler commands under a Linux environment are similar (if Intel's oneAPI ifort is installed). Also, if necessary, modify the path for the Dislin library on your system.

```
gfortran -o Prog_BAT.out -O3  chap3_alg1_Mod_NumPrec_types.f90
    chap3_alg7_Mod_NumMethods.f90 chap3_alg8_Mod_MINPACK.f90
    chap3_alg2_Mod_BAT_model.f90 chap3_alg5_Mod_Dislin_plots.f90
    chap3_alg3_BAT_Example_1.f90 chap3_alg6_BAT_Example_2.f90
    chap3_alg4_Prog_BAT_Examples.f90  -ldislin_d
    -I/usr/local/dislin/gf/real64
```

**Listing 3.13** Command line for Example 2 when using gfortran from a Linux terminal.

When instead using the provided Visual Studio solution/project with the Intel® Fortran compiler on Windows from within Visual Studio, simply select `example_no = 2` in `Prog_BAT_Examples` prior to building and running the project. Alternatively, you could also compile and link the program from a dedicated "Intel oneAPI command prompt for Intel 64 Visual Studio" (on Windows) using the command shown in Listing 3.14 (see also the provided file chap3_alg4c_Example2_command_line.txt for instructions). Then, the program can be run by typing ./Prog_BAT.exe.

```
ifort /o Prog_BAT.exe /O3  chap3_alg1_Mod_NumPrec_types.f90
    chap3_alg7_Mod_NumMethods.f90 chap3_alg8_Mod_MINPACK.f90
    chap3_alg2_Mod_BAT_model.f90 chap3_alg5_Mod_Dislin_plots.f90
    chap3_alg3_BAT_Example_1.f90 chap3_alg6_BAT_Example_2.f90
    chap3_alg4_Prog_BAT_Examples.f90  /I"c:\dislin\ifc\real64"
    c:\dislin\disifl_d.lib c:\dislin\disifd_d.lib user32.lib
    gdi32.lib
```

**Listing 3.14** Command line for Example 2 when using Intel's ifort compiler on Windows from a dedicated "Intel oneAPI command prompt for Intel 64 Visual Studio."

Back to the system shown in Figure 3.4: how can we determine the phase separation region of a binary system? Before we discuss the specifics of a method to do so, let us generate an additional plot of the (normalized) molar Gibbs energy of mixing of our system. This is based on code sections (2.5 and 2.6) in subroutine BAT_example_2. Since we are interested in the nonideal mixing effects and the phase separation region, all we need to know is the change in the Gibbs energy due to mixing; here normalized by $R\,T$ and the total molar amounts $n = \sum n_j$ of components in the system, resulting in a dimensionless quantity. Since the standard chemical potentials (see Equation (3.6)) are unaffected by mixing, they constitute a constant Gibbs energy contribution to the system, making them irrelevant in the context of understanding the effects of mixing. The relevant expression for a single liquid phase containing water and organics is given by

$$\frac{\Delta G_{\mathrm{mix}}}{nRT} = \sum_j x_j \ln\left(a_j\right), \tag{3.20}$$

where the sum over $j$ covers all system components and $x_j = n_j/n$. See the code snippet in Listing 3.15 for the binary mixture case.

```
!(2.5)  Compute normalized molar Gibbs energy of mixing from
!   the computed activity data.
allocate( delta_Gmix_by_nRT(np), delta_Gmix_ideal_by_nRT(np) )
where (x_org_vec(:) > 0.0_wp .AND. (1.0_wp - x_org_vec(:)) &
& > 0.0_wp)
    delta_Gmix_by_nRT = &
        & (1.0_wp - x_org_vec)*log(activities(:,1)) &
        & + x_org_vec*log(activities(:,2))
    !ideal mixing case for comparison:
    delta_Gmix_ideal_by_nRT = &
        & (1.0_wp - x_org_vec)*log((1.0_wp - x_org_vec)) &
        & + x_org_vec*log(x_org_vec)
elsewhere
    delta_Gmix_by_nRT = 0.0_wp
    delta_Gmix_ideal_by_nRT = 0.0_wp
endwhere
```

**Listing 3.15** Example 2 subroutine (part 2); code for the calculation of the normalized Gibbs energy change due to mixing (excerpt from file chap3_alg6_BAT_Example_2.f90).

The where–elsewhere–endwhere construct in that listing allows for concurrent (vectorized) processing of the tested conditions operating over the whole array or array slices (since the shapes of the array slices are compatible). Figure 3.5 shows the normalized Gibbs energy change due to mixing as generated by the code in Steps (2.5) and (2.6)

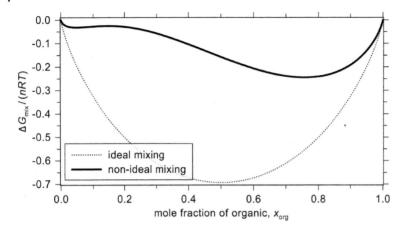

**Figure 3.5** Predicted normalized Gibbs energy change due to mixing in the binary water + 1-pentanol system (single-phase case).

of subroutine `BAT_Example_2` based on the activity and composition data shown in Figure 3.4.

The curvature of $\Delta G_{mix}/(nRT)$ for the nonideal mixing case differs in an important way from that of an ideal mixture: the former curve has two inflection points, leading to a bulge in the curve toward higher values. The resulting curvature means that there exist two points on the curve which share a common tangent, along which the normalized Gibbs energy would be lower than that of the single-phase curve bulging above it (see also Figure 3.6). The same construction cannot be done along the ideal mixing curve. This "common tangent" geometric aspect of the nonideal mixing curve in Figure 3.5 is a well-known necessary (and sufficient) condition for LLPS to occur [21, 22]. The contact points on the curve sharing the common tangent indicate the range of phase separation. In an LLPS state, phase $\alpha$ will have composition $x_{org}^{\alpha}$, and phase $\beta$ composition $x_{org}^{\beta}$. Associated with those compositions are the Gibbs energy of mixing contributions from phases $\alpha$ and $\beta$. From Equation (3.20) we can derive

$$\frac{\Delta G_{mix}}{RT} = \sum_j n_j^{\alpha} \ln (a_j^{\alpha}) + n_j^{\beta} \ln (a_j^{\beta}). \tag{3.21}$$

Introducing $f^{\alpha}$, the fraction of the cumulative molar amounts residing in phase $\alpha$, $f^{\alpha} = \sum_j n_j^{\alpha}/n$, we can express the normalized Gibbs energy of mixing as

$$\frac{\Delta G_{mix}}{nRT} = f^{\alpha} \cdot \sum_j \left[ x_j^{\alpha} \ln(a_j^{\alpha}) \right] + (1 - f^{\alpha}) \cdot \sum_j \left[ x_j^{\beta} \ln(a_j^{\beta}) \right]; \tag{3.22}$$

that is,

$$\frac{\Delta G_{mix}}{nRT} = f^{\alpha} \cdot \frac{\Delta G_{mix}^{\alpha}}{nRT} + (1 - f^{\alpha}) \cdot \frac{\Delta G_{mix}^{\beta}}{nRT}. \tag{3.23}$$

Considering the fact that at equilibrium $a_j^{\alpha} = a_j^{\beta}, \forall j$ (see Section 3.4.1.1), within the miscibility gap, Equation (3.22) can be expressed equivalently by

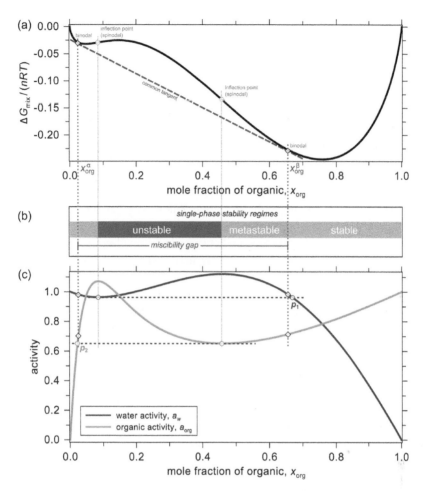

**Figure 3.6** Predicted binary water + 1-pentanol mixing properties. (a) Common tangent (dashed red) to the normalized Gibbs energy of mixing curve (black), revealing the liquid–liquid equilibrium composition range, which is denoted by $x^\alpha_{org}$, $x^\beta_{org}$. (b) Characterization of single-phase stability regimes: a single (fully mixed) phase is stable in the gray ranges, metastable in the pale red ranges, and intrinsically unstable in the red range. (c) Schematic construction of an approximation of the miscibility gap by points $p_1$, $p_2$ (they serve as an initial guess). Points $p_1$, $p_2$ are determined via the local minima on the activity curves.

$$\frac{\Delta G^{LLE}_{mix}}{nRT} = \sum_j \ln(a^{LLE}_j) \left[ f^\alpha x^\alpha_j + (1 - f^\alpha) x^\beta_j \right]. \tag{3.24}$$

The value of $f^\alpha$ varies with the overall input composition. In the binary system case, for $x^{init}_{org} = x^\alpha_{org}$, $f^\alpha = 1.0$, while for $x^{init}_{org} = x^\beta_{org}$, $f^\alpha = 0.0$. Also, $f^\beta = 1 - f^\alpha$.

What is the interpretation of Equations (3.22)–(3.24)? These equations state that in the LLPS case the $\Delta G_{mix}/(nRT)$ value of the stable equilibrium configuration depends on a weighted linear combination of the $\Delta G_{mix}/(nRT)$ values at the (binodal) limits of phase separation. Graphically, this linear scaling is represented by the common tangent "tunneling" the miscibility gap. It is relatively easy to grasp this geometrically for a binary system (see Figure 3.6(a)); analogously, a tangent (hyper)plane criterion also applies to more complex multicomponent systems, just in a higher dimensional composition space. Beyond three dimensions, this will be difficult to graphically visualize and imagine. The higher dimensional parameter space of a multicomponent mixture is a key reason why computing the most stable LLPS state is a hard problem [23]. It is far less daunting to work with the binary mixture case, which is what we do in our Example 2.

Considering our binary water + 1-pentanol system, if we choose an initial mixture composition, $x_{\text{org}}^t$, that falls within the miscibility gap, for example, $x_{\text{org}}^{\text{init}} = 0.3$, the actual normalized Gibbs energy of mixing (after LLPS has occurred) will be represented by the point on the common tangent line for $x_{\text{org}}^{\text{init}} = 0.3$. The corresponding $G_{\text{mix}}/(nRT)$ value of about −0.118 is clearly smaller than the corresponding value ($\sim$ −0.057) on the solid black curve representing the single-phase solution. This means that the LLPS state is favorable and, thus, the equilibrium state of the system.

Figure 3.6 further illustrates the relationship between the normalized $\Delta G_{\text{mix}}$ curve and the mixture activities. Equation (3.20) already indicated a direct relationship. Figure 3.6(c) also suggests that the local maximum on one of the activity curves is perfectly aligned with the local minimum on the other curve in terms of $x_{\text{org}}$. This is not a coincidence. One can confirm that this alignment is a consequence of the BAT model equations (Equations (3.11) and (3.12)). In fact, any binary activity coefficient model that fulfills the Gibbs–Duhem relation will have this property. Because of this alignment of extrema, it also follows that the inflection points on the normalized $\Delta G_{\text{mix}}$ curve (Figure 3.6(a)) occur at the same $x_{\text{org}}$ coordinates as the local extrema on the activity curves. Those inflection points mark the so-called spinodal limits of intrinsic (single-phase) stability [e.g., 5]. This means that in the $x_{\text{org}}$ range between the inflection points, an initially well-mixed solution will spontaneously undergo phase separation (see Figure 3.6(b)). Given these geometrical relationships, in case of a binary system, one only needs to look at one of the component activity curves to judge whether LLPS may occur. When the curve exhibits a local minimum/maximum for any $x_{\text{org}}$ (where $0.0 < x_{\text{org}} < 1.0$), LLPS will certainly occur in some range (the extent of this range remains to be determined), otherwise LLPS will be impossible.

### 3.4.1.1 Isoactivity Condition

There is one additional equilibrium condition that will prove useful in devising a method for determining the exact extent of the phase separation composition range. In Section 3.1.1.1, we had discussed that at thermodynamic equilibrium the Gibbs energy of a system attains a global minimum and also that the chemical potential of any individual species must be equivalent across the coexisting phases. That is, $\mu_j^\alpha = \mu_j^\beta$, for a two-phase equilibrium (Equation (3.4)). Using the definition of the chemical potential (Equation (3.6)), at LLE between liquid phases $\alpha$ and $\beta$, this means

$$\mu_j^{\alpha,\ominus} + RT \ln\left(a_j^\alpha\right) = \mu_j^{\beta,\ominus} + RT \ln\left(a_j^\beta\right). \tag{3.25}$$

Without loss of generality, we can apply the same standard state to both phases (e.g., the chemical potential of the pure component at temperature $T$ and pressure $p$), such that $\mu_j^{\alpha,\ominus} = \mu_j^{\beta,\ominus}$. It then follows from Equation (3.25) that $a_j^\alpha = a_j^\beta$ at equilibrium. This is known as the "isoactivity condition." In turn, if an activity curve has more than one point with a certain activity value, for example, refer to the horizontal line for point $p_1$ in Figure 3.6(c), it must mean that the mixture exhibits phase separation at least within the composition range where such isoactivity tie-lines are possible. For mole-fraction-based activities, the isoactivity condition can be expressed as $x_j^\alpha \gamma_j^\alpha = x_j^\beta \gamma_j^\beta$, which implies the following equivalence at LLE:

$$\frac{x_j^\alpha}{x_j^\beta} = \frac{\gamma_j^\beta}{\gamma_j^\alpha}. \tag{3.26}$$

Here, the left-hand side represents a composition ratio, sometimes denoted as the equilibrium partitioning constant of component $j$ between the two phases. In the case of a binary mixture at constant temperature, as discussed in Example 2, the two LLE compositions, $x_{org}^\alpha$ and $x_{org}^\beta$, represent fixed compositions, namely the end points of the miscibility gap (=points on binodal). If we generate an initial mixture of mole fraction $x_{org}^{init}$ located anywhere between those two mole fraction limits (i.e., inside the miscibility gap), after undergoing phase separation, the system will be split into two phases. One will have composition $x_{org}^\alpha$ and the other $x_{org}^\beta$; refer to the red labels in Figure 3.6(a). Importantly, whether $x_{org}^{init}$ was chosen closer to $x_{org}^\alpha$ or $x_{org}^\beta$ does not affect the LLE phase compositions, it only affects the absolute component amounts in each phase, that is, whether phase $\alpha$ is bigger or smaller than phase $\beta$, which we have expressed by $f^\alpha$ (see Equation (3.23)). Note that this is only generally true for binary systems because the phase separation occurs along a single tie-line. In multicomponent systems, the initial (overall) mixture composition, when chosen within the miscibility gap, will impact both the phase size ratio as well as the compositions of the stable phases at LLE [e.g., 23].

### 3.4.1.2 Generating a Good Initial Guess for LLPS

At this point, we are ready to focus on the computation of the phase separation limits in a binary solution. Many methods have been developed for phase separation computations in multicomponent systems, because each approach is accompanied by advantages and drawbacks in terms of reliability and computational efficiency [e.g., 23, 24]. The existence of undesired trivial solutions—and risk of convergence to those solutions—make efficient phase separation computations often a challenging numerical problem. The situation is not quite as intricate when constrained to the case of a binary system. In the following, we will outline and implement an approach for binary systems that is almost always successful and reasonably efficient.

Generally, computing the phase compositions of coexisting phases in an LLE case follows one of the two approaches: (1) solving directly the global minimization of the Gibbs energy of mixing function, or (2) solving a system of nonlinear equations (derived from the isoactivity condition) [24]. We will implement a method based on the latter approach in our BAT Example 2. In either approach, due to the nonlinear nature of the problem, it is highly advantageous to first generate a good initial guess of the two LLE phase compositions. This is key, because otherwise there is a good chance that our calculation will converge to a trivial solution, one where the two phases in equilibrium are of exactly same composition (which fulfills the isoactivity condition). On the flip side, if we are able to generate an initial guess close enough to the final solution of the LLE problem, our chances for convergence to the desired nontrivial solution are excellent (but not strictly guaranteed).

Figure 3.6(c) outlines schematically the construction of an approximation of the LLPS composition range by means of the $x_{org}$ coordinates of points $p_1$ and $p_2$. This method is motivated by the approach introduced by Gorkowski et al. [7] for estimating the phase separation region and the water activity of phase separation in binary aqueous systems. Here, we will use this approach to determine a good initial guess, followed by solving the LLE problem numerically to sufficiently high accuracy.

---

**Outline for computing the LLE state of a binary system (as implemented in subroutine** `BAT_example_2`**):**

1. Characterize the activity curves by computing a few points over the full $x_{org}$ range (Step 2.2 in subroutine `BAT_example_2`).
2. Determine the approximate coordinates of local minima/maxima from the array of activity "curve" data—or absence of local extrema (Step (2.7)). If phase separation is not detected, skip the remaining steps (we are done).
3. If phase separation is detected, numerically resolve the locations of the two local minima.
4. Numerically estimate the $x_{org}$ values of points $p_1$, $p_2$, located on the opposite sides of the local maxima of the activity curves (step 2.8); see Figure 3.6.
5. Use $x_{org}(p_1)$, $x_{org}(p_2)$ to generate an initial guess of the equilibrium partitioning constant; then solve a system of nonlinear equations by improving $x_{org}^{\alpha}$, $x_{org}^{\beta}$ under consistent molar balance constraints attempting to fulfill the isoactivity condition (Step 2.9 and separate subroutines).

---

In the following, we will look at snippets of the Fortran code associated with the outlined steps. You will find the complete code and further details in the provided .f90 files.

When computational efficiency is of concern, one needs to find a good balance between the total number of `BAT_light` subroutine calls necessary to solve the LLPS problem, while managing the risk of missing detection of an LLPS case due to insufficient resolution of the activity curves. Therefore, in Step (2.2) of subroutine `BAT_example_2`, for the initial characterization of the activity curves, we suggest to usually use somewhere between 10 and 30 points. For our example, we set `np = 15` on line 52 in file chap3_alg6_BAT_Example_2.f90, meaning 15 initial calls of `BAT_light`. The activity "curve" data are then only known at the symbol positions shown in Figure 3.7. If we were to use very few points, say 4, it would become likely that we will miss a present local minimum or maximum, especially if the LLPS region is small (which depends on the organic/system considered).

Listing 3.16 shows the code for determining the array indices of the local minimum and/or maximum on each activity curve based on only the low-resolution activity data. We simply compare the activity of each point to its two immediate neighbors (since our array data is already ordered from low to high corresponding $x_{org}$). Because of the use of low-resolution data, we carry out this detection process for both curves. Also, because of the functional form of the BAT model, we know that only a single local min/max may be present on a curve (note that other binary activity coefficient models may potentially allow more, requiring further considerations). Once we find both extrema, the search can be stopped (listing code line 23).

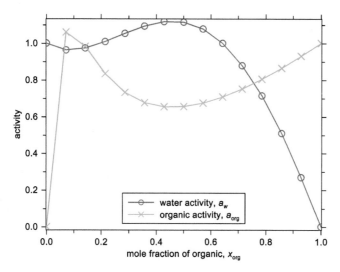

**Figure 3.7** BAT-predicted activity data evaluated for `np` = 15 points equally spaced over the full $x_{org}$ range (water + 1-pentanol system).

```
!(2.7) find a local minimum and a local maximum of the activity
!   curves -- if bracketing intervals are present.
!   first determine approximate index locations of extrema
!   for both water and organic activity curves:
phase_sep_detected = .false.
imin = 0
imax = 0
do k = 1,2     !loop over the two mixture components
    do i = 2,np-1
        if ( activities(i,k) < activities(i-1,k) ) then
            if ( activities(i,k) < activities(i+1,k) ) then
                !found a local minimum
                imin(k) = i
                phase_sep_detected = .true.
            endif
        else
            if ( activities(i,k) > activities(i+1,k) ) then
                !found a local maximum
                imax(k) = i
                phase_sep_detected = .true.
            endif
        endif
        if (imin(k) > 0 .AND. imax(k) > 0) then
            exit
        endif
    enddo
enddo !k
!Note: if any local extremum is found for any curve, then both
!a local max and min must be present.  Due to the functional
!relation between aw and aorg, we know that the x-coordinate of
!a local min of one curve is the same as the local max x_org
!for the other curve and analogous for a local maximum.
!Because limited curve resolution was used, make sure that all
!indices are correctly set when local extrema are present:
if (phase_sep_detected) then
    !possibly adjust based on local min/max from other curve:
    if (imin(1) /= imax(2)) then
        imin(1) = max(imin(1), imax(2))
```

```
            imax(2) = imin(1)
    endif
    if (imin(2) /= imax(1)) then
        imin(2) = max(imin(2), imax(1))
        imax(1) = imin(2)
    endif
    !also consider a local min/max being near the activity =
    !1.0 end point (one index off) due to low resolution of
    !curve (missing extrema) and knowing that x_org_vec values
    !increase with increasing index number.
    if (imin(1) == 0 .AND. imax(2) == 0) then
        imin(1) = 2
        imax(2) = 2
    endif
    if (imin(2) == 0 .AND. imax(1) == 0) then
        imin(2) = np-1
        imax(1) = np-1
    endif
endif
```

**Listing 3.16** Example 2 subroutine (part 3); determine array indices associated with local extrema from low-resolution activity data array (excerpt from file chap3_alg6_BAT_Example_2.f90).

As noted in the comments on lines 28–34 of Listing 3.16, in the case of a detected phase separation (recorded by `phase_sep_detected = .true.`), we make sure (listing code lines 35–57) that the array indices of both the local minima (`imin(1:2)`) and maxima (`imax(1:2)`) are correctly determined for both curves.

Next, we numerically solve for the $x_{org}$ values of the local minima on both activity curves as shown in Listing 3.17. Achieving a moderate precision (`tol_init`, see listing line 2) is sufficient here, because this is only serving our construction of a "good" initial guess for the LLPS solver. However, it is not advised to completely skip this step. When using a low initial curve resolution in Step (2.2), our present guess for the minima (`x_org_vec(imin(k))`) could be rather poor, making this numerical minimum detection necessary.

```
if (phase_sep_detected) then
    tol_init = max( 1.0E-4_wp, sqrt(epsilon(1.0_wp)) )
    write(*,'(A,ES12.5)') 'set tolerance for local_min: ', tol_init
    !set general BAT wrapper function options via elements
    !of custom derived type fopt:
    fopt%M_org = M_org
    fopt%OtoC = OtoC
    fopt%density_org = density_org
    do k = 1,2
        !determine local minimum within interval [ax, bx]:
        ax = x_org_vec(imin(k) -1)
        bx = x_org_vec(imin(k) +1)
        !set fopt type components for finding a local minimum
        !in water (k=1) or organic (k=2) activity:
        fopt%determine_local_max = .false.
        fopt%component_no = k
        !use local_min method:
        flocmin = local_min(ax, bx, tol_init, fnc_BAT_activity, &
            & fopt, x_org_min(k), nfcalls)
        !save activity value at determined local min:
        activity_locmin(k) = flocmin
        write(*,'(A,I0,A,ES13.6,A,ES13.6)') 'found local min &
            & of activity curve at x_org_min(',k,') = ', &
```

```
              & x_org_min(k), ', flocmin = ', flocmin
         write(*,'(A,I0,/)') 'number of BAT_light calls during &
              & local min search = ', nfcalls
    enddo  !k
```

**Listing 3.17** Example 2 subroutine (part 4); use function `local_min` from module `Mod_NumMethods` to solve for the precise $x_{org}$ values of the local minima within the set numerical tolerance (excerpt from file chap3_alg6_BAT_Example_2.f90).

The function `local_min` seeks a local minimum of a user-supplied function of the form `fnc(x, fopt)` in an *x*-value interval $[a, b]$. In our case, we have written a wrapper function, `fnc_BAT_activity(x_org, fopt)` that matches the function interface required by `local_min`. Function `fnc_BAT_activity` simply calls the subroutine `BAT_light` with a certain value of `x_org`, while setting all the other `BAT_light` interface arguments based on those stored in a customized derived type variable `fopt`; see the implementation in `Mod_BAT_model`. The implementation details of function `local_min` are provided in module `Mod_NumMethods` (file chap3_alg7_Mod_NumMethods.f90). This numerical minimum-finding method uses a combination of the "successive inverse parabolic interpolation" and "golden section search" approaches. It does not require derivative information from the targeted function and offers superlinear convergence. Our implementation is a slightly modified version based on the Fortran 90 version by John Burkardt (https://people.sc.fsu.edu/~jburkardt/f_src/), which is based on the algorithm and original FORTRAN 77 implementation by Richard Brent [25]. Our modifications include the use of set real(kind = wp) working precision and the optional passing of additional function parameters via derived type `foptions` declared in module `Mod_NumPrec_Types` (i.e., passing variable fopt in our example). While other numerical methods could be used, this approach works well and determines the minimum to set precision typically within 5–8 `BAT_light` subroutine calls. As a result, we obtain the $x_{org}$ coordinate of the minimum in the *k*th activity curve (saved as `x_org_min(k)`) as well as the activity at the minimum (saved as activity_locmin(k); line 21 of Listing 3.17).

In Step (2.8) of subroutine `BAT_Example_2`, we determine the $x_{org}$ coordinates of the points $p_1$ and $p_2$ shown in Figure 3.6. This is achieved by finding the approximate intersection on a component's activity curve where "activity" ≈ "activity_locmin(k)" on the opposite side of the local maximum of the curve (relative to the minimum location). See details in the code snippet of Listing 3.18.

```
!(2.8) Determine approximate x_org coordinates of the points p1 and
    p2.
do k = 1,2
    !determine approx. array index value (iloc) of the p1
    !or p2 point:
    if (imin(k) <= imax(k)) then
        iloc1 = imax(k)-1 &
            & + minloc( abs(activities(imax(k):np,k) &
            & -activity_locmin(k)), DIM = 1 )
    else
        iloc1 = minloc( abs(activities(1:imax(k),k) &
            & -activity_locmin(k)), DIM = 1 )
    endif

    !determine curve data interval indices [iloc1, iloc2]
```

```
!to bracket the point:
if (iloc1 > 1 .AND. iloc1 < np) then
    if (activities(iloc1,k) > activity_locmin(k)) then
        if (activities(iloc1+1, k) < activity_locmin(k)) then
            iloc2 = iloc1 +1
        else
            iloc2 = iloc1 -1
        endif
    else
        if (activities(iloc1+1, k) > activity_locmin(k)) then
            iloc2 = iloc1 +1
        else
            iloc2 = iloc1 -1
        endif
    endif
else
    !iloc1 is first or last index, so iloc2 must be
    !the only neighbor
    if (iloc1 == 1) then
        iloc2 = 2
    else
        iloc2 = np-1
    endif
endif
write(*,'(A,I0,A,ES12.5,/)') 'approx x_org of p',k,' &
    & point is: ', 0.5_wp*( x_org_vec(iloc1) + &
    & x_org_vec(iloc2) )

!use Ridders' method to determine the x_org value of the
!p1 or p2 points (within set tolerance):
tol_init = max( 1.0E-3_wp, sqrt(epsilon(1.0_wp)) )
fopt%component_no = k
fopt%refval = activity_locmin(k)
fnc_loc1 = (activities(iloc1,k) / activity_locmin(k)) -1.0_wp
fnc_loc2 = (activities(iloc2,k) / activity_locmin(k)) -1.0_wp

x_org_p1p2(k) = Ridders_zero(fnc_BAT_act_dev, fopt, &
    & x_org_vec(iloc1), x_org_vec(iloc2), fnc_loc1, &
    & fnc_loc2, tol_init, nfcalls)

!for debugging output to screen:
write(*,'(A,I0,A,ES12.5)') 'determined x_org of p',k,' &
    & point is: ', x_org_p1p2(k)
write(*,'(A,ES12.5)') 'relative deviation from targeted &
    & activity: ', fnc_BAT_act_dev(x_org_p1p2(k), fopt)
write(*,'(A,I0,/)') 'number of BAT_light function calls &
    & during Ridders_zero = ', nfcalls
enddo !k
```

**Listing 3.18** Example 2 subroutine (part 5); use function `local_min` from module `Mod_NumMethods` to solve for the precise $x_{org}$ values of the local minima within the set numerical tolerance (excerpt from file chap3_alg6_BAT_Example_2.f90).

If we were using a large number of data points to characterize the activity curves (e.g., np > 500), we could simply use the determined, best-matching point `iloc` found from the `minloc` function evaluation of the array data–or perhaps apply linear interpolation. However, when using very few initial curve data points, we typically need to better resolve the location of the $p_1$ or $p_2$ point. As shown in Listing 3.18 (lines 43–53), we use Ridders' method (function `Ridders_zero` from module `Mod_NumMethods`) based on the Fortran version by Press et al. [26]. Ridders' method efficiently computes a root (zero) of a function in one unknown "x" when provided with an input interval $[x_1, x_2]$ in which the function value changes sign. Our implementation of this method allows for passing of optional user-function parameters via variable `fopt`, which we need for setting additional BAT_light parameters and to indicate for which component the activity deviation should be computed. Note, instead one could use a global (public) module variable to pass additional parameters to `fnc_BAT_act_dev` (and BAT_light), yet the use of `fopt` is perhaps more transparent from a code-readability point of view. Similar to the minimum finding problem, we generate a wrapper function `fnc_BAT_act_dev` for our BAT_light-based calculation that complies with the function interface demanded by function `Ridders_zero`. You find function `fnc_BAT_act_dev` in module `Mod_BAT_model`; it returns for any $x_{\text{org}}$ input the relative deviation between the BAT-computed activity and a given reference value (here the value at the corresponding local activity minimum).

The function `Ridders_zero` returns the $x_{\text{org}}$ coordinates of the points $p_1$ and $p_2$, which we store in array `x_org_p1p2`. These two $x_{\text{org}}$ values provide us with the good initial guess we were aiming for.

### 3.4.1.3 Determination of the LLPS Limits

In Step (2.9) of subroutine `BAT_Example_2`, subroutine `BAT_miscibility_gap` is called, which implements a method for the computation of the LLPS composition limits. The goal is to numerically solve for the (binodal) points, $x_{\text{org}}^{\alpha}$, $x_{\text{org}}^{\beta}$, defining the limits of the miscibility gap. The method applied here is based on the theory and notation outlined by Zuend and Seinfeld [23]. Consider an initially completely mixed solution of any number of nondissociating components, each of total molar amount $n_j^t$. Note that Zuend and Seinfeld [23] also consider electrolytes, which we will ignore here. The associated initial (total solution) mole fraction is then given by $x_j^t = n_j^t / \sum_k n_k^t$, where $k$ covers all components. In a case where this mixture will undergo LLPS, the total molar amounts remain conserved quantities and we can define the fraction of $j$ in phase $\alpha$ as $q_j^{\alpha} = n_j^{\alpha} / n_j^t$. Similarly, it follows that $q_j^{\beta} = n_j^{\beta} / n_j^t = 1 - q_j^{\alpha}$. We can use $q_j^{\alpha}$ and $x_j^t$ to express the mole fractions in phases $\alpha$ and $\beta$ by

$$x_j^{\alpha} = \frac{q_j^{\alpha} x_j^t}{\sum_k q_k^{\alpha} x_k^t} \quad \text{and} \quad x_j^{\beta} = \frac{(1 - q_j^{\alpha}) x_j^t}{\sum_k (1 - q_k^{\alpha}) x_k^t}. \tag{3.27}$$

Based on the isoactivity condition (Equation (3.26)), the following equation must hold at LLE for all components [23]:

$$q_j^{\alpha} = \frac{\frac{\gamma_j^{\beta}}{\gamma_j^{\alpha}}}{\omega + \frac{\gamma_j^{\beta}}{\gamma_j^{\alpha}}}. \tag{3.28}$$

Here, $\omega$ is the molar $\beta$-to-$\alpha$ phase size ratio defined by

$$\omega = \frac{\sum_k n_k^{\beta}}{\sum_k n_k^{\alpha}}; \quad \text{also,} \quad \omega = \frac{x_j^{\alpha} - x_j^t}{x_j^t - x_j^{\beta}}. \tag{3.29}$$

The equation on the right shows that $\omega$ is related to the ratio of geometric distances between the initial composition point and the LLE phase composition points bounding the miscibility gap. Moreover, from our definition of $f^{\alpha}$ (Equation (3.21)), it follows that $\omega = (1 - f^{\alpha}) / f^{\alpha}$.

Equations (3.27–3.29) form a coupled system of equations, since $x_j^\alpha$ depends on $q_j^\alpha$, which depends on $\omega$ and the activity coefficient ratio $\gamma_j^\beta/\gamma_j^\alpha$, which we also call the phase "affinity of $j$." This name is appropriate since any ratio $\gamma_j^\beta/\gamma_j^\alpha > 1.0$ indicates that $j$ has a higher affinity for phase $\alpha$, while a value less than 1.0 denotes preference of phase $\beta$. In addition, $\omega$ also depends on $q_j^\alpha$ via $x_j^\alpha$ and $x_j^\beta$. For the general multicomponent system case, solving those coupled equations can be done in various ways, with additional challenges when dissociating electrolyte components are also considered; see discussion by Zuend and Seinfeld [23].

However, in the binary system case, given a good initial guess, we can take a few shortcuts, since we are not primarily interested in finding a consistent set of values for $q_j^\alpha$ and $\omega$. Instead, our goal is to find a set of $q_j^\alpha$ values (and associated $x_j^\alpha, x_j^\beta$) such that the following system of relative activity deviations (one equation per component) is solved within a set tolerance level:

$$\frac{a_j^\alpha - a_j^\beta}{\frac{1}{2}(a_j^\alpha + a_j^\beta)} = 0, \quad \forall j. \tag{3.30}$$

Solving Equation (3.30) will mean that we found a liquid–liquid partitioning solution that fulfills the isoactivity condition—and if it is a nontrivial solution, we have successfully solved the LLPS problem.

Given that $q_j^\alpha$ is constrained to the range $0 \leq q_j^\alpha \leq 1$, we could use the set of $q_j^\alpha$ values directly as the unknowns (solver variables) with a nonlinear equation solver, constrained to find solutions of Equation (3.30) while maintaining $q_j^\alpha$ within the [0, 1] interval. However, in this case, our solver of choice in Fortran is the modified "Powell's hybrid method" from the MINPACK software package developed at Argonne National Laboratory by Moré et al. [27, 28], which is an unconstrained solver (see the description in module `Mod_MINPACK`; file chap3_alg8_Mod_MINPACK.f90). Therefore, we take inspiration from Equation (3.28), noticing that it forms a sigmoidal map between the unconstrained real number line, when represented by $\ln(\gamma_j^\beta/\gamma_j^\alpha)$, and the constrained $0 \leq q_j^\alpha \leq 1$ range [23]. We choose $\omega = 1.0$ as a fixed mapping function parameter (in the following approach). This particular value is a good choice because $\omega = 1.0$ is also consistent with selecting $x_{\mathrm{org}}^t = \frac{1}{2}\left[x_{\mathrm{org}}(p_1) + x_{\mathrm{org}}(p_2)\right]$ from our initial guesses for $x_{\mathrm{org}}^\alpha$ and $x_{\mathrm{org}}^\beta$. $x_{\mathrm{org}}^t$ represents an initial overall solution (input) composition that we estimate to be within the miscibility gap, such that LLPS is expected to occur. Thus, our unknowns to be determined by the solver will be $\ln(F_j)$, where $F_j$ represents a form of affinity of $j$ that will be mapped to $q_j^\alpha$ by function

$$q_j^\alpha = \frac{F_j}{1 + F_j}. \tag{3.31}$$

The outlined LLPS computation approach is implemented by means of two subroutines within module `Mod_BAT_model`. Subroutine `BAT_miscibility_gap` is shown in Listing 3.19.

```fortran
subroutine BAT_miscibility_gap(x_org_p1p2, M_org, OtoC, &
            & density_org, x_org_LLE_limits)

use Mod_MINPACK, only : hybrd1

implicit none
!Interface arguments:
real(wp),dimension(2),intent(in) :: x_org_p1p2
real(wp),intent(in) :: M_org, OtoC, density_org
real(wp),dimension(2),intent(out) :: x_org_LLE_limits
!local variables:
integer :: info
real(wp),parameter :: tol_LLE = max(1.0E-5_wp, &
  & sqrt(epsilon(1.0_wp)))
real(wp) :: sum_qx, xtot_org
real(wp),dimension(2) :: aff, lnaff, fvec, qA
real(wp),dimension(2,2) :: activities, ln_actcoeff
!...............................................

!omega = 1.0 initially (and fixed for sigmoidal mapping)
xtot_org = 0.5_wp*(x_org_p1p2(1) + x_org_p1p2(2))

!set BATopt values to be passed via private module variable
!Mod_BAT_model to the 'calc_BAT_LLE_dev' subroutine:
BATopt%xtot = [1.0_wp - xtot_org, xtot_org]     !x_w, x_org
BATopt%M_org = M_org
BATopt%OtoC = OtoC
BATopt%density_org = density_org

!evaluate BAT at initial guess points for phases alpha 'A'
!and beta 'B';
!activities(:,1) = val. for water and org. of phase A (= 1):
write(*,'(A,2(ES13.6,1X),/)') 'initial guess for xA_org, xB_org = ', &
    x_org_p1p2(2), x_org_p1p2(1)
call BAT_light(x_org_p1p2(1), M_org, OtoC, density_org, &
  & ln_actcoeff(1,:), activities(1,:))
call BAT_light(x_org_p1p2(2), M_org, OtoC, density_org, &
  & ln_actcoeff(2,:), activities(2,:))
ncalls_LLE = 2      !count BAT_light calls

!initialize lnaff(:) as unknown (solver variable):
lnaff = ln_actcoeff(1,:) - ln_actcoeff(2,:)

!use Powell's hybrid method from module Mod_MINPACK to solve
!the system of equations of the LLE isoactivity conditions
!(implemented in subroutine calc_BAT_LLE_dev below), starting
!with the guess for lnaff:
call hybrd1(calc_BAT_LLE_dev, 2, lnaff, fvec, tol_LLE, info)

!for information, report diagnostics and relative deviations
!from isoactivity condition:
write(*,'(A,I0)') 'number of BAT_light calls during LLE calc.: ', &
    ncalls_LLE
write(*,'(A,I0)') 'after hybrd1 solver; info = ', info
write(*,'(A,2(ES13.6,1X))') 'after hybrd1 solver; fvec = ', fvec

if (info < 0) then
    !terminated due to convergence to trivial solution
    !based on test within calc_BAT_LLE_dev
    qA = 0.5_wp
else
    aff(:) = exp(lnaff)
    qA = aff/(1.0_wp + aff)
endif
```

```
!mole fractions in phases alpha and beta from determined qA:
sum_qx = sum(qA*BATopt%xtot)
x_org_LLE_limits(1) = qA(2)*BATopt%xtot(2)/sum_qx
sum_qx = sum((1.0_wp - qA)*BATopt%xtot)
x_org_LLE_limits(2) = (1.0_wp - qA(2))*BATopt%xtot(2)/sum_qx
!make x_org_LLE_limits(1) (= xA_org) the water-rich phase
!(by choice):
if (x_org_LLE_limits(1) > x_org_LLE_limits(2)) then
    sum_qx = x_org_LLE_limits(1)
    x_org_LLE_limits(1) = x_org_LLE_limits(2)
    x_org_LLE_limits(2) = sum_qx
endif
write(*,'(A,2(ES13.6,1X),/)') 'after hybrd1 solver; xA_org, xB_org =
    ', x_org_LLE_limits(1:2)

end subroutine BAT_miscibility_gap
```

**Listing 3.19** Code for computation of LLE phase compositions based on an initial guess (excerpt from file chap3_alg2_Mod_BAT_model.f90).

Subroutine `calc_BAT_LLE_dev` provides the calculation of the relative activity deviations (Equation (3.30)) for given compositions of phases $\alpha$ and $\beta$ computed via the $\ln(F_j)$ variables (`lnaff` in the code) proposed by the employed solver. This subroutine has the standardized interface required by the `hybrd1` solver (Powell's hybrid method). See the implementation details of `calc_BAT_LLE_dev` in file chap3_alg2_Mod_BAT_model.f90. Due to the restricted interface, additional BAT model parameters are transferred to `calc_BAT_LLE_dev` via the private module variable `BATopt`—one advantage of placing related subroutines into the same module. Powell's hybrid method from MINPACK, subroutine `hybrd1`, is called on line 45 of Listing 3.19. Briefly, the task of `hybrd1` is to find a zero (vector) of $N$ nonlinear functions in $N$ unknowns; in this case the associated Jacobian is computed by a forward-difference approximation. The MINPACK module also offers other, more general solver variants, including ones where the Jacobian is evaluated/provided by the user-supplied subroutine, as well as methods for minimizing $M$ functions in $N$ unknowns using a version of the Levenberg–Marquardt algorithm. Details for each method and its application are stated in the header to each subroutine in chap3_alg8_Mod_MINPACK.f90.

Subroutine `BAT_miscibility_gap` returns the compositions of the detected phases, that is, the extent of the miscibility gap. If the calculation failed due to a poor initial guess, the subroutine will return two phases of identical compositions (the trivial solution). In a binary system that was determined to exhibit LLPS, we need to call `BAT_miscibility_gap` only once (when successful), since the phase separation limits will be known thereafter. Those values (array variable `x_org_LLE_limits`) can be used to compute the relative phase splits for any input solution composition that falls within the miscibility gap. In a more general case of a multicomponent system, one needs to recalculate the LLPS compositions for every input composition, because the miscibility gap is higher than 1-dimensional and the inputs rarely fall on the same (previous) tie-line [23], while the latter is always the case for binary mixtures (at isothermal, isobaric conditions).

When you run Prog_BAT_Examples for Example 2, you will find output similar to Listing 3.20 for an LLPS computation.

```
## Example 2 -- consideration of LLPS ##
number of BAT_light calls in step (2.3): 15

set tolerance for local_min: 1.00000E-04
found local min  of activity curve at x_org_min(1) =  8.519337E-02, flocmin =  9.624295E-01
number of BAT_light calls during  local min search = 7

found local min  of activity curve at x_org_min(2) =  4.592518E-01, flocmin =  6.518775E-01
number of BAT_light calls during  local min search = 7

approx x_org of p1  point is:  6.78571E-01

determined x_org of p1  point is:  6.68032E-01
relative deviation from targeted  activity: 1.33058E-08
number of BAT_light calls during  Ridders_zero = 3

approx x_org of p2  point is:  3.57143E-02

determined x_org of p2  point is:  2.24158E-02
relative deviation from targeted  activity: 3.17415E-11
number of BAT_light calls during  Ridders_zero = 5

initial guess for xA_org, xB_org =  2.241582E-02  6.680316E-01

number of BAT_light calls during LLE calc.: 16
after hybrd1 solver; info = 1
after hybrd1 solver; fvec =  1.209536E-09  3.142757E-09
after hybrd1 solver; xA_org, xB_org =  2.591991E-02  6.572402E-01
```

**Listing 3.20** Snippet of BAT Example 2 output for LLPS computation.

In this example, our initial guesses for the LLPS limits $[x_{org}^\alpha, x_{org}^\beta]$ were quite good indeed; guesses [0.02242, 0.6680] versus final [0.02592, 0.6572] determined by the solver, that is, absolute deviations of −0.0035 and +0.011. By varying the properties of the organic component (e.g., the O:C ratio or molar mass), you can confirm that our approach for generating initial guesses is generally very good. Listing 3.20 also reports the number of BAT_light subroutine calls during the construction of the initial guess and the equation solving with Powell's hybrid method: 15 calls for curve characterization, a combined 14 during local_min searches, 8 during Ridders' method, and 16 during Powell's method. That is, 37 calls to generate a very good initial guess followed by 16 during the process of solving the LLE equations (a total of 53 BAT_light calls).

### 3.4.1.4 Computation of Equilibrium Phase Compositions and Phase Amounts

Now that we have determined the extent of the miscibility gap and associated binodal compositions, we can run an equilibrium BAT model calculation covering the full $x_{org}$ composition range. The goal of such a calculation is to determine the equilibrium-state phase composition(s), related system properties, such as activities and activity coefficients, and, in case of an LLPS, the fractional molar amounts in each phase (e.g., quantified by $f^\alpha$). Note that we can express $f^\alpha$ based on the mole fractions of a single component as follows:

$$f^\alpha = \frac{\sum_j n_j^\alpha}{\sum_j n_j^t} = \frac{x_j^t - x_j^\beta}{x_j^\alpha - x_j^\beta}. \tag{3.32}$$

The expression on the right is known as the "lever rule," since geometrically, along a tie-line the numerator represents the distance between the (initial) overall mole fraction of $j$ (somewhere within the miscibility gap) and the opposite LLPS limit (i.e., $\beta$ for $f^\alpha$), while the denominator represents the maximum extent of the miscibility gap. When outside the phase separation region, where only a single liquid phase exists, we arbitrarily define $f^\alpha = 1$ for $x_j^t < x_{org}^\alpha$ and $f^\alpha = 0$ for $x_j^t > x_{org}^\beta$. Related to $f^\alpha$, the phase partitioning of individual components, $q_j^\alpha$, can also be expressed using determined mole fractions,

$$q_j^\alpha = f^\alpha \frac{x_j^\alpha}{x_j^t}. \tag{3.33}$$

The equilibrium-state BAT calculation follows similar steps as that of the (forced) single-phase computation we had carried out initially—except that one has to consider whether an input $x_{org}^t$ value falls outside or within the LLPS range. We also need to compute (and save) the phase composition and activity coefficient data for the two phases in case of LLPS. Listing 3.21 shows the implementation located in subroutine `BAT_example_2` under the Step (2.10) code section.

```
!(2.10) Generate equilibrium-state phase fractions and activity
!       data with consideration of potential presence of LLPS.
deallocate ( x_org_vec, ln_actcoeff, activities, delta_Gmix_by_nRT )

np = 1001
allocate( x_org_vec(np), ln_actcoeff(np,2), activities(np,2), &
    & ln_actcoeff_B(np,2), activities_B(np,2), &
    & delta_Gmix_by_nRT(np), fracA(np), qA_org(np), qA_w(np), &
    & x_org_A(np), x_org_B(np) )

xinc = 1.0_wp/real(np-1, kind=wp)
x_org_vec = [(i*xinc, i = 0, np-1)]        !the input (total) x_org

if (phase_sep_detected) then
    !compute activity coeff. and activities for LLE compositions;
    !phase alpha:
    call BAT_light(x_org_LLE_limits(1), M_org, OtoC, &
        & density_org, ln_ac_LLE_A, act_LLE_A)
    !phase beta:
    call BAT_light(x_org_LLE_limits(2), M_org, OtoC, &
        & density_org, ln_ac_LLE_B, act_LLE_B)
endif

!For equilibrium computations, call the BAT_light model to
!compute activities *only* when outside the miscibility gap:
do i = 1,np
    if ( x_org_vec(i) < x_org_LLE_limits(1) .OR. &
        & x_org_vec(i) > x_org_LLE_limits(2) ) then
        ! --> outside miscibility gap:
        call BAT_light(x_org_vec(i), M_org, OtoC, &
            & density_org, ln_actcoeff(i,:), activities(i,:))
        if (x_org_vec(i) < x_org_LLE_limits(1)) then
            fracA(i) = 1.0_wp
            qA_org(i) = 1.0_wp
            qA_w(i) = 1.0_wp
        else
            fracA(i) = 0.0_wp
            qA_org(i) = 0.0_wp
            qA_w(i) = 0.0_wp
        endif
        x_org_A(i) = x_org_vec(i)
        x_org_B(i) = x_org_vec(i)
        ln_actcoeff_B(i,:) = ln_actcoeff(i,:)
        activities_B(i,:) = activities(i,:)
    else
        ! --> initial input x_org is within miscibility gap:
        x_org_A(i) = x_org_LLE_limits(1)
```

```
      x_org_B(i) = x_org_LLE_limits(2)
      activities(i,:) = act_LLE_A
      activities_B(i,:) = act_LLE_B
      ln_actcoeff(i,:) = ln_ac_LLE_A
      ln_actcoeff_B(i,:) = ln_ac_LLE_B

      !use lever rule to determine total fraction in phase A:
      fracA(i) = (x_org_vec(i) - x_org_B(i)) / &
          & (x_org_A(i) - x_org_B(i))

      !calculate fractions qA_org and qA_w:
      qA_org(i) = fracA(i) * x_org_A(i) / x_org_vec(i)
      qA_w(i)   = fracA(i) * (1.0_wp - x_org_A(i)) &
          & / (1.0_wp - x_org_vec(i))
  endif
enddo

!compute equilibrium-state (normalized) Gibbs energy of mixing:
where (activities(:,1) > 0.0_wp .AND. activities(:,2) > 0.0_wp)
  !the generally valid version based on Eq. (3.22):
  delta_Gmix_by_nRT = fracA*( (1.0_wp - x_org_A) * &
      & log(activities(:,1)) + x_org_A*log(activities(:,2)) ) &
      & + (1.0_wp - fracA)*( (1.0_wp - x_org_B) * &
      & log(activities_B(:,1)) + x_org_B * &
      & log(activities_B(:,2)) )
elsewhere
  delta_Gmix_by_nRT = 0.0_wp
endwhere
```

**Listing 3.21** Example 2 subroutine (part 6); code for BAT model equilibrium-state calculations with consideration of LLPS (excerpt from file chap3_alg6_BAT_Example_2.f90).

Figure 3.8 shows several results from the equilibrium calculation. Comparing panel (a) of Figure 3.8 with that from Figure 3.6 reveals that the equilibrium curve of the normalized Gibbs energy of mixing follows the common tangent path within the miscibility gap denoted by the diamond symbols. Outside of the LLPS range the equilibrium composition ($x_{org}$) is equivalent to the input composition, $x_{org}^t$. Panel (b) quantifies how the total molar partitioning shifts from a larger water-rich phase ($f^\alpha > 0.5$) to a larger organic-rich phase as $x_{org}^t$ increases over the LLPS range, as illustrated in panel (c). The activity curves (panel e) show horizontal curve sections within the LLPS range, consistent with the expectation from the isoactivity condition. Panel (d) shows that there are distinct phase partitioning preferences by the two components for inputs that fall within the LLPS range, with water (blue curve) showing a higher affinity for phase $\alpha$ than 1-pentanol (as expected).

The code for the generation of the different Figure 3.8 panels is provided under Step (2.11) in subroutine BAT_example_2 (file chap3_alg6_BAT_Example_2).

In the atmospheric aerosol context, one aspect of great interest is the water uptake (hygroscopicity) behavior of particles as a function of relative humidity (RH), especially at high RH. Ignoring for a moment particle size effects and absolute aerosol mass and volume changes due to gas–particle partitioning of both water and the organic when RH changes, we can use the BAT equilibrium prediction to depict how a particle's overall water mole fraction would increase as a function of water activity and, hence, RH. Recall that water activity is equivalent to RH under bulk conditions, as expressed by modified Raoult's law (Equation (3.9)).

Starting at low RH, Figure 3.9 shows that water uptake by a binary water + 1-pentanol particle is relatively modest up to a water activity of $a_w = 0.979$, the onset of phase

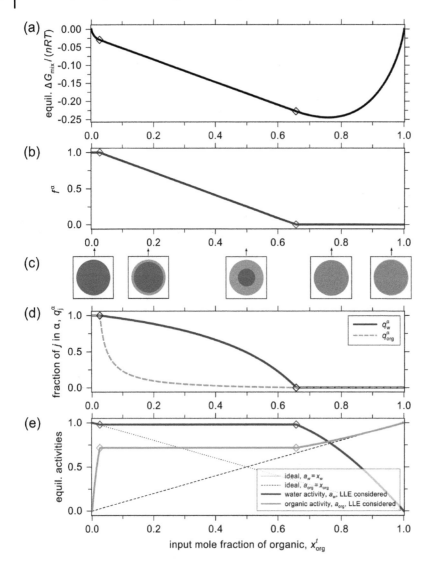

**Figure 3.8** Equilibrium-state BAT predictions for the water + 1-pentanol system with consideration of LLPS. The $x$-axis refers to the total (input) mole fraction of 1-pentanol. The symbols (diamonds) mark the phase separation limits (miscibility gap between them). (a) Normalized equilibrium-state Gibbs energy of mixing (Equation (3.22)); (b) fraction of cumulative molar amounts in phase $\alpha$ (Equation (3.32)); (c) illustration of equilibrium-state phase compositions and, in LLPS range, relative phase sizes at selected $x^t_{org}$; (d) individual component fractions in phase $\alpha$ (Equation (3.33)); (e) equilibrium-state water and 1-pentanol activities.

separation. Reaching that phase separation limit from the water activity perspective, any further increase in RH will lead to substantial water uptake by our particle, since the equilibrium $x_w$ instantly jumps to the upper branch of the blue curve. Note that there exists no single-phase equilibrium $x_w$ between the branches marked by the open diamonds. As a feature of LLPS in binary systems, the upper and lower diamonds are at exactly the same water activity; thus, there is an apparent discontinuity in $x_w$ (the

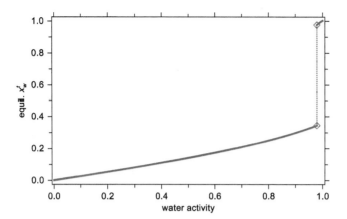

**Figure 3.9** Equilibrium predictions of the total mole fraction of water ($x_w^t$) versus water activity of the water + 1-pentanol system, accounting for LLPS. The symbols (diamonds) mark the phase separation limits.

miscibility gap) when comparing equilibrium RH conditions just below and above the LLPS $a_w$. In practice, a particle equilibrated at a water activity just below the discontinuity, when exposed to higher RH, will experience continuous net water uptake by water vapor condensation (gas–particle partitioning). It will transition smoothly from an overall water-poor to a water-rich particle composition at the phase separation $a_w$ by means of substantial water uptake into a new (nucleated) water-rich liquid phase (of composition of the upper curve branch in Figure 3.9). This water-rich phase will temporarily coexist with the water-poor one and grow on the expense of the latter as more water is absorbed. During this process, the particle's $a_w$ will not increase until the $x_w^t$ value has surpassed that of the water-rich equilibrium state (upper branch) at which point only a single, water-rich liquid phase will remain. This liquid–liquid phase transition is analogous to the isothermal phase transition from ice to liquid water at the melting point: one needs to supply energy (latent heat) to the system (ice + liquid water) to melt the ice, yet the temperature will remain unchanged at the melting point until all ice has melted. In multicomponent systems, such a water-poor-to-water-rich transition with increasing RH would typically span a range in particle water activities (no jump discontinuity) since the mixture will offer many degrees of freedom in terms of LLPS phase compositions [7].

Since the phase transition in our water + 1-pentanol system occurs at quite high water activity, an aerosol particle composed of only these components would likely have a higher effective hygroscopicity under supersaturated conditions (i.e., $\kappa_{CCN}$ prior to cloud droplet activation) than its hygroscopic growth factor measured, say at 85 % RH, may suggest [20]. Activity coefficient models like BAT and other, more advanced models, can aid in understanding and predicting the expected water uptake behavior as well as the nonideal mixing effects on gas–particle partitioning.

## 3.5 Multicomponent Aerosol Thermodynamics Models

In the previous sections of this chapter, we have explored the details and applications of a binary mixture activity coefficient model. It is illustrative for the understanding of the principles of activity coefficient and phase separation computations. However, one key issue is the limitation to two mixture components. It is well known that atmospheric aerosol particles consist of a multitude of components of both organic and inorganic nature. The mixture of particle components tends to contain molecules of a relatively wide range in polarities, sizes, volatilities, and shapes. Hence, the need for methods capable of representing nonideal mixing and related effects on gas–particle partitioning, phases, viscosities and multiphase chemistry. There exist ways to use the BAT model or other binary activity coefficient models to approximately compute, using a mixing rule, the water uptake of aqueous organic mixtures containing several different organic compounds for given RH conditions; however, one will need to ignore organic–organic interactions. Such an approach has been outlined and implemented by Gorkowski et al. [7]. In this section, we will turn our focus to community models that treat multicomponent mixing directly and therefore offer advanced options for the modeling of aerosol thermodynamics.

Over the past decades, many models have been developed and parameterized for the treatment of chemical thermodynamics targeting various applications within the physical sciences and engineering disciplines. Several types of models have been introduced specifically for the treatment of atmospheric aerosol thermodynamics, finding application in box models and/or large-scale atmospheric models. Some of those models focus on the nonideal mixing of atmospherically relevant ions in aqueous inorganic (electrolyte) solutions, including equilibrium particle water uptake—and in some cases formation of solid salts. Examples for such models are the ISORROPIA II model [29], the EQUISOLV II model [30], and the Extended Aerosol Inorganics Model (E-AIM), which offers several model variants, including some with treatment of organic mixture components [31–34]. Other multicomponent models focus on (aqueous) organic mixtures only, for example, the Universal Quasi-Chemical Functional group Activity Coefficients (UNIFAC) model by Fredenslund et al. [35]. The Model for Simulating Aerosol Interactions and Chemistry (MOSAIC) by Zaveri et al. [36] includes activity coefficient treatments for aqueous electrolytes [37] and organic compounds, although organic–inorganic interactions are not explicitly accounted for. Only a few aerosol thermodynamic models exist for the treatment of multicomponent organic–inorganic mixtures. One of those models is the Aerosol Inorganic–Organic Mixtures Functional Groups Activity coefficient model (AIOMFAC) by Zuend et al. [38, 39], which we will consider in more detail in Section 3.6. The review article by Pye et al. [3] provides an overview and brief characterization of the above-mentioned models. Some of these models also provide more details on dedicated websites and some offer a web-interface for custom model calculations (e.g., E-AIM: http://www.aim.env.uea.ac.uk/aim/aim.php; AIOMFAC: https://aiomfac.lab.mcgill.ca).

## 3.6 Activity and LLE Computations with the AIOMFAC Model

The AIOMFAC model is a versatile thermodynamic mixing model with explicit consideration of interactions among water, organic compounds, and more than 12 inorganic ions in multicomponent phases. Yin et al. [40] provide details on the set of ions and organic functional groups covered; see also https://aiomfac.lab.mcgill.ca for recent updates. At its core, AIOMFAC calculates the activity coefficients of components/species for a given input mixture composition and temperature. That is, the basic AIOMFAC model does not calculate liquid–liquid, solid–liquid, or gas–particle partitioning equilibria, although it includes partial dissociation equilibria of certain ions like bisulfate ($HSO_4^-$). Additional AIOMFAC-based models have been developed for the computation of coupled liquid–liquid and gas–particle (gas–liquid) equilibria [e.g., 3, 5, 6]. In this section, we will learn how to use and customize a model variant termed AIOMFAC-LLE. This model enables the computation of activities, yet also conducts the detection and numerical solution of the phase compositions for cases where a mixtures undergoes LLPS at equilibrium.

Unlike the expressions for the BAT (light) model discussed earlier, the equations for activity coefficients as computed by AIOMFAC are far more involved. Those expressions and the related theory underpinning the AIOMFAC model are described in detail elsewhere [38, 39]. Briefly, the activity coefficient expressions are derived from a combination of (1) a local composition model (a variant of UNIFAC) for the treatment of interactions among nonionic components, (2) a Pitzer-like ion-interaction model for mixtures of aqueous inorganic ions, and (3) a middle-range interaction treatment for the interactions among ions and organic groups. We will mainly focus on the practical use and related code, rather then the equations and implementation details. The basic model design was strongly influenced by the group-contribution model LIFAC, developed by Yan et al. [41] for chemical engineering applications. Like UNIFAC and LIFAC, AIOMFAC makes use of a group-contribution concept for the representation of organic compounds. This means that organic molecules are described as an assembly of organic (functional) groups (called subgroups). For example, the molecule methylsuccinic acid, SMILES: O=C(O)C(C)CC(O)=O, can be described in "AIOMFAC notation" as $(CH_3)(CH_2)(CH)(COOH)_2$ since it consists of 1 $CH_3$, 1 $CH_2$, 1 CH, and 2 COOH (carboxyl) subgroups. The first three subgroups are part of the alkyl $CH_n$ main group, with the main groups referring to a lower level of structural detail used for certain expressions. Within AIOMFAC, a multicomponent mixture is actually described as a solution of groups, with interactions among groups computed first, followed by mapping of the subgroup and main group activity coefficient contributions in the correct proportions for associated molecular compounds (see also https://aiomfac.lab.mcgill.ca/about.html). In this method, water and different inorganic ions are described by their own subgroups (e.g., ion $NH_4^+$ is assigned subgroup number 204). In the aerosol context, an important advantage of the group-contribution method is its versatility in describing countless possible organic structures by a relatively modest number of subgroups. The group-contribution concept, making use of a set of AIOMFAC group–group interaction parameters, offers a level of predictability for the activity coefficients of molecular structures and mixtures that have never been experimentally determined, as discussed in Zuend et al. [39]. Still, a higher level of structural information about organic mixture components is required than with the use of reduced-complexity models like BAT.

As mentioned above, different equilibrium models have been built around the core AIOMFAC module. AIOMFAC-LLE (version 3.0) provides a sophisticated method for computing the LLE state of a given input mixture composition using AIOMFAC for the composition-dependent activity coefficients in different phases. The AIOMFAC-LLE implementation allows for LLPS calculations with up to two liquid phases while covering mixtures of any number of components (in principle), including consideration of multiple electrolytes and ions that only partially dissociate. This AIOMFAC-LLE model represents an extended and improved version of the method by Zuend and Seinfeld [23]. Since this model needs to solve the multicomponent case, some of the shortcuts we were able to take with the BAT model for LLPS are no longer applicable.

For example, it is not straightforward to determine whether a mixture will undergo LLPS or not [23]. Extensive initial exploration of the multidimensional curvature of the Gibbs energy function for that purpose is computationally far more costly than in the binary mixture case and impractical. Overall, the goal remains to determine a nontrivial LLPS state that simultaneously fulfills the isoactivity condition for all components (Equation (3.26)) and represents a (global) minimum in the Gibbs energy of mixing. As in the case of the BAT model example, generating a good initial guess for a potential LLPS state is a key feature for a reliable and efficient method. Because this in itself is a challenging problem, a set of several distinct initial guesses is constructed, increasing the chances that at least one will be sufficiently good. The main steps of the method (designed by Andreas Zuend) implemented in the AIOMFAC-LLE program are outlined in the following.

---

**Method outline for liquid–liquid equilibrium computations as implemented in the AIOMFAC-LLE program (version 3.0):**

1. Generate several initial guesses for the two liquid phase compositions (an empirical procedure). For this, the physicochemical properties of the mixture components are used to characterize them in terms of a proxy metric for polarity. We use a combination of the O:C and H:C ratios (for organics)—or an O:C-equivalent assigned to inorganic components/ions. This is useful because the most polar and the least polar nonelectrolyte components are expected to exhibit affinities for different liquid phases at equilibrium, should a LLE prevail. The initial guesses are then distributed along a designated polarity axis, with scaling of the individual component amounts present in each phase as a function of component polarity. The number of initial guesses generated is empirically determined based on the present spread in polarities. For implementation details, see submodule `SubModinitGuessesLLE` of the AIOMFAC-LLE code.

2. Use a damped fix-point method to improve each initial guess to a set threshold accuracy in fulfilling a coupled system of nonlinear algebraic equations (Equations (3.28) and (3.29)) and associated equations for electrolyte components while maintaining charge neutrality. Activity coefficients for each phase composition will be computed/updated multiple times using calls to AIOMFAC. This step will either lead to a highly improved candidate solution for LLPS, convergence toward a previously determined candidate solution, or convergence to a trivial solution (i.e., two phases of the same composition). For implementation details, see submodule `SubModPhaseSeparation` and methods in module `ModPhaseSep`.

3. If LLPS candidate solutions were detected, numerically resolve those to higher accuracy solving Equation (3.30) (using Powell's hybrid method). For implementation details, see submodule `SubModLLEsolver`.

4. Compute the normalized Gibbs energy of mixing for the (potentially several) LLPS state solutions as well as the (forced) single-phase state solution. Determine the equilibrium state of lowest Gibbs energy and return it as the solution for the given inputs. For implementation details, see submodule `SubModPhaseSeparation`.

We note that the implementation details of Steps 1 and 2 in the outlined method differ from those described in the related work by Zuend and Seinfeld [23]. Improvements were made to the efficiency and reliability of the method as well as additions to correctly treat the multi-ion case.

### 3.6.1 Customizing and Running AIOMFAC-LLE

Given the complexity of the AIOMFAC-LLE model implementation, we will focus on an example program providing insights into how to build and run the AIOMFAC-LLE program. For this purpose, we will consider the different top-level steps necessary in any AIOMFAC program run and treat most other Fortran modules simply as available "black boxes." The example discussed below will also allow you to modify the model for use in other applications, for example, when AIOMFAC-LLE is used as a module of a box model.

Let us first cover some file logistics for this example. The complete AIOMFAC-LLE program code and required folders for input and output of the example system are provided in a zip-archive, file chap3_alg9_AIOMFAC-LLE_Fortran_code.zip. After unzipping it, that code version can be compiled and linked from a command terminal with any recent Fortran compiler. Moreover, we also provide a complete MS Visual Studio project/solution for this program in zip-file chap3_alg10_AIOMFAC-LLE_VS_proj.zip. For MS Windows users, the Visual Studio project is likely the most convenient way for exploring, modifying, and running this example program. For those working from a command terminal environment, the command lines for ifort and gfortran compilations of the Fortran code are listed in file chap3_alg10_AIOMFAC-LLE_command_line.txt. Listing 3.22 shows the compilation example for gfortran on Linux, including linking to the Dislin plotting library (which needs to be installed separately, as for the BAT model examples above).

```
gfortran -o Prog_AIOMFAC-LLE.out -O3 -ffree-line-length-none
   Mod_NumPrec.f90 ModTransformations.f90 zerobracket_inwards.f90
   ridderzero.f90 brent.f90 ModSystemProp.f90 ModAIOMFACvar.f90
   ModMRpart.f90 ModCompScaleConversion.f90 Mod_MINPACK.f90
   ModSubgroupProp.f90 Mod_Dislin_plots.f90 Mod_InputOutput.f90
   ModPhaseSep.f90 ModComponentNames.f90 SubModinitGuessesLLE.f90
   SubModLLEsolver.f90 Mod_PureViscosPar.f90 ModSRunifac.f90
   SubModDefSystem.f90 ModCalcActCoeff.f90
   SubModPhaseSeparation.f90 SubModGibbsEnergyLiq.f90
   AIOMFAC_LLE_inout.f90 Main_AIOMFAC_LLE_prog.f90  -ldislin_d
   -I/usr/local/dislin/gf/real64
```

Listing 3.22 Command line for building the AIOMFAC-LLE program using gfortran from a Linux terminal, including linking to the Dislin library for plotting examples included in this model version.

After successful compilation, the program can be run using a command like ./Prog_AIOMFAC-LLE.out ./Inputfiles/input_0404.txt, in which the second command-line argument specifies the path to an input file characterizing the system components and mixture compositions for which calculations should be made. The AIOMFAC-LLE (version 3.0) program uses the same input file structure and a similar core code structure as contemporary versions of the AIOMFAC-web model (see https://aiomfac.lab.mcgill.ca and related source code at https://github.com/andizuend/AIOMFAC). The main difference between our AIOMFAC-LLE example program and AIOMFAC-web (version 2.32) is that the latter does not (yet) support LLE computations; plus, our program includes a few plotting examples aside from text-file output. In future editions, the AIOMFAC website model will likely also provide phase separation computations as a standard feature. Given this similarity, one convenient option for generating valid input files is by using the AIOMFAC-web input form on the "Run Model" page of the AIOMFAC website (https://aiomfac.lab.mcgill.ca). We will revisit the topic of input files with a specific example further below (Listing 3.24).

The main program procedure includes a few key code sections covering program initialization, calculations, and output processing, as outlined in the following:

---

**Key code sections of Main_AIOMFAC_LLE_prog.f90:**

1. Program initialization, reading of an input file and associated data processing.
2. Loading of AIOMFAC model parameters and initialization of a specific system of components (from input file information).
3. Performing AIOMFAC-LLE calculations for one or multiple data points stating mixture composition and temperature (calling subroutine `AIOMFAC_LLE_inout`).
4. Output of computed activity coefficients, equilibrium-phase compositions, and other information about the system and specific mixture data points.
5. Finalization of the program (de-allocation of data arrays, etc.).

---

Here these different tasks are combined into a single main program procedure (and .f90 file) for ease of use. However, we note that for other applications, such as condensed-phase diffusion computations requiring repeated calls of AIOMFAC for various phase compositions, it can be advantageous to place the actual AIOMFAC-LLE mixture calculation task into a separate subroutine.

The Fortran implementation of the first item from the above list of key sections is shown in Listing 3.23.

```fortran
program Main_AIOMFAC_LLE_prog

use Mod_NumPrec, only : wp
use ModSystemProp, only : errorflagmix, nindcomp, NKNpNGS, SetSystem
    , topsubno, waterpresent
use ModSubgroupProp, only : SubgroupAtoms, SubgroupNames
use ModMRpart, only : MRdata
use ModSRunifac, only : SRdata
use Mod_InputOutput, only : OutputLLE_TXT, OutputLLE_plots,
    ReadInputFile, RepErrorWarning

implicit none
!set preliminary input-related parameters:
integer,parameter :: maxpoints = 101
integer,parameter :: ninpmax = 51          !set the maximum number of
    mixture components allowed (arbitrary parameter)
!local variables:
character(len=4) :: VersionNo
character(len=200) :: filename
character(len=3000) :: filepath, folderpathout, fname, txtfilein
character(len=200),dimension(:),allocatable :: cpnameinp
character(len=200),dimension(:),allocatable :: outnames, &
```

```fortran
      & name_species_TeX
integer :: allocstat, errorflag, errorind, i, nc, ncp, npoints, &
    & nspecies, nspecmax, pointi, unito, warningflag, &
    & warningind, watercompno
integer,dimension(:,:),allocatable :: cpsubg    !list of input
    component subgroups and corresponding subgroup quantities
real(wp),dimension(:),allocatable :: T_K
real(wp),dimension(:),allocatable :: inputconc
real(wp),dimension(:,:),allocatable :: composition, compos2, &
    & out_LLEprop
real(wp),dimension(:,:,:),allocatable :: out_data_A, out_data_B
real(wp),dimension(:),allocatable :: LLEprop        !selected
    properties of predicted LLE; structure ( #-phases, omega, phi,
    Gibbs-E_diff, LstarA, LstarB, reladiffbest )
real(wp),dimension(:,:,:),allocatable :: LLEoutvars  !3-D array
    !with computed compositions and activities for each species
    !and phase; structure is: (| input_mass-frac, mass-frac,
    !mole-frac, molality, act.coeff., activity, ion-indicator |
    !species-no |phase-no |)
logical :: ignore_LLPS, filevalid, verbose, xinputtype
!..................................................................


!
!==== initialization section =====================================
!
VersionNo = "3.0"
verbose = .true.    !if true, some debugging information will be
                    !printed to the unit "unito" (errorlog file)
nspecmax = 0
errorind = 0
warningind = 0
!
!==== input data section =========================================
!
!read command line for text-file name (which contains the input
!parameters to run the AIOMFAC program):
call get_command_argument(1, txtfilein)
if (len_trim(txtfilein) < 4) then
    !no command line argument stated; use specific input file
    !for tests:
    txtfilein = './Inputfiles/input_0404.txt'
endif
filepath = adjustl(trim(txtfilein))
write(*,*) ""
write(*,'(A,A)') "MESSAGE from AIOMFAC-LLE: program started, command
    line argument 1 = ", trim(filepath)
write(*,*) ""
allocate(cpsubg(ninpmax,topsubno), cpnameinp(ninpmax), &
    & composition(maxpoints,ninpmax), T_K(maxpoints), &
    & STAT=allocstat)
!--
call ReadInputFile(filepath, folderpathout, filename, ninpmax, &
    & maxpoints, unito, verbose, ncp, npoints, warningind, &
    & errorind, filevalid, cpnameinp, cpsubg, T_K, composition, &
    & xinputtype)
```

**Listing 3.23** AIOMFAC-LLE main program code: variable declarations and input file processing (snippet from file Main_AIOMFAC_LLE_prog.f90).

On code line 57 of above listing, the relative path and name of a (default) input file (input_0404.txt) is stated, which will be used if an input file name is not specified as a command-line argument during program execution (line 53). The input file will be read and processed by the subroutine `ReadInputFile` from module `Mod_InputOutput`. Let us look at the input file of for our example system, which contains the four components water, isobutanol, methylsuccinic acid, and ammonium bisulfate ($NH_4HSO_4$); see Listing 3.24.

```
Input file for AIOMFAC-web model

mixture components:
....
component no.:  01
component name: 'Water'
subgroup no., qty:  016, 01
....
component no.:  02
component name: 'Isobutanol'
subgroup no., qty:  145, 02
subgroup no., qty:  147, 01
subgroup no., qty:  150, 01
subgroup no., qty:  153, 01
....
component no.:  03
component name: 'Methylsuccinic_acid'
subgroup no., qty:  001, 01
subgroup no., qty:  002, 01
subgroup no., qty:  003, 01
subgroup no., qty:  137, 02
....
component no.:  04
component name: 'NH4HSO4'
subgroup no., qty:  204, 01
subgroup no., qty:  248, 01
....
++++
mixture composition and temperature:
mass fraction?  0
mole fraction?  1
....
point, T_K,  cp02, cp03, cp04
1 298.15   3.333E-04 3.333E-04 3.333E-04
2 298.15   3.333E-03 3.333E-03 3.333E-03
```

**Listing 3.24** Snippet of an AIOMFAC-web input file.

Such input files state the AIOMFAC subgroups which characterize each molecule or ion by means of a subgroup identifier and corresponding quantity of entities in the compound. For example, subgroup no. 137 refers to the carboxyl group. A table of supported subgroups is provided on https://github.com/andizuend/AIOMFAC. The input information will be used to determine all possible components/species in the system considered. In this case, the number of species differs from those stated in the file due to the presence of an electrolyte component ($NH_4HSO_4$) and the partial dissociation of the bisulfate ion into additional ion species ($H^+$, $SO_4^{2-}$) in aqueous solution. Therefore, after input processing, AIOMFAC will consider three nonelectrolyte components and four ions during the subsequent computations. Temperature and mixture compositions for one or more data points can be stated in the last section of an input file. If desired we could ignore those data points and generate other temperature and mixture data for customized calculations directly within the AIOMFAC-LLE main program, while the system components/species remain fixed as defined by the input file. Also note that the input file does not list the mole fraction of component 01 (here water), since it will be calculated within the AIOMFAC-LLE program from the sum of all other component mole fractions (as difference from 1.0). Input data for mole fractions is defined on the basis of undissociated (input) electrolyte components.

```
!load the MR and SR interaction parameter data:
call MRdata()
call SRdata()
call SubgroupNames()
call SubgroupAtoms()

!set system properties based on the data from the input file:
call SetSystem(1, .true., ncp, cpnameinp(1:ncp), &
    & cpsubg(1:ncp,1:topsubno) )

!check whether water is present in the mixture and as which
!component number:
watercompno = 0
if (waterpresent) then
    watercompno = findloc(cpsubg(1:ncp,16), value=1, dim=1)
endif
!transfer composition data to adequate array size:
allocate(compos2(npoints,ncp), STAT=allocstat)
do nc = 1,ncp
    compos2(1:npoints,nc) = composition(1:npoints,nc)
enddo
deallocate(cpsubg, composition, STAT=allocstat)

if (errorflagmix /= 0) then  !a mixture-related error occurred:
    call RepErrorWarning(unito, errorflagmix, warningflag, &
        & errorflag, i, errorind, warningind)
endif
```

**Listing 3.25** AIOMFAC-LLE main program code: model and system initialization (snippet from file Main_AIOMFAC_LLE_prog.f90).

The code section covering the initialization of AIOMFAC interaction parameters and the setting of properties of the system components/species is shown in Listing 3.25. The call of subroutine `SetSystem` from module `ModSystemProp` defines a large number of component and system properties stored in public and private variables of that module for use elsewhere. For use cases in which AIOMFAC-LLE is integrated into a parent program, make sure that model parameter and system initialization procedures are only called once for the same system (e.g., as part of a separate AIOMFAC initialization procedure).

```
!==== AIOMFAC-LLE calculation section ==========================
!
if (errorind == 0) then !perform AIOMFAC-LLE calculations
                        !; else jump to termination section;
allocate(inputconc(nindcomp), outnames(NKNpNGS), &
& out_data_A(8,npoints,NKNpNGS), out_data_B(8,npoints,NKNpNGS), &
& out_LLEprop(7,npoints), LLEprop(8), LLEoutvars(8,NKNpNGS,2), &
& name_species_TeX(NKNpNGS), STAT=allocstat)
inputconc = 0.0_wp
out_data_A = 0.0_wp
out_data_B = 0.0_wp

!set AIOMFAC input and call the main AIOMFAC-LLE subroutine for
```

```
!all composition points;
ignore_LLPS = .false.    !(set .true. to force a single-phase
                         !calculation)

do pointi = 1,npoints    !loop over mixture points, changing
                         !composition and/or temperature
    inputconc(1:ncp) = compos2(pointi,1:ncp)
    !..
    call AIOMFAC_LLE_inout(inputconc, xinputtype, T_K(pointi), &
    & ignore_LLPS, nspecies, LLEprop, LLEoutvars, outnames, &
    & errorflag, warningflag)
    !..
    if (warningflag > 0 .OR. errorflag > 0) then
        call RepErrorWarning(unito, errorflagmix, warningflag, &
        & errorflag, pointi, errorind, warningind)
    endif
    !..
    !save properties of this input point:
    out_LLEprop(:,pointi) = LLEprop
    !out_data_A array structure: | LLEoutvars data columns <1:7>
    ! | data point no. | species no.|
    do nc = 1,nspecies
        out_data_A(1:7,pointi,nc) = LLEoutvars(1:7,nc,1)
        out_data_B(1:7,pointi,nc) = LLEoutvars(1:7,nc,2)
        out_data_A(8,pointi,nc) = real(errorflag, kind=wp)
        out_data_B(8,pointi,nc) = real(errorflag, kind=wp)
        if (errorflag == 0 .AND. warningflag > 0) then
            !do not overwrite an errorflag if present
            if (warningflag == 16) then
                !a warning that only affects viscosity calc.
                !(here not used)
            else
                out_data_A(8,pointi,nc) = real(warningflag, &
                & kind=wp)
                out_data_B(8,pointi,nc) = real(warningflag, &
                & kind=wp)
            endif
        endif
    enddo !nc
    !..
enddo !pointi
```

**Listing 3.26** Call of the main AIOMFAC-LLE activity coefficient and phase separation calculations for individual mixture composition points (snippet from file Main_AIOMFAC_LLE_prog.f90).

Listing 3.26 shows the code section that contains the call to subroutine AIOMFAC_LLE_inout for each mixture data point. AIOMFAC_LLE_inout provides a relatively user-friendly interface to coordinate the calculation of activity coefficients and the potential detection and determination of LLPS (via call to subroutine PhaseSeparation). It also processes the internal calculation outputs and returns phase- and species-specific output data of typical interest. In the parent procedure (Main_AIOMFAC_LLE_prog), those outputs are then stored in arrays out_LLEprop, out_data_A and out_data_B. Note that when a single liquid phase is the stable equilibrium state for a given mixture point, the corresponding phase $\beta$ data in array out_data_B will simply show the single-phase data (identical to that in out_data_A for that data point). Line 15 of Listing 3.26 indicates that one could switch off LLPS computations (set ignore_LLPS = .true.), which means that the model would be forced to assume a single liquid phase regardless of whether LLPS would actually occur or not. This leads to significantly faster program execution and could be a desired option when calculations are performed for systems that have been determined to form only a single liquid phase as stable equilibrium state; for example, organic-free aqueous electrolyte solutions (when ignoring formation of solids).

In the last sections of the main program, the subroutines `OutputLLE_TXT` and (optionally) `OutputLLE_plots` are called to write the output data to text files and to generate a selection of plots. This section could be skipped if the goal is simply to return output from a single data point calculation (when part of a larger program). For our water + isobutanol + methylsuccinic acid + $NH_4^+$ + $HSO_4^-$ + $H^+$ + $SO_4^{2-}$ system, defined by input file 0404, calling subroutine `OutputLLE_TXT` will generate AIOMFAC-LLE_output_0404.txt in folder "Outputfiles." This file lists selected properties of the system and each equilibrium mixture state. For example, it lists properties such as number of liquid phases, the fraction of mass in phase $\alpha$ (useful in LLPS case), activity coefficients, activities, and different concentration metrics for all species in all liquid phases. You can regenerate this file by compiling and running the provided code.

For the chosen example, the input composition data were generated somewhat arbitrarily by fixing the molar mixing ratios of the three nonwater (input) components to a 1:1:1 ratio. The molar fraction of water was then changed from high to low values (in slightly irregular steps), leading to 30 data points representative of humid to dry conditions. This could be understood as conducting an experiment in which the dry mixture composition and mass of components stays constant (treated as nonvolatile) while the amount of water is varied. Obviously, with seven mixture species, many other mixing ratios would be possible. Thus, the data generation approach applied for this example only yields a single mixture "dehydration" trace through a multidimensional mixture composition space at a temperature of 298.15 K. Figure 3.10 shows several properties of this set of AIOMFAC-LLE calculations.

Panel (a) of Figure 3.10 indicates that the activity coefficients of the two organic components are relatively large and increase toward lower water mass fractions. Clearly, in the composition range where LLPS is present, the organic components have substantially lower activity coefficients in phase $\beta$ (panel b), indicating that this is their preferred phase. For such graphs, using mass fraction (or water activity) for the $x$-axis is usually preferred over mole fraction. This is the case because the established equilibrium-state mole fractions will depend on the degree of dissociation of the ions, which varies in this system, while the mass fraction of water and the organics is not affected by the degree of ion dissociation and therefore a robust composition metric less likely to be misinterpreted. Panel (c) shows also that the mass fraction of water in phase $\beta$ is lower than that in coexisting phase $\alpha$, as well as that of the overall (input) mixture composition. This suggests that phase $\beta$ is likely water-poor and organic-rich, while phase $\alpha$ is expected to be ion- and water-rich. We will confirm this below. Panel (d) indicates that with decreasing water activity (and related water mass fraction), the dominant phase in terms of mass shifts from phase $\alpha$ to phase $\beta$. At a water activity of 0.79, both phases are of about the same mass; thus, likely also of similar volume. The code for the plots shown in Figure 3.10 is provided in subroutine `OutputLLE_plots`.

In addition, to visualize the relative phase compositions at different water activity levels, we have made stacked bar graphs as shown in Figure 3.11 using an external, Dislin-based program. This figure confirms that at water activities below 0.6, phase $\alpha$ is predominantly an aqueous electrolyte-rich phase, while phase $\beta$ is organic-rich, yet also contains some amounts of bisulfate, ammonium, and sulfate. You can run the AIOMFAC-LLE program with modified molar (or mass) mixing ratios of the input components, for example, using a copy of our example input file, to explore the impact on equilibrium phase compositions and the distribution of mass among coexisting phases.

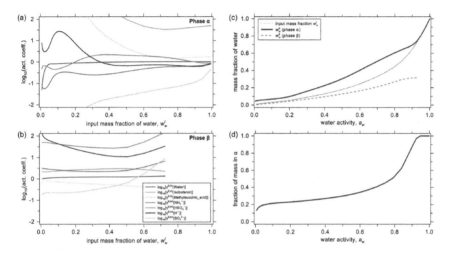

**Figure 3.10** AIOMFAC-LLE calculation outputs for the water + isobutanol + methylsuccinic acid +NH$_4$HSO$_4$ system at $T = 298.15$ K, for a 1:1:1 dry-mixture mixing ratio. (a, b) Predicted log$_{10}$-scale activity coefficients of the mixture species in phases $\alpha$ (top) and $\beta$ (bottom) versus the input mass fraction of water (i.e., that in the overall mixture, not a specific phase). The $y$-axis range was limited to the range from 0.01 to 100. (c) Comparison of the mass fractions of water for the overall mixture as well as each phase in the LLPS range, which starts at a water activity of $\sim 0.94$. (d) Fraction of the total mass residing in phase $\alpha$ as function of water activity.

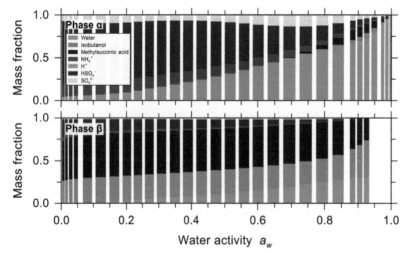

**Figure 3.11** Predicted equilibrium phase mass fractions (stacked bars) for the water + isobutanol + methylsuccinic acid + NH$_4$HSO$_4$ example system (see Section 3.6.1).

The AIOMFAC-LLE main program code and calculation example discussed in this section has provided you with a practical introduction to the use of a sophisticated community thermodynamics model for nonideal mixing in aerosols (and liquid solutions in general). For most applications and customized modifications of this model, it will be sufficient to work with (and understand) the code of file Main_AIOMFAC_LLE_prog.f90 and perhaps file AIOMFAC_LLE_inout.f90. These two files provide the main interface to the underlying core AIOMFAC and LLE procedures carrying out the activity coefficient, dissociation equilibrium, and phase separation calculations. The AIOMFAC-LLE code can also be implemented as a module contained within a gas–particle partitioning box model, which may further also include thermodynamic solid–liquid equilibrium computations. Such coupled gas-particle and LLE models based on AIOMFAC have been used in various studies, including for quantifying the impact of LLPS on gas–particle partitioning [5, 6], impacts on surface tension and cloud droplet activation [19], or to study the water uptake and phase behavior (hygroscopicity) of laboratory systems mimicking mixed organic–inorganic aerosols [e.g., 42, 43].

## Bibliography

[1] J. M. Prausnitz, R. N. Lichtenthaler, and E. G. de Azevedo. *Molecular Thermodynamics of Fluid-Phase Equilibria. Physical and Chemical Engineering Sciences.* Prentice-Hall PTR, 1999.

[2] J. H. Seinfeld and S. N. Pandis. *Atmospheric Chemistry and Physics: From Air Pollution to Climate Change.* J. Wiley & Sons, New York, USA, 1998.

[3] H. O. T. Pye, A. Nenes, B. Alexander, A. P. Ault, M. C. Barth, S. L. Clegg et al. The acidity of atmospheric particles and clouds. *Atmospheric Chemistry and Physics,* 20(8):4809–4888, 2020.

[4] Terry Renner. *Quantities, Units and Symbols in Physical Chemistry.* The Royal Society of Chemistry, 2007.

[5] A. Zuend, C. Marcolli, T. Peter, and J. H. Seinfeld. Computation of liquid–liquid equilibria and phase stabilities: implications for RH-dependent gas/particle partitioning of organic–inorganic aerosols. *Atmospheric Chemistry and Physics,* 10(16):7795–7820, 2010.

[6] A. Zuend and J. H. Seinfeld. Modeling the gas-particle partitioning of secondary organic aerosol: the importance of liquid–liquid phase separation. *Atmospheric Chemistry and Physics,* 12(9):3857–3882, 2012.

[7] K. Gorkowski, T. C. Preston, and A. Zuend. Relative-humidity-dependent organic aerosol thermodynamics via an efficient reduced-complexity model. *Atmospheric Chemistry and Physics,* 19(21):13383–13407, 2019.

[8] M. L. McGlashan. Deviations from Raoult's law. *Journal of Chemical Education,* 40(10):516–518, 1963.

[9] Gregory S. Girolami. A simple "back of the envelope" method for estimating the densities and molecular volumes of liquids and solids. *Journal of Chemical Education,* 71(11):962, 1994.

[10] Windows Subsystem for Linux website. https://docs.microsoft.com/en-us/windows/wsl/, 2021 (accessed 26 July 2021).

**[11]** DISLIN website. https://www.dislin.de, 2021 (accessed 26 July 2021).

**[12]** Helmut Michels. *The Data Plotting Software DISLIN - Version 11*. Shaker Media GmbH, 2017.

**[13]** A. K. Bertram, S. T. Martin, S. J. Hanna, M. L. Smith, A. Bodsworth, Q. Chen et al. Predicting the relative humidities of liquid–liquid phase separation, efflorescence, and deliquescence of mixed particles of ammonium sulfate, organic material, and water using the organic-to-sulfate mass ratio of the particle and the oxygen-to-carbon elemental ratio of the organic component. *Atmospheric Chemistry and Physics*, 11(21):10995–11006, 2011.

**[14]** M. Song, C. Marcolli, U. K. Krieger, A. Zuend, and T. Peter. Liquid-liquid phase separation in aerosol particles: Dependence on o:c, organic functionalities, and compositional complexity. *Geophysical Research Letters*, 39, 2012.

**[15]** Yuan You, Mackenzie L. Smith, Mijung Song, Scot T. Martin, and Allan K. Bertram. Liquid-liquid phase separation in atmospherically relevant particles consisting of organic species and inorganic salts. *International Reviews in Physical Chemistry*, 33(1):43–77, 2014.

**[16]** Yuanzhou Huang, Fabian Mahrt, Shaun Xu, Manabu Shiraiwa, Andreas Zuend, and Allan K. Bertram. Coexistence of three liquid phases in individual atmospheric aerosol particles. *PNAS*, 118(16):e2102512118, 2021.

**[17]** H. O. T. Pye, A. Zuend, J. L. Fry, G. Isaacman-VanWertz, S. L. Capps, K.W. Appel et al. Coupling of organic and inorganic aerosol systems and the effect on gas-particle partitioning in the southeastern us. *Atmospheric Chemistry and Physics*, 18(1):357–370, 2018.

**[18]** L. Renbaum-Wolff, M. Song, C. Marcolli, Y. Zhang, P. F. Liu, J.W. Grayson et al. Observations and implications of liquid-liquid phase separation at high relative humidities in secondary organic material produced by $\alpha$-pinene ozonolysis without inorganic salts. *Atmospheric Chemistry and Physics*, 16(12):7969–7979, 2016.

**[19]** Jurgita Ovadnevaite, Andreas Zuend, Ari Laaksonen, Kevin J. Sanchez, Greg Roberts, Darius Ceburnis et al. Surface tension prevails over solute effect in organic-influenced cloud droplet activation. *Nature*, 546(7660):637–641, 2017.

**[20]** N. Rastak, A. Pajunoja, J. C. Acosta Navarro, J. Ma, M. Song, D. G. Partridge et al. Microphysical explanation of the RH-dependent water affinity of biogenic organic aerosol and its importance for climate. *Geophysical Research Letters*, 44(10):5167–5177, 2017.

**[21]** M. L. Michelsen. The isothermal flash problem. part i. stability. *Fluid Phase Equilibria*, 9(1), 1982.

**[22]** L. E. Baker, A. C. Pierce, and K. D. Luks. Gibbs energy analysis of phase equilibria. *Society of Petroleum Engineers Journal*, 22(5), 1982.

**[23]** Andreas Zuend and John H. Seinfeld. A practical method for the calculation of liquid–liquid equilibria in multicomponent organic-water-electrolyte systems using physicochemical constraints. *Fluid Phase Equilibria*, 337:201–213, 2013.

**[24]** Y. S. Teh and G. P. Rangaiah. A study of equation-solving and Gibbs free energy minimization methods for phase equilibrium calculations. *Chemical Engineering Research and Design*, 80(7):745–759, 2002.

**[25]** R. P. Brent. *Algorithms for Minimization without Derivatives*. Prentice-Hall, Prentice-Hall, Englewood Cliffs, New Jersey, 1973, 195 pp. ISBN 0-13-022335-2.

**[26]** W.H. Press, S.A. Teukolsky, W.T. Vetterling, B.P. Flannery, and M. Metcalf. *Numerical Recipes in Fortran 90: Volume 2, Volume 2 of Fortran Numerical Recipes: The Art of Parallel Scientific Computing.* Cambridge CB2 8RU, United Kingdom also:Published in the United States of America by Cambridge University Press, New York 1996.

**[27]** J. J. Moré, B. S. Garbow, and K. E. Hillstrom. *User Guide for MINPACK-1.* Argonne National Laboratory Report ANL-80-74, 1980.

**[28]** J. J. Moré, D. C. Sorensen, K. E. Hillstrom, and B. S. Garbow. *The MINPACK project, in Sources and Development of Mathematical Software.* Englewood Cliffs, New Jersey, United States, Prentice-Hall, Inc., 1984. Wayne R. Cowell, ISBN 0-13-823501-5.

**[29]** C. Fountoukis and A. Nenes. Isorropia II: a computationally efficient thermodynamic equilibrium model for $K^+$–$Ca^{2+}$–$Mg^{2+}$–$NH_4^+$–$Na^+$–$SO_42$––$NO_3^-$–$Cl^-$–$H_2O$ aerosols. *Atmospheric Chemistry and Physics*, 7(17):4639–4659, 2007.

**[30]** Mark Z. Jacobson. *Fundamentals of Atmospheric Modeling.* Cambridge University Press, New York, 1999.

**[31]** Simon Clegg and Kenneth Pitzer. Thermodynamics of multicomponent, miscible, ionic solutions: generalized equations for symmetrical electrolytes. *Journal of Physical Chemistry*, 96(8):3513–3520, 1992.

**[32]** S. L. Clegg, P. Brimblecombe, and A. S. Wexler. Thermodynamic model of the system $H^+$–$NH_4^+$–$SO_4^{2-}$–$NO_3^-$–$H_2O$ at tropospheric temperatures. *Journal of Physical Chemistry A*, 102(12):2137–2154, 1998.

**[33]** S. L. Clegg, J. H. Seinfeld, and P. Brimblecombe. Thermodynamic modelling of aqueous aerosols containing electrolytes and dissolved organic compounds. *Journal of Aerosol Science*, 32(6):713–738, 2001.

**[34]** Anthony S. Wexler and Simon L. Clegg. Atmospheric aerosol models for systems including the ions $H^+$, $NH_4^+$, $Na^+$, $SO_4^{2-}$, $NO_3^-$, $Cl^-$, $Br^-$, and $H_2O$. *Journal of Geophysical Research: Atmospheres*, 107(D14):ACH 14–1–ACH 14–14, 2002.

**[35]** A. Fredenslund, R. L. Jones, and J. M. Prausnitz. Group-contribution estimation of activity coefficients in nonideal liquid mixtures. *AICHE J*, 21(6):1086–1099, 1975.

**[36]** Rahul A. Zaveri, Richard C. Easter, Jerome D. Fast, and Leonard K. Peters. Model for simulating aerosol interactions and chemistry (mosaic). *Journal of Geophysical Research: Atmospheres*, 113(D13), 2008.

**[37]** Rahul A. Zaveri, Richard C. Easter, and Anthony S. Wexler. A new method for multicomponent activity coefficients of electrolytes in aqueous atmospheric aerosols. *Journal of Geophysical Research: Atmospheres*, 110(D2), 2005.

**[38]** A. Zuend, C. Marcolli, B. P. Luo, and T. Peter. A thermodynamic model of mixed organic-inorganic aerosols to predict activity coefficients. *Atmospheric Chemistry and Physics*, 8(16):4559–4593, 2008.

**[39]** A. Zuend, C. Marcolli, A. M. Booth, D. M. Lienhard, V. Soonsin, U. K. Krieger et al. New and extended parameterization of the thermodynamic model AIOMFAC: calculation of activity coefficients for organic-inorganic mixtures containing carboxyl, hydroxyl, carbonyl, ether, ester, alkenyl, alkyl, and aromatic functional groups. *Atmospheric Chemistry and Physics*, 11(17):9155–9206, 2011.

**[40]** H. Yin, J. Dou, L. Klein, U. K. Krieger, A. Bain, B. J.Wallace et al. Extension of the AIOMFAC model by iodine and carbonate species: applications for aerosol acidity and cloud droplet activation. *Atmospheric Chemistry and Physics Discussions*, 2021:1–57, 2021.

**[41]** W. D. Yan, M. Topphoff, C. Rose, and J. Gmehling. Prediction of vapor–liquid equilibria in mixed-solvent electrolyte systems using the group contribution concept. *Fluid Phase Equilibria*, 162(1-2):97–113, 1999.

**[42]** N. Hodas, A. Zuend, W. Mui, R. C. Flagan, and J. H. Seinfeld. Influence of particle-phase state on the hygroscopic behavior of mixed organic-inorganic aerosols. *Atmospheric Chemistry and Physics*, 15(9):5027–5045, 2015.

**[43]** Hichem Bouzidi, Andreas Zuend, Jakub Ondráek, Jaroslav Schwarz, and Vladimir dímal. Hygroscopic behavior of inorganic-organic aerosol systems including ammonium sulfate, dicarboxylic acids, and oligomer. *Atmospheric Environment*, 229:117481, 2020.

**[44]** J. D. Surratt, A. W. H. Chan, N. C. Eddingsaas, M. N. Chan, C. L. Loza, A. J. Kwan et al. Reactive intermediates revealed in secondary organic aerosol formation from isoprene. *PNAS*, 107(15):6640–6645, 2010.

# 4

# Chemical Mechanisms and Pure Component Properties

In all other chapters, we discuss how representations of chemical and physical state space of aerosol particles are manifest through bulk, modal, sectional, and single particle representations. Knowledge about the composition space of aerosol particles can be known a-priori through a series of direct and indirect observations [1, 2]. In this chapter, we discuss mechanisms that describe the creation and evolution of individual compounds through gas phase reactions. These mechanisms contain the rules that dictates which compounds react with each other along with any new compounds that are produced and the rate at which this all happens. With the information held in text files, we take a short detour in this chapter and explore methods to extract this information and subsequently set up the differential equations for solution. This process leads to the automatic generation of aerosol models driven by a single chemical mechanism file and we provide examples of two existing automatic aerosol model generation schemes provided in both Python and Julia. In reviewing these existing frameworks we have to design an appropriate mapping of compounds in the gas and condensed phases to a set of arrays continuing the developments covered in Chapter 2. Whilst we do not provide examples of condensed phase reactions in the proceeding text, we re-iterate that the purpose of this book is to demonstrate and practice mechanisms for solving generic problems in aerosol science. With this in mind, methods for extracting information from chemical mechanism files and creating a dynamic simulation could be translated to a system with reactions taking place in multiple phases.

In any aerosol model that captures a particulate composition, we need to define values of pure components and mixture properties. Whilst we might use experimental values directly, for the many thousands of possible organic compounds in atmospheric aerosols, manual calculation becomes both laborious and largely infeasible. To align with the automatic generation of equations that dictate the chemical evolution of a gas phase, here we discuss methods used to predict the properties of individual molecules. We again take a little detour and look at common formats used to represent molecular structures and tools that take that information to extract specific functional groups. There are multiple methods used to predict pure component properties, and each will require its own combination of specific functional groups. Whilst this chapter focuses on pure component vapor pressures and density, the tools we present are generic and can be used in any workflow that requires moving from a chemical structure to a particular property.

*Introduction to Aerosol Modelling: From Theory to Code.*
First Edition. Edited by David Topping and Michael Bane.
© 2022 John Wiley & Sons Ltd. Published 2022 by John Wiley & Sons Ltd.

## 4.1 Chemical Mechanisms

A chemical mechanism defines a set of rules by which a series of chemical reactions take place. From a programming perspective, these mechanisms are typically encapsulated in a text file that defines each individual chemical reaction and the rate at which it occurs. For example, the master chemical mechanism (MCM) [3–8] describes the detailed gas-phase chemical processes involved in the tropospheric degradation of a series of primary emitted volatile organic compounds (VOCs). The MCM protocol is divided into a series of subsections dealing with initiation reactions, the reactions of the radical intermediates and the further degradation of first and subsequent generation products [8]. Once the complexity of the parent molecules and reaction pathways grows and potentially millions of compounds are generated, this far exceeds the size of mechanisms that can be written manually. With this in mind, The Generator for Explicit Chemistry and Kinetics of Organics in the Atmosphere (GECKO-A) has been developed for the automatic writing of explicit chemical schemes of organic species and their partitioning between the gas and condensed phases [9, 10].

There have been many research papers on detailed chemical kinetics in the gas phase, and we direct the reader to the academic literature for more information on these [1]. In the proceeding sections, we focus on how we use information held within chemical mechanisms to develop a model that not only captures the evolving chemistry of the gas phase but allows us to develop models that couple gas and condensed phase processes.

For atmospheric aerosol, implementing this coupling is essential to capture the dynamic change in aerosol size and composition. For example, as the degradation of a parent VOC proceeds, production of compounds with a low enough vapor pressure could lead to condensational growth of pre-existing aerosol particles depending on the composition and phase state of said particles (see Chapters 2 and 3, respectively). Likewise, a changing gas phase environment could lead to evaporative loss and shrinkage from existing particles depending on the difference between the equilibrium pressure at the surface of the particle and the partial pressure in the gas phase. We discuss this within the context of solving the droplet growth equation introduced in Chapter 2. In this chapter, we extend those developments to include an evolving gas phase and explicitly track the concentration of each component in a "reacting" gas phase and through partitioning to the aerosol phase.

### 4.1.1 Gas Phase Only Model

Given chemical mechanism files contain information on reactants, products, and reaction rates, we need some way of extracting that information and simulating evolving concentrations according to some initial conditions. In the following, we will demonstrate the general approach taken by two existing community models, PyBox [11] and JlBox [12], both of which are designed to automatically generate a model using Python and Julia, respectively. This provides us with a useful comparison between two models that rely on the same model structure whilst using specific language dependent features. First, we review their structure and the method for automated parsing of a chemical mechanism file. Following [12], the gas phase reaction of chemicals in atmosphere follows the gas kinetics equation:

$$\frac{d}{dt}[C_{g,i}] = -\sum_{l} r_l S_{i,l} \tag{4.1}$$

$$r_l = k_l \prod_{\forall i, S_{i,l} > 0} [C_{g,i}]^{S_{i,l}} \tag{4.2}$$

where $[C_{g,i}]$ is the concentration of compound $i$ in the gas phase (molecules $\cdot$ cm$^{-3}$), $r_l$ is the reaction rate of reaction $l$, $k_l$ is the corresponding reaction rate coefficient of reaction $l$, and $S_{i,l}$ is the value of the stoichiometry matrix for compound $i$ in this reaction. The algebraic sum of components $S_{i,l}$ in equation is the total order of the reaction [13]. It is worthwhile looking at how units for $k_l$ can subsequently vary, according to the order of reaction taking place. For example, imagine component $C_{g,i}$ is produced through a first-order reaction with component $C_{g,j}$. We can write the reaction as follows, noticing the positive sign:

$$\frac{d}{dt}[C_{g,i}] = k[C_{g,j}] \tag{4.3}$$

where $k$ is given in units of s$^{-1}$. Next we consider a scenario where $C_{g,i}$ is lost through a second-order reaction with $C_{g,j}$ to produce $C_{g,k}$. We can write the relevant equations as

$$\frac{d}{dt}[C_{g,i}] = -k[C_{g,i}][C_{g,j}] \tag{4.4}$$

$$\frac{d}{dt}[C_{g,k}] = k[C_{g,i}][C_{g,j}] \tag{4.5}$$

where $k$ is now given in units of cm$^3 \cdot$ molecule$^{-1} \cdot$ s$^{-1}$. Two examples of third-order reactions are given below:

$$\frac{d}{dt}[C_{g,i}] = -k[C_{g,i}]^2[C_{g,k}] \tag{4.6}$$

$$\frac{d}{dt}[C_{g,z}] = k[C_{g,i}][C_{g,j}][C_{g,k}] \tag{4.7}$$

where $k$ is now given in units of cm$^6 \cdot$ molecule$^{-2} \cdot$ s$^{-1}$. From a programming perspective this tells us that, as we use arrays to track how the concentrations of $C_{g,i}$, $C_{g,j}$, $C_{g,k}$, etc. evolve over time, we also need to multiply concentrations of these components with rate coefficients $k$ according to the stoichiometry of reactions $S_{i,l}$ we are simulating. The above ordinary differential equations (ODEs), for example Equation (4.1), fully determines the concentrations of gas phase chemicals at any time given reaction coefficients $k_j$, a stoichiometry matrix $\{S_{ij}\}$ and initial values.

We can use a more general expression in Equation (4.8) [12] that we will expand on when coupling to a condensed phase in Section 4.1.2:

$$\frac{dy}{dt} = f(y; \vec{p}), y = (C_{g,1}, C_{g,2}, \ldots, C_{g,n_{chem}}) \tag{4.8}$$

where $y$ represents the states of the ODE, $n_{chem}$ is the number of chemicals, $\vec{p}$ is a vector of parameters of the ODE, and $f(y)$ is the right-hand side (RHS) function implicitly defined by Equation (4.1). For example, the gas phase simulation of a mechanism with $n_{chem} = 400$ chemicals has to solve an ODE with 400 states. Meanwhile, the Jacobian (required by implicit ODE solvers) will require an $400 \times 400$ matrix [12]. We do not delve into the advantages and disadvantages of individual ODE solvers in this book, but we can provide general guidance based on the domain-based knowledge that has accrued from aerosol model developments and known best practices. The Jacobian is a matrix that contains the partial derivatives of each RHS function with respect to all components, as expressed in Equation (4.9).

$$J = \begin{bmatrix} \frac{\partial f(y_1)}{\partial y_1} & \cdots & \frac{\partial f(y_1)}{\partial y_n} \\ \cdots & \cdots & \cdots \\ \frac{\partial f(y_3)}{\partial y_1} & \cdots & \frac{\partial f(y_3)}{\partial y_1} \end{bmatrix} \tag{4.9}$$

An analytical form of the Jacobian is beneficial when using an implicit or semi-implicit ODE solver since this can be one of the most expensive steps. Implicit methods are used for solving problems based around the chemical kinetics of atmospheric composition due to their *stiff* nature. Loosely defined, a stiff system is one in which there are widely varying timescales of change between individual components. As Huang et al. [12] note, the Jacobian matrix of the RHS is needed in implicit ODE solvers as well as in adjoint sensitivity analysis. The accuracy of the Jacobian matrix, however, has variable requirements in each case. For implicit ODE solvers, when doing forward simulations, the accuracy of the matrix only affects the rate of convergence instead of the accuracy of the result. Some methods can tolerate inaccurate Jacobian matrices, whilst for adjoint sensitivity analysis, accurate Jacobian matrices are needed [12].

For gas phase chemical kinetics, an analytical form of $J$ is relatively easy to generate by applying the rules of differentiation to the expressions provided in Equations (4.1)–(4.8). For coupling between the gaseous and condensed phase, this becomes more complex to derive manually as the total loss and gain terms include the droplet growth equation and any associated representations of condensed phase properties. Nonetheless, if an analytical form cannot be constructed then, typically, there would be the option to use a finite difference approach. For example, the first expression $\frac{\partial f(y_1)}{\partial y_1}$ in (4.9) could be calculated using two calls to $f(y_1)$ as follows:

$$\frac{\partial f(y_1)}{\partial y_1} \approx \frac{(f(y_1 + \Delta y_1) - f(y_1))}{\Delta y_1} \tag{4.10}$$

where $\Delta y_1$ needs to be defined by the user. Once we know how many components we need to track in the gas phase, we can initialize an array that holds the evolving concentration of each compound. If we wish to build a model of an evolving gas phase by solving $n$ ODEs in Python, we might follow the structure followed in code Listing 4.1, where we implement a set of $n$ ODEs defined by Equation (4.1). If we are dealing with a relatively *small* chemical mechanism with, for example, <20 compounds we might write the relevant code by hand. Indeed, we can use the template already provided in Listing 4.1 and simulate a relatively simple gas phase mechanism. To do this, we modify the volatility distribution, or VBS, example given in Chapter 2. To recap, in the VBS each volatility bin is linearly separated in log space. Representing the concentration of condensed mass that would lead to 50% of the mass in each bin condensing as a volatility metric, these bins typically range from $10^{-6}$ to $10^3$ µg · m$^{-3}$. In solving the droplet growth equation, we did not allow our representative compounds, which keep a fixed volatility, to move between each bin. In the following code demonstration, we build a gas phase only simulation where material is redistributed between each volatility bin through second-order reactions involving an oxidant as described by Equation (4.4).

**Table 4.1** Initial conditions to simulate transfer of mass across each volatility bin through second-order reactions with a fixed oxidant. In this simulation, we assume our oxidant remains at a concentration of 10 parts-per-billion (ppb), the temperature is 298.15 K, and the molecular weight of each representative compound, thus volatility bin, is 200 $g \cdot mol^{-1}$. We also assume the rate coefficient for each second-order reaction is $10^{-15}$ $cm^3 \cdot molecule^{-1} \cdot s^{-1}$.

**Partitioning parameters**

| $Log10(C_i^*)$ | $C_i$ ($\mu g \cdot m^{-3}$) |
| --- | --- |
| −6 | 0.1 |
| −5 | 0.1 |
| −4 | 0.1 |
| −3 | 0.3 |
| −2 | 0.6 |
| −1 | 1.0 |
| 0 | 1.5 |
| 1 | 3.0 |
| 2 | 2.0 |
| 3 | 1.0 |

We specify a fixed abundance of our oxidant, in this example ozone, in parts-per-billion [ppb] and use the same rate coefficient for each volatility bin. Table 4.1 outlines the initial conditions used in this simulation whist Figure 4.1 illustrates a schematic of our model. Note that, for the highest volatility bin ($Log10(C_i^*) = 3$), our model only simulates loss from this bin, described by the reaction with our oxidant to generate mass that moves mass into the next lowest volatility bin ($Log10(C_i^*) = 2$). Likewise, for the lowest volatility bin ($Log10(C_i^*) = -6$), our model only simulates an increase of mass into this bin depending on the concentration in the next highest volatility bin. For all other volatility bins, there are loss and gain routes from neighboring bins. We should point out that the first exercise at the end of this chapter is to create this model yourself. If you would like to try this yourself, please do so before reading on.

```
# Import modules
import numpy as np
from scipy.integrate import odeint


# -- define some initial conditions --
# .....
# ----------------------------------------

def dydt(input):

    # -- perform some calculations --
    # .....
    # ----------------------------------------

    return dydt_array
```

```
# Define the time over which the simulation will take place
t = np.linspace(0, 10000, num=1000)

# Call the ODE solver with reference to our function, dydt, setting
    the
# absolute and relative tolerance, atol and rtol respectively.
solution = odeint(dy_dt, array, t, rtol=1.0e-6, atol=1.0e-4, tcrit=
    None)

# Do something with the output contained within variable 'solution
    '..
```

**Listing 4.1** Template structure for solving ODEs using the Scipy Python package. After importing both Numpy and the **odeint** function from the Scipy module, we define a function **dydt** that will return the RHS function values when called by **odeint**. Following this, we define an array t that holds the points in time at which we would like the values in y to be passed back to us. *Associated code - chap4_alg1_ODE_template_example.py.*

Code Listing 4.2 provides the full example for comparison with the template given in Listing 4.1. As you read through this code listing, take some time to understand how the competing reactions are contributing to the rates of change of concentrations in each bin according to the governing equations given by (4.1). We specify a fixed abundance of our oxidant, in this example ozone, in parts-per-billion [ppb] and convert this to molecules · cm$^{-3}$ using a constant conversion factor of $2.4631010^{10}$ [13]. In our Python script, as this is defined before the call to the ODE solver and is within the same Python *.py* file then we can access this value from our RHS definition. Please

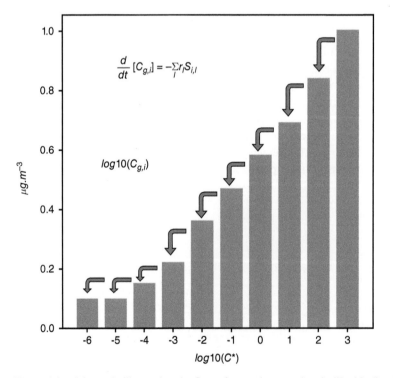

**Figure 4.1** Schematic illustrating the flow of mass down each volatility bin (Log10($C_i^*$)) through a second-order reaction with an oxidant.

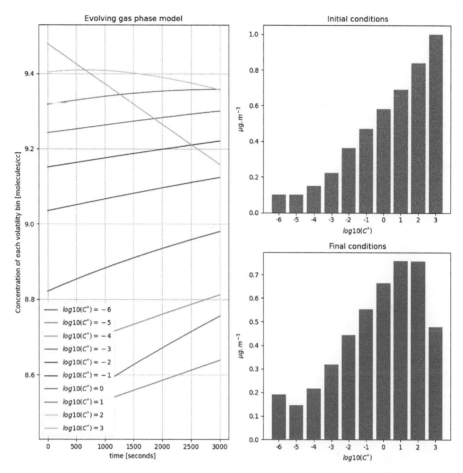

**Figure 4.2**  Evolution of gas-phase concentrations through simulation of second-order reactions between each volatility bin and fixed concentration of oxidant. The left hand figure plots the change in concentrations ($\log10(\text{molecules} \cdot \text{cm}^{-3})$) over time. The top right barplot illustrates the initial volatility distribution in $\mu g \cdot m^{-3}$, whilst the bottom right plot illustrates the change in the volatility distribution at the end of the simulation.

note in this example we thus assume that our oxidant concentration remains constant, whereas in simulating a chemical mechanism with multiple production and loss terms, this would be considered an nonphysical representation. Nonetheless, this demonstration helps us build on our existing frameworks. To supplement this, Figure 4.2 not only plots the change in concentration in each bin as a function of time through the simulation but also compares the initial VBS distribution with the final distribution following the transfer of mass between each bin.

```
# Import the relevant libraries, including an ODE solver from Scipy
# We only want to use the 'odeint' [Solve Initial Value Problems]
    from the
# scipy.integrate package
import numpy as np
from scipy.integrate import odeint
import matplotlib.pyplot as plt # Import Matplotlib so we can plot
    results
```

```python
# Ideal gas constant [m3 atm K-1 mol-1]
R_gas_other=8.2057e-5
# Avogadro's number
NA=6.0221409e+23

# Set the ambient temperature
Temp_K=298.15

# Defining the gas phase components
num_species = 10 # Number of condensing species from the gas phase

# The molecular weight of each condensing specie [g/mol]
mw_array=np.zeros((num_species), dtype=float)
mw_array[:]=200.0 # Assuming a constant value for all components

# The volatility of each specie, using the mass based C* convention
# Here we assuming a linear seperation in Log10 space
log_c_star = np.linspace(-6, 3, num_species)
Cstar = np.power(10.0,log_c_star)
# Convert C* to a pure component saturation vapour pressure [atm]
P_sat = (Cstar*R_gas_other*Temp_K)/(1.0e6*mw_array)

# Initialise an abundance of material in each volatility bin [
    micrograms/m3]
abundance = np.zeros((num_species), dtype = float)
abundance[0] = 0.1
abundance[1] = 0.1
abundance[2] = 0.15
abundance[3] = 0.22
abundance[4] = 0.36
abundance[5] = 0.47
abundance[6] = 0.58
abundance[7] = 0.69
abundance[8] = 0.84
abundance[9] = 1.0

# Initialise abundance of a gas phase oxidant in ppb
ozone = 10.0
Cfactor= 2.55e+10 #ppb-to-molecules/cc
# convert abundance of oxidant to molecules / cc
ozone_abundance = ozone*Cfactor
# Define a rate coefficient for second order reactions
rate_coefficient = 1.0e-15

# Unit conversion of gas abundance to molecules / cc
gas_abundance = ((abundance*1.0e-6)/(mw_array))*1.0e-6*NA

# New define an array that holds the molecular abundance
array = np.zeros((num_species), dtype=float)
array[0:num_species] = gas_abundance


###################################################################


# Define RHS function
def dy_dt(array,t):

    dy_dt_array = np.zeros((num_species), dtype=float)

    # Calculate the rate of each reaction explicitly
    rate_1 = ozone_abundance*array[1]*rate_coefficient
    rate_2 = ozone_abundance*array[2]*rate_coefficient
    rate_3 = ozone_abundance*array[3]*rate_coefficient
    rate_4 = ozone_abundance*array[4]*rate_coefficient
```

```
    rate_5 = ozone_abundance*array[5]*rate_coefficient
    rate_6 = ozone_abundance*array[6]*rate_coefficient
    rate_7 = ozone_abundance*array[7]*rate_coefficient
    rate_8 = ozone_abundance*array[8]*rate_coefficient
    rate_9 = ozone_abundance*array[9]*rate_coefficient

    # Write RHS for each volatility bin according to our simple
    # mechanism
    dy_dt_array[0] = rate_1
    dy_dt_array[1] = rate_2 - rate_1
    dy_dt_array[2] = rate_3 - rate_2
    dy_dt_array[3] = rate_4 - rate_3
    dy_dt_array[4] = rate_5 - rate_4
    dy_dt_array[5] = rate_6 - rate_5
    dy_dt_array[6] = rate_7 - rate_6
    dy_dt_array[7] = rate_8 - rate_7
    dy_dt_array[8] = rate_9 - rate_8
    dy_dt_array[9] = -1.0*rate_9

    return dy_dt_array

t = np.linspace(0, 3000, num=1000)
```

Listing 4.2 Python code to simulate a relatively simply evolving gas phase mechanism through second-order reactions with a fixed concentration of oxidant. *Associated code - chap4_alg2_gas_only.py.*

Whilst the previous example is a useful demonstrator of simulations that are amenable to manual code generation, as our mechanisms grow beyond hundreds if not thousands of differential equations, it becomes advantageous to create an automated model generation process [14]. There are additional benefits associated with automatic model generation, including the ability to make functional changes that automatically propagate through the entire code structure as well as reduced chance of errors entered in small regions of code that might be otherwise difficult to track down. In addition, as parameters including reaction rates might be updated, then an automated file generation could adapt to the change in scientific literature. To create a model structure and then solve Equation (4.1) we need to perform a number of steps as clarified in Algorithm 4.1.

In the following, we will develop a strategy to perform each point in turn. Regarding point (3), code Listing 4.3 provides a template of how this may look in Python. In the explicit case, each reaction and thus $f(y_i)$ is written line by line, effectively providing a code variant of Equations (4.3)–(4.6) depending on the order of the reaction. The alternative is to pass information on the stoichiometry of each compound in each reaction, $S_{i,l}$ from Equation (4.1). The former will allow the user to interpret the specific reactions that contribute to $\frac{d}{dt}[C_{g,i}]$ line by line, whilst the latter is done through a generic set of operations that may be carried out using loops or vectorized array operations. The first method is intended to statically figure out the symbolic expressions of the loss and gain for each species as combinations of rate coefficients and gas concentrations and generate the RHS function line by line from the relevant expressions. Whilst this method is straightforward, especially for small cases, it may consume lots of memory and time for compiling when the mechanism file is large (i.e., $>1000$ equations). In this instance, another approach is to store the data on stoichiometry and reactant/product combinations as integer matrices and the RHS function loops through the data to calculate the result.

---

**Algorithm 4.1:** Automated model generation procedure to solve equations defined by (4.1).

---

1 **Result:** Code base for solving Equation (4.1)

2 1. Parse mechanism file;

3 2. Extract reactants, products, stoichiometry, and reaction rates;

4 3. Write RHS function (equations defined by 4.8) as an explicit set of procedures OR as generic function that implements the form of Equation (4.1);

5 **if** *Analytical Jacobian can be derived* **then**

6 | Write Jacobian function to be called by ODE solver;

7 **else**

8 | Specific the use of finite difference Jacobian;

9 **end**

---

```python
# A demonstration of two different approaches to solving
# the ODES for a gas phase simulation.
import numpy as np
# The function that is called by the ODE solver and accepts
# time and concentration arrays as inputs as follows:
# input:
# t - time variable [internal to solver]
# y - concentrations of all compounds in both phases [molecules/cc]
# output:
# dydt - the dy_dt of each compound in each phase [molecules/cc.sec]

def dydt_func(t,y):

    # make sure the y array is not a list.
    y_asnumpy=np.array(y)

    # Initialise the output array
    dy_dt=np.zeros((num_species,1),)

    # Calculate time of day
    time_of_day_seconds=start_time+t

    # Calculate the concentration of RO2 species, using an index
    file created
    # during parsing
    RO2=np.sum(y_asnumpy[RO2_indices])

    # Call a separate function to evaluate reaction coefficients as
    a function
    # of temperature and other variables.
    rates=evaluate_rates(time_of_day_seconds,RO2,H2O,temp)

    # Call a separate function to calculate product of all reactants
    and
    # stoichiometry for each reaction [A^a*B^b etc]
    reactants=reactant_product(y_asnumpy)

    # Multiply product of reactants with rate coefficient to get
    reaction rate
    reactions = np.multiply(reactants,rates)
```

```
# Option (1):
# Use an explicit set of procedures
dydt_gas=dydt_eval(np.zeros((num_species)),reactions)
dydt_gas[0]=-1.0*reactions[0]-reactions[1]\
    -reactions[2]-reactions[3]
dydt_gas[1]=-1.0*reactions[7]+reactions[0]\
    +reactions[41]
dydt_gas[2]=-1.0*reactions[12]-reactions[13]
dydt_gas[4]=-1.0*reactions[2]-reactions[3]\
    +reactions[178]+reactions[593]
dydt_gas[5]=-1.0*reactions[18]-reactions[19]\
    +reactions[2]
# .......

# Option (2):
# Use generic function that multiplies stoichiometric matrix by the
# variables in reactions array
dydt_gas=np.dot(stoich,reactions)
# .......

dy_dt[0:num_species,0]=dydt_gas

return dy_dt
```

**Listing 4.3** Options for defining the Python implementation of ODEs from Equations (4.3) to (4.8). *Associated code - chap4_alg3_dydt_example.py.*

In this specific code example, our function takes a concentration array y and time t of the simulation (s) as input whilst relying on a global variable that defines the number of gas phase components involved in the simulation (num_species). This example is built on the use cases of the PyBox framework, where we also have an array defined as RO2_indices that tracks which compounds in our mechanism are defined as radicals. These are subsequently used to calculate the reaction rates on line 29, passed as input to a function **evaluate_rates**. The list of radicals, and thus relevant indices of our concentration array y, have been automatically generated in PyBox. Specifically, the compound names of each radical are known in advance and, when the mechanism file is parsed, the associated index in the concentration array is stored. On line 36, we multiply the reaction coefficients rates with the combined reactant concentrations (reactants) to arrive at an array that holds the rate of each reaction. In the proceeding code, we then see two options for using this information to evaluate the RHS expressions. In option (1), we provide an example set of lines that explicitly define the reactions involved in the loss and gain terms for each compound, which have indices 0, 1, 2, and so on. In option (2), we use an additional matrix stoich that contains the stoichiometry of each compound in each reaction. Atmospheric chemical mechanisms can have a sparse structure and so we expect stoich to be a sparse matrix. For $n_{chem}$ compounds (ODE states) and $L$ reactions, stoich would have the dimensions $(n_{chem}, L)$. In both cases, we return an array of size $n_{chem}$ which contains the evaluated RHS for each compound at a specific set of concentrations and ambient conditions. In other words, we return an array that stores the value for $\frac{dy}{dt}$ for each compound.

Whether the first or second option is selected, we need to work through points 1 and 2 in Algorithm 4.1 to extract the relevant information from our mechanism file. Following two existing community box-models, PyBox [11] and JlBox [12], we can use regular expressions in Python to extract all the relevant information we need. A regular expression is a set of characters that define a search pattern. As stated in the official Python documentation [15], regular expressions represent *a tiny, highly specialized programming language embedded inside Python and made available through the remodule. Using this little language, you specify the rules for the set of possible strings that you want to match; this set might contain English sentences, or e-mail addresses, or TeX commands, or anything you like.* There is extensive literature and indeed training courses on how to use regular expressions, but let us demonstrate the application on a subset of reactions taken from the MCM mechanism for $\alpha$-pinene, a biogenic atmospheric VOC, as provided in Text Box 4.1. You may wish to create your own set of tools using regular expressions, and hopefully this acts as a useful starting point. In Text Box 4.1, we also add three hypothetical reactions at the end of this example for the purposes of illustrating extracting information for a range of chemical kinetic formats, as demonstrated shortly.

---

**Text Box 4.1: Example chemical mechanism file.**

{1.} APINENE + NO3 = NAPINAO2 : 1.2D-12*EXP(490/TEMP)*0.65 ;
{2.} APINENE + NO3 = NAPINBO2 : 1.2D-12*EXP(490/TEMP)*0.35 ;
{8.} NAPINAO2 + HO2 = NAPINAOOH : KRO2HO2*0.914 ;
{11.} NAPINAO2 = APINBNO3 : 6.70D-15*0.1*RO2 ;
{12.} NAPINAO2 = NAPINAO : 6.70D-15*0.9*RO2 ;
{13.} NAPINBO2 + HO2 = NAPINBOOH : KRO2HO2*0.914 ;
{14.} NAPINBO2 + NO = NAPINBO + NO2 : KRO2NO ;
{16.} NAPINBO2 = APINANO3 : 2.50D-13*0.1*RO2 ;
{17.} NAPINBO2 = NAPINBO : 2.50D-13*0.8*RO2 ;
{18.} NAPINBO2 = NC101CO : 2.50D-13*0.1*RO2 ;
{19.} APINOOA = C107O2 + OH : KDEC*0.55 ;
{20.} APINOOA = C109O2 + OH : KDEC*0.45 ;
{21.} APINOOB = APINBOO : KDEC*0.50 ;
{43.} NAPINAOOH = NAPINAO + OH : J(41) ;
{44.} NAPINAO = PINAL + NO2 : KDEC ;
{45.} APINBNO3 + OH = APINBCO + NO2 : 3.64D-12 ;
{46.} NAPINBOOH + OH = NAPINBO2 : 1.90D-12*EXP(190/TEMP) ;
{47.} NAPINBOOH + OH = NC101CO + OH : 1.23D-11 ;
{48.} NAPINBOOH = NAPINBO + OH : J(41) ;
{60.} 3.0A + B = 2C : 1.2D-13 ;
{61.} C + 2.0D = E : 1.8D-13 ;
{62.} D + A = 1.0E : 1.2D-12 ;

---

Here we see a variety of string expressions used to represent compound names within a chemical equation format. For example, the first line describes the reaction between

$\alpha$-pinene and $NO_3$ to produce a new compound NAPINAO2. The syntax you see in this box has been generated to work with the Kinetic PreProcessor (KPP) [16]. The KPP tool was designed for use by the atmospheric science community. Similar to the example we are about to construct, KPP translates a chemical mechanism into Fortran77, Fortran90, C, or Matlab simulation code together with a suitable numerical integration scheme.

The choice of naming convention is one that has been dictated to us through the mechanism generator we have at hand, where each name is also associated to a chemical structure as discussed in Section 4.2 when we need to predict pure component and molecular properties. For now, we focus on extracting all unique compound names and the structure of the chemical equations that dictate the evolving concentrations over time.

The equation number identifiers wrapped in the {} brackets are generated by the MCM extraction facility and do not increase linearly in this example as we have selected only a subset from the entire file. Likewise, for our additional examples, we simply prescribe numbers above the previous subset. Clearly, the format does not strictly matter provided our regular expressions match the relevant features to search for. Indeed, it quickly becomes obvious that even though we might create a parsing function to extract all the information we need to construct our model, we need to be careful that all the relevant components are captured. This is certainly true when our mechanism files contain a large number of equations that become infeasible to count and check manually. In this case, evaluating the correctness of our method can be done by manually checking a subset that represents all potential combinations of string-numeric expressions.

Code Listing 4.4 is adapted from a set of parsing functions from PyBox. This code snippet reads in the mechanism file outline earlier, saved as a text file we name `example_mechanism.txt`. Once opened, we first strip any newline or tab characters using the **.replace** operator. The variable `equation_numbers` in this particular example is a list of all numbers found between the {} separators with the knowledge that these are only used for this purpose. In this case, the regular expression $\setminus\{ (.*?).\setminus\}$ used within the **findall** function of the Python module **re** finds all equation numbers. In a similar fashion, we can extract a list that contains the text entry for every reaction using the regular expression $\setminus\}(.*?)\setminus;$ where we have now defined our boundaries using the final equation number identifier bracket } and semi-colon ";" that signals the end of a given reaction expression.

```python
# Import the relevant modules
import re
import pdb
import collections

# First read in the text file and print extracted equations to
    screen
filename= 'example_mechanism.txt'
text=open(filename,'rU')
with open (filename, "r") as myfile:
    data=myfile.read().replace('\n', '').replace('\t', '')

equation_numbers = re.findall(r"\{(.*?).\}",data)
print("Equation numbers = ",equation_numbers)
```

```
print("Total number of equations = ",len(equation_numbers))
eqn_list=re.findall(r"\}(.*?)\;",data)
print("Equation list = ",eqn_list)
```

**Listing 4.4** Example of parsing routine to extract the equation numbers and total number of equations in a chemical mechanism file. *Associated code - chap4_alg4_Parsing_example.py.*

If you run this code using the Python interpreter, you would see output provided in Listing 4.5:

```
$ Equation numbers =  ['1', '2', '8', '11', '12', '13', '14', '16',
  '17', '18', '19', '20', '21', '43', '44', '45', '46', '47', '48'
  ]
$ Total number of equations =  19
$ Equation list =  ['  APINENE + NO3 = NAPINAO2 : 1.2D-12*EXP(490/
  TEMP)*0.65 ', '  APINENE + NO3 = NAPINBO2 : 1.2D-12*EXP(490/TEMP)
  *0.35 ', '  NAPINAO2 + HO2 = NAPINAOOH : KRO2HO2*0.914 ', '
  NAPINAO2 = APINBNO3 : 6.70D-15*0.1*RO2 ', '  NAPINAO2 = NAPINAO :
  6.70D-15*0.9*RO2 ', '  NAPINBO2 + HO2 = NAPINBOOH : KRO2HO2
  *0.914 ', '  NAPINBO2 + NO = NAPINBO + NO2 : KRO2NO ', '
  NAPINBO2 = APINANO3 : 2.50D-13*0.1*RO2 ', '  NAPINBO2 = NAPINBO :
  2.50D-13*0.8*RO2 ', '  NAPINBO2 = NC101CO : 2.50D-13*0.1*RO2 ',
  '  APINOOA = C10702 + OH : KDEC*0.55 ', '  APINOOA = C10902 + OH
  : KDEC*0.45 ', '  APINOOB = APINBOO : KDEC*0.50 ', '  NAPINAOOH =
  NAPINAO + OH : J(41) ', '  NAPINAO = PINAL + NO2 : KDEC ', '
  APINBNO3 + OH = APINBCO + NO2 : 3.64D-12 ', '  NAPINBOOH + OH =
  NAPINBO2 : 1.90D-12*EXP(190/TEMP) ', '  NAPINBOOH + OH = NC101CO
  + OH : 1.23D-11 ', '  NAPINBOOH = NAPINBO + OH : J(41) ']
```

**Listing 4.5** Output from running the file *chap4_alg4_Parsing_example.py.*

We can expand this short code example a little further and extract the reactants and products whilst identifying the form of our reaction rate coefficient from every entry in the list `eqn_list`. Code Listing 4.6 is again adapted from PyBox and is designed to split the string entry for each reaction into each component according to key symbols. The aim is to record the reactants and products of each reaction, along with their stoichiometry and form of the reaction rate, so we can build a unique list of all interacting components and build a structure that we can use as a basis to solve our ODEs given by Equation (4.1).

```
# Initialise dictionaries that will store information on the loss
    and gain
# reactions for each compound. Also create dictionaries that store
    the
# stoichiometry and rate coefficient expressions as strings. Finally
# we create a dictionary that will allow us to map a compound name
    to
# an index that is used in the numerical arrays of our simulation.

rate_dict=collections.defaultdict(
        lambda: collections.defaultdict())
loss_dict=collections.defaultdict(
        lambda: collections.defaultdict())
gain_dict=collections.defaultdict(
        lambda: collections.defaultdict())
stoich_dict=collections.defaultdict(
        lambda: collections.defaultdict())
rate_dict_reactants=collections.defaultdict(
        lambda: collections.defaultdict())
species_dict=collections.defaultdict()
```

```
species_dict2array=collections.defaultdict()

#Create an integer that stores number of unique species
species_step=0

# In this loop we interrogate each line that has been extracted from
    our
# mechanism file and is now stored as a list
for equation_step in range(len(equation_numbers)):
    equation_full=eqn_list[equation_step]
    #split the line into reactants and products
    equation=equation_full.split(':',1)[0].split('=',1)
    # extract content to the left of the previous split [reactants]
    reactants=equation[0].split('+')
    #strip away all whitespace
    reactants= [x.strip(' ') for x in reactants]
    # extract content to the right of the previous split [products]
    products=equation[1].split('+')
    #strip away all whitespace
    products = [x.strip(' ') for x in products]
    #At the moment, we have not separated the reactant/product from
    #its stoichiometric value
    # Now extract the reaction rate expression
    rate_full=equation_full.split(':',1)[1]
    #strip away all whitespace
    rate_full=rate_full.strip()
    rate_dict[equation_step]="".join(rate_full.split())

    #used to identify reactants by number, for any given reaction
    reactant_step=0
    product_step=0
```

Listing 4.6 Extracting unique compounds and nature of products and reactants. *Associated code - chap4_alg4_Parsing_example.py.*

Now we can proceed to extract the unique reactants and products, we also need to decide on how they are represented in our numerical model. We normally create a structure where the index of any given numerical arrays corresponds with a known component of a model. In this case, this would be a chemical component within a specific phase or the property of an individual component or ensemble. For example, we might specify that the third component of array x, which we would access through the syntax x[2] in Python (recall Python indexing starts at 0), represents the concentration of Ozone in the gas phase. In some cases, it is convenient to store and retrieve information from a non-numerical identifier, including a compound name or objects. Indeed this would improve readability of the code, since an index would have context. We give specific examples around objects in Section 4.3 when calculating fundamental properties and storing information in dictionaries. So far we haven't used dictionaries in Python. It is best to think of a dictionary as a set of key: value pairs, with the requirement that the keys are unique (within one dictionary) [16]. In code Listing 4.7, we create a simple dictionary and manually create key:value pair entries. We then cycle through each pair and print the values to screen.

```
#First we define the dictionary by name
#with no entries to start with
aerosol_dict = {}

#Now we can add a few entries
aerosol_dict['type'] = 'Sodium Chloride'
aerosol_dict['size [microns]'] = 10.56
```

```
aerosol_dict['Measured by'] = 'Owen'
aerosol_dict['values'] = [103,304,112,33]

# Cycle through each key:value pair and
# print to the screen
for key, value in aerosol_dict.items():
    print("Key   = ", key)
    print("Value = ", value)

# We can also check whether a key exists
if 'type' in aerosol_dict.keys():
  print("It is in there!")
else:
  print("Not there")
```

Listing 4.7 Example dictionary in Python. *Associated code - chap4_alg5_dict_example.py.*

We can also create multidimensional dictionaries and append to key: value pairs using the **collections** module available in Python as demonstrated in code Listing 4.8. In this example, we initialize a dictionary to expect lists as value entries.

```
#We can also create 2D dictionaries which starts
#to become useful when embedding a structure to
#our key:value pairs. For example, in the following
#we manually create a 2D dictionary of aerosol type
#(by composition) and concentrations measured in
#a particular bin structure
aerosol_records = {'Sodium Chloride' : {'size [microns]':10.56, \
  'place':'coast', 'values':[103,304,112,33]}, \
  'Ammonium Sulphate' : {'size [microns]':0.345, \
  'place':'city', 'values':[450,1500,2003,579]}}
#We can then 'look up' and print the concentration array
#for ammonium sulphate as follows:
print(aerosol_records['Ammonium Sulphate']['values'])

# An alternative approach is to use the collections
#module. This has advantages over the previous approach
#if you want to avoid error messages with missing values
#or retain the order in which entries were added to
#the dictionary. For example the following takes an
#existing list and is able to append values to existing
#keys
aerosol_list = [('NaCl', 10.3), ('NH4NO3', 0.45), \
    ('NaCl', 23.5), ('NH4NO3', 0.34), ('NaCl', 12.1)]
aerosol_dict2 = collections.defaultdict(list)
for type, value in aerosol_list:
    aerosol_dict2[type].append(value)
print(aerosol_dict2.items())
```

Listing 4.8 Example dictionary in Python that stores lists as entries. *Associated code - chap4_alg5_dict_example.py.*

Turning back to our specific problem, we need to extract information to translate Equation (4.1) into the function **dydt**, in code Listing 4.3, that our choice of ODE solver will integrate. The mechanism file as it is currently presented does not give this directly; rather we need to extract the following information:

- Unique species involved in the chemical mechanism.
- A record of each equation that each specie is involved in.

- The rate coefficient, whether constant or variable, for each equation.
- A mapping of each unique specie to a number identifier, such that we can construct our array *y* that represents the states [concentrations] of the ODE.

We can also use this information to construct a Jacobian using simple rules of differentiation. Code Listing 4.9 provides an example solution for combining the required operations.

```python
# In this loop we interrogate each line that has been extracted from
    our
# mechanism file and is now stored as a list
for equation_step in range(len(equation_numbers)):
    equation_full=eqn_list[equation_step]
    #split the line into reactants and products
    equation=equation_full.split(':',1)[0].split('=',1)
    # extract content to the left of the previous split [reactants]
    reactants=equation[0].split('+')
    #strip away all whitespace
    reactants= [x.strip(' ') for x in reactants]
    # extract content to the right of the previous split [products]
    products=equation[1].split('+')
    #strip away all whitespace
    products = [x.strip(' ') for x in products]
    #At the moment, we have not separated the reactant/product from
    #its stoichiometric value
    # Now extract the reaction rate expression
    rate_full=equation_full.split(':',1)[1]
    #strip away all whitespace
    rate_full=rate_full.strip()
    rate_dict[equation_step]="".join(rate_full.split())

    #used to identify reactants by number, for any given reaction
    reactant_step=0
    product_step=0

    # Loop through 'reactants' in this equation, extracting the
    # stoichiometry
    for reactant in reactants:
        reactant=reactant.split()[0]
        # - Extract stoichiometry and unique identifier
        try: #Extract a stoichiometric coefficient, if given
            temp=re.findall(r"[-+]?\d*\.\d+|\d+|\d+",reactant)
            #This extracts all numbers either side
            stoich=temp[0] #Select the first number extracted.
            # Check if this value is before the variable
            # If after, we ignore this. EG. '2NO2'
            # or just 'NO2'
            # If len(temp)==1 then we only have one number
            # and can proceed with the following
            if len(temp)==1:
                if reactant.index(stoich) == 0 :
                    reactant=reactant.split(stoich,1)[1]
                    stoich=float(stoich)
                else:
                    stoich=1.0
            elif len(temp)>1:
                # If this is the case, ensure the reactant
                # extraction is unique. If string is '2NO2'
                # the above procedure extracts 'NO'.
                # We need to ensure the reactant is 'NO2'.
                # We cut the value in temp[0] away from the
                # original string. We can attach the first
```

```
                    # part with the second number. Thus
                    if reactant.index(stoich) == 0 :
                        reactant=reactant.split(stoich,1)[1]+temp[1]
                        stoich=float(stoich)
                    else:
                        stoich=1.0
            except:
                stoich=1.0
            # Store stoichiometry and species flags in dictionaries
            if reactant not in ['hv']:
                stoich_dict[equation_step][reactant_step]=stoich
                rate_dict_reactants[equation_step][reactant_step]=
reactant
                # -- Update species dictionaries --
                #check to see if entry already exists
                if reactant not in species_dict.values():
                    species_dict[species_step]=reactant
                    species_dict2array[reactant]=species_step
                    species_step+=1
                # -- Update loss dictionaries --
                if equation_step in loss_dict[reactant]:
                    loss_dict[reactant][equation_step]+=stoich
                else:
                    loss_dict[reactant][equation_step]=stoich
```

**Listing 4.9** Interrogating each reaction string to identify the reactants, products, and stoichiometry. *Associated code - chap4_alg4_Parsing_example.py.*

In this somewhat large example, we loop through each reaction expression, separating the reactants, products, and rate coefficients. We then loop through each reactant and extract the stoichiometric coefficient. This is again achieved through the use of regular expressions, where the function call defined in code Listing 4.10 extracts all integer and floating point numbers associated with a string expression.

```
temp=re.findall(r"[-+]?\d*\.\d+|\d+|\d+",reactant)
```

**Listing 4.10** Regular expressions used to extract all integer and floating point numbers associated with a string expression. *Associated code - chap4_alg4_Parsing_example.py.*

We also must ensure that we only extract a number that precedes a compound identifier (e.g., $NO_2$). As we encounter unique reactants and products, we create a mapping between compound name, as a dictionary key, and a numerical index for arrays modified in our simulation through, in this case, the dictionary `species_dict2array`.

Once this information is collected we can print to screen, for each compound, the reactions involved in dictating production and loss. For example, we can manually select specie identifiers "NAPINAOOH," "NAPINAO," "NAPINBO2" and "A" and access the identifier and stoichiometry of each reaction associated with production and loss captured in the 2D dictionaries `gain_dict` and `loss_dict`, respectively (see code Listing 4.9). By running the Python script *chap4_alg4_Parsing_example.py*, you should see the same output given in code Listing 4.11.

```
Gain dictionary entries for NAPINAOOH
Equation number:   8  with stoichiometry 1.0
Gain dictionary entries for NAPINAO
Equation number:  12  with stoichiometry 1.0
Equation number:  43  with stoichiometry 1.0
Loss dictionary entries for NAPINBO2
Equation number:  13  with stoichiometry 1.0
Equation number:  14  with stoichiometry 1.0
Equation number:  16  with stoichiometry 1.0
Equation number:  17  with stoichiometry 1.0
Equation number:  18  with stoichiometry 1.0
Loss dictionary entries for A
Equation number:  60  with stoichiometry 3.0
Equation number:  62  with stoichiometry 1.0
```

Listing 4.11 Output from example script to extract reactant and product information. This displays the equation number, as defined in our mechanism file, and stoichiometric information for specific species.

In addition, the sample script loops through each species identifier held in the dictionary `species_dict2array` and prints the list of all other species that dictate both production and loss as gathered from the 2D dictionaries `gain_dict` and `loss_dict`, as shown in Listing 4.12.

```
Species dependent on  APINENE  are: ['APINENE', 'NO3']
Species dependent on  NO3  are: ['APINENE', 'NO3']
Species dependent on  NAPINAO2  are: ['HO2', 'APINENE', 'NAPINAO2', '
    NO3']
Species dependent on  NAPINBO2  are: ['HO2', 'NAPINBO2', 'NAPINBOOH',
    'OH', 'NO3', 'NO', 'APINENE']
Species dependent on  HO2  are: ['HO2', 'NAPINAO2', 'NAPINBO2']
Species dependent on  NAPINAOOH  are: ['NAPINAOOH', 'HO2', 'NAPINAO2'
    ]
Species dependent on  APINBNO3  are: ['OH', 'NAPINAO2', 'APINBNO3']
...
```

Listing 4.12 Output from example script to extract reactant and product information. This displays the relationships between specific compounds in our example mechanism file.

Listing 4.12 serves as a useful small demonstrator of the potential sparsity of chemical mechanisms used in atmospheric aerosol studies. Depending on the ecosystem of tools provided by a programming environment, this might also influence the structure of the code developed. For example, the increased diversity in solver options provided by Julia is also reflected in the ability to benefit from the use of sparse linear solvers. In JlBoxv1.1 [12], the accumulated rate of change of each compound can be expressed as a sparse matrix-vector product of the stoichiometry matrix and the rates of equations vector: a design choice that could be integrated into other language bases.

Whilst we do not build a complete end–end code base here, our demonstration of these parsing tools will allow you to extract information from a mechanism file and translate Equation (4.1) into the function **dydt**, in code Listing 4.3. Of course, there are existing community models that do this. It is worthwhile briefly reviewing these here, where we use those models to conduct a gas phase simulation from a mechanism file. PyBoxv1.0 [11] and JLBoxv1.0 [12] both automatically create a model framework and are written in Python and Julia, respectively. There are some important differences between model architectures, stemming largely from the desire for greater computational performance and the ecosystem of features that Julia currently provides. This presents an interesting set of decisions that you, the reader, may come across as the landscape of scientific knowledge and software evolves. At the time of writing, Python is now a mature language and has a significant support base and wide ecosystem that enables developers to connect, for example, informatics suites with numerical methods in a one-language solution. Julia on the other hand is relatively new, has a smaller support base, and is driven heavily by applications in data science and numerical simulations.

Written in Python, PyBoxv1.0 uses Numba [17] or the Fortran-to-Python-Generator f2py [18] to define both the RHS function **dydt** and the Jacobian within a library of ODE solvers provided by the Assimulo package [19]. The Assimulo module combines a variety of different solvers written in Fortran, C, and even Python via a common high-level interface. Whilst PyBox uses a multilanguage approach, JLBoxv1.1 retains pure Julia implementations throughout.

JlBox uses the DifferentialEquations.jl library to solve the relevant ODEs, which presently includes over 100 solvers. For aerosol simulations, we generally choose semi-implicit/implicit solvers including Rosenbrock, SDIRK, and BDF types of solvers as our problem is numerically stiff. In both PyBox and JlBox, the default solvers use an adaptive time-step and reach a solution according to an absolute and relative error provided by the user. Huang et al. [12] notes that higher error tolerance allows larger time steps, resulting in faster simulation time and versa. However, the error tolerance could also influence the convergence of fully implicit ODE solvers due to the nonlinear nature of the ODE, so it may fail to converge if the tolerance is too high [12]. Whilst these general rules allow a developer and then user to select relevant simulation options, the most appropriate choice will often require an investigation into the trade off between, say, run-time, and relative tolerance. Nonetheless, the Julia documentation associated with the DifferentialEquations.jl library provides guidance for stiff and nonstiff systems. The following informal guidelines on choice of solver are extracted from the Julia documentation (https://julialang.org/):

> For stiff problems at high tolerances (>1e-2?) it is recommended that you use Rosenbrock23 or TRBDF2. These are robust to oscillations and massive stiffness is needed, though are only efficient when low accuracy is needed. Rosenbrock23 is more efficient for small systems where re-evaluating and re-factorizing the Jacobian is not too costly, and for sufficiently large systems TRBDF2 will be more efficient. At medium tolerances (>1e-8?) it is recommended you use Rodas5, Rodas4P (the former is more efficient but the later is more reliable), Kvaerno5, or KenCarp4. As native DifferentialEquations.jl solvers, many Julia numeric types (such as BigFloats, ArbFloats, or DecFP) will work. When the equation is defined via the @ode_def macro, these will be the most efficient. For faster solving at low tolerances (<1e-9) but when VectorFloat64 is used, use radau. For asymptotically large systems of ODEs (N>1000?) where f is very costly and the complex eigenvalues are minimal (low oscillations), in that case CVODE_BDF will be

the most efficient but requires VectorFloat64. CVODE_BDF will also do surprisingly well if the solution is smooth. However, this method can handle less stiffness than other methods and its Newton iterations may fail at low accuracy situations. Another good choice for this regime is lsoda.

There may be some terminology in this statement that is confusing, but the guidance is of general use. When referring to methods such as "Rosenbrock23 or TRBDF2" that is of course referring to the programmatic routine that translates a given ODE solver option into code within Julia. Likewise, you may come across language-dependent references to the same schemes, if available. For example, the same DifferentialEquations.jl documentation (https://diffeq.sciml.ai/stable/) provides a guide for cross referencing the same solvers between Matlab, Python, and Julia, respectively, as

- ode23 -> BS3()
- ode45/dopri5 -> DP5(), though in most cases Tsit5() is more efficient
- ode23s -> Rosenbrock23(), though in most cases Rodas4() is more efficient
- ode113 -> VCABM(), though in many cases Vern7() is more efficient
- dop853 -> DP8(), though in most cases Vern7() is more efficient
- ode15s/vode -> QNDF(), though in many cases CVODE_BDF(), Rodas4(), KenCarp4(), TRBDF2(), or RadauIIA() are more efficient
- ode23t -> Trapezoid()
- ode23tb -> TRBDF2()
- lsoda -> lsoda(), though AutoTsit5(Rosenbrock23()) or AutoVern7(Rodas5()) may be more efficient. Note that lsoda() requires the LSODA.jl extension, which can be added via ]add LSODA; using LSODA.
- ode15i -> IDA(), though in many cases Rodas4() can handle the DAE and is significantly more efficient

### 4.1.2 Coupling the Gaseous and Condensed Phase Using a Fully Moving Sectional Approach

In Section 4.1, we have discussed methods for extracting information from a chemical mechanism file to simulate the evolving chemistry in the gas phase. For this, we had to consider the structure of function definitions required by available ODE solvers. Depending on the volatility of compounds in the gas phase, the ambient conditions and phase state of any available aerosol particles, gas-to-particle partitioning may change the composition of both the gaseous and condensed phases. In Chapter 2, we introduced the droplet growth equation and presented implementations that allowed us to simulate the growth of either a fully moving sectional or modal distribution. However, in those examples we did not allow our gas phase to evolve other than through condensational loss. In this section, we can combine the developments in Chapter 2 with Section 4.1 presented above and develop an approach to create a model that simulates the evolving gas phase chemistry whilst simultaneously capturing the gas-to-particle partitioning process; thus condensational growth and evaporative loss in a mixed phase model. We restrict our model to representing an ideal liquid state for our condensed phase [see Chapter 3], with no condensed phase reactions. It is worthwhile reminding ourselves of the structure of arrays we can use to track the concentration in the gaseous and condensed phases. We show this in Figure 4.3 with an additional visual reference to the coupling of gas phase chemical kinetics and the droplet growth equations.

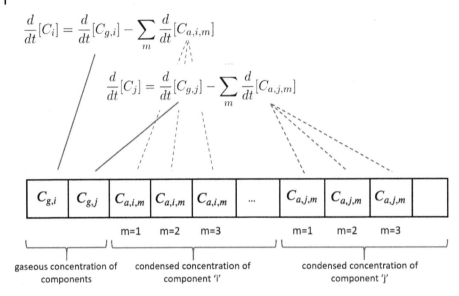

$$\frac{d}{dt}[C_i] = \frac{d}{dt}[C_{g,i}] - \sum_m \frac{d}{dt}[C_{a,i,m}]$$

$$\frac{d}{dt}[C_j] = \frac{d}{dt}[C_{g,j}] - \sum_m \frac{d}{dt}[C_{a,j,m}]$$

| $C_{g,i}$ | $C_{g,j}$ | $C_{a,i,m}$ | $C_{a,i,m}$ | $C_{a,i,m}$ | ... | $C_{a,j,m}$ | $C_{a,j,m}$ | $C_{a,j,m}$ | |

m=1  m=2  m=3          m=1  m=2  m=3

gaseous concentration of        condensed concentration of        condensed concentration of
components                      component 'i'                     component 'j'

**Figure 4.3** Schematic example of representing concentrations in both the gaseous ($C_{g,i}$) and condensed phases ($C_{a,i,m}$) array of concentrations of species $i$ and $j$ in the gas phase, via a state vector, or array, $y$.

Of course, one could use another structure but we still need to ensure the relevant coupling is capture in the RHS function **dydt**. Specifically, we need to expand Equation (4.1) to (4.11) through (4.12).

$$\frac{d}{dt}[C_i] = \frac{d}{dt}[C_{g,i}] - \sum_m \frac{d}{dt}[C_{a,i,m}] \tag{4.11}$$

$$\frac{d}{dt}[C_i] = \left|\sum_l r_l S_{il}\right|_{production} - \left|\sum_l r_l S_{il}\right|_{loss} - \sum_m \frac{d}{dt}[C_{a,i,m}] \tag{4.12}$$

where $C_i$ is the total abundance of compound $i$, split between the gaseous and condensed phases according to Equation (4.13), where $C_{a,i,m}$ is the abundance of compound $i$ in the condensed phase, specifically in size "bin" $m$.

$$C_i = C_{g,i} + \sum_m C_{a,i,m} \tag{4.13}$$

If we combine Equation (2.44) from Chapter 2 that described the rate of change in gas phase concentrations from gas-to-particle partitioning, with Equation (4.1) we arrive at Equation (4.14):

$$\frac{d}{dt}[C_i] = \left|\sum_l r_l S_{il}\right|_{production} - \left|\sum_l r_j S_{il}\right|_{loss} - \sum_m k_{i,m}\left(C_{g,i} - C_{a,i,m}^{eq}\right) \tag{4.14}$$

where $k_{i,m}$ is the first-order mass transfer coefficient for species $i$ to size bin $m$ (please see Chapter 2) and $C_{a,i,m}^{eq}$ is the equivalent equilibrium concentration of species $i$ according to the current composition and phase state of particles in bin $m$. We have also explicitly separated the gain and loss terms from gas phase reactions, specified through the terms $\left|\sum_l r_l S_{il}\right|_{production}$ and $\left|\sum_l r_j S_{il}\right|_{loss}$, respectively. Equation (4.14) is the ODE that described the change in total concentration of each compound involved in the simulation. However, we need to separate the concentrations in both the gaseous and condensed phases such that we can track the relative change in composition, size, and properties of the condensed phase and ensure mass conservation. Now we need to design an appropriate state vector (array) $y$. To repeat this, we track the concentration of each compound $C_i$ in both the gas phase and in each size bin, hence both $C_{g,i}$ and $C_{i,a,m}$, as particles grow and shrink according to the droplet growth equation REF. This defines our state vector $y$, as illustrated in Figure 4.3. Therefore, each cell in our state vector requires us to translate the relevant components of Equation (4.14). For the gas phase components, our production and loss terms represent the set of reactions, as defined by our chemical mechanism file, that lead to an increase and decrease in the concentration of the component $i$. In these expressions, $r_l$ is the reaction rate of reaction $l$ and $S_{il}$ is the stoichiometry of compound $i$ in reaction $l$. We have not included any other production and loss terms, such as primary emissions or deposition, in our work thus far. We also account for any gas-to-particle loss through a difference in partial and effective equilibrium concentrations, summed across all size bins, through the final term. For the cells in our state vector that track the concentration of compound $i$ in each size bin, we implement Equation (4.15):

$$\frac{d}{dt}[C_{a,i,m}] = k_{i,m}\left(C_{g,i} - C_{a,i,m}^{eq}\right) \tag{4.15}$$

Again, our current framework assumes only absorptive partitioning dictates the change in condensed concentrations, but we could add additional production and loss terms to Equation (4.15) including nucleation [Chapter 6], coagulation [Chapter 5], and surface phase reactions, and so on. For the first two, we would need to alter how we represent our size distribution in a sectional approach and we refer the reader to Chapter 7 where we discuss various strategies to do this.

Given this new structure, we can now expand our existing gas phase only model provided in code Listing 4.2 through additions to the RHS function **dydt**. Specifically, we need to integrate the code structure that enabled us to solve the Droplet Growth equation in Chapter 2. If you have not read the previous Section 4.1.1 in this chapter, or Chapter 2, you may find it useful to do so before proceeding. The first exercise of this chapter in particular will help you practice creating your own simple gas phase only model which we expand on in the following section to couple with solving the droplet growth equation.

In this expanded structure, we retain the same initial conditions defined in Table 4.1, whilst also keeping the same polydisperse sectional distribution used in Chapter 2. Code Listings 4.13 and 4.14 present a subset of the modified form of a RHS function that now couples changes in gas phase components due to second-order reactions with a fixed concentration of oxidant and partitioning to a sectional size distribution. In Listing 4.13, we explicitly account for gas phase reactions and add the production and loss terms to array `dy_dt_array` prior to calculating the additional rate of change through solving the droplet growth equation through the loop identified at the end of this listing. We have written the rate of each reaction explicitly, reflecting the format used in Equation (4.1). In each instance, we access the value of our ozone concentration and rate coefficient through variables `ozone_abundance` and `rate_coefficient`, whilst multiplying by the concentration of material in the appropriate volatility bin according to the schematic in Figure 4.1. Following this, we then evaluate the loss and gain for each volatility bin according to the first two terms of Equation (4.12). In the proceeding portion of code (not shown here), we cycle through each size bin and evaluate Equation (4.15) which is then subtracted from the rate of change for each gas phase component as per the third term in Equation (4.12).

In Listing 4.14, we remind ourselves that the RHS function returns the total contribution to the rate of change of all compounds in the gas and condensed phases, where the matrix `dy_dt_gas_matrix` stores the contributions from condensational growth or evaporative loss. Representing each size bin as column in this matrix, we subtract the total loss/gain for each gas phase component. The entire code structure can be found in file *chap4_alg6_droplet_growth_coupled.py*. When examining this file, you will find we provide two RHS functions named **dy_dt** and **dy_dt_ozone**. These are useful for identifying the required changes, but also so we can solve independently of each other and compare the impact on the growth of our sectional size distribution as shown in Figure 4.4.

```python
# Version of the RHS function with ozone effects
def dy_dt_ozone(array,t):

    Cg_i_m_t = array[0:num_species]

    # We are working with 8 size bins, each of which has an
    involatile core
    size_array = np.zeros((num_bins), dtype=float)
    dy_dt_array = np.zeros((num_species+num_species*num_bins), dtype
    =float)
    dy_dt_gas_matrix = np.zeros((num_species,num_bins), dtype=float)

    # Calculate the rate of each reaction explicitly
    rate_1 = ozone_concentration*array[1]*rate_coefficient
    rate_2 = ozone_concentration*array[2]*rate_coefficient
    rate_3 = ozone_concentration*array[3]*rate_coefficient
    rate_4 = ozone_concentration*array[4]*rate_coefficient
    rate_5 = ozone_concentration*array[5]*rate_coefficient
    rate_6 = ozone_concentration*array[6]*rate_coefficient
    rate_7 = ozone_concentration*array[7]*rate_coefficient
    rate_8 = ozone_concentration*array[8]*rate_coefficient
    rate_9 = ozone_concentration*array[9]*rate_coefficient

    # Now write out the RHS for each volatility bin according to our
    simple
    # mechanism
    dy_dt_array[0] = rate_1
    dy_dt_array[1] = rate_2 - rate_1
    dy_dt_array[2] = rate_3 - rate_2
    dy_dt_array[3] = rate_4 - rate_3
    dy_dt_array[4] = rate_5 - rate_4
    dy_dt_array[5] = rate_6 - rate_5
    dy_dt_array[6] = rate_7 - rate_6
    dy_dt_array[7] = rate_8 - rate_7
    dy_dt_array[8] = rate_9 - rate_8
    dy_dt_array[9] = -1.0*rate_9

    # Now cycle through each size bin
    for size_step in range(num_bins):
```

**Listing 4.13** Accounting for gas phase reactions in new form of RHS prior to implementing the droplet growth equation. *Associated code - chap4_alg6_droplet_growth_coupled.py.*

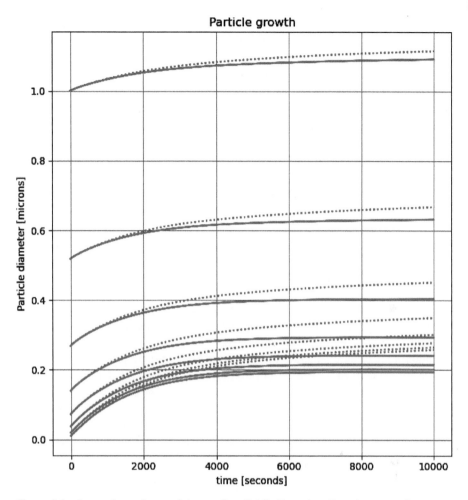

**Figure 4.4** Comparison of an evolving section distribution when there is no gas phase reactions taking place [solid line], versus a simulation that accounts for simple second-order reactions [dashed line].

```
# Add contributory loss from the gas phase to particle phase
dy_dt_gas_matrix[0:num_species,size_step]=dm_dt
# Add a contributory gain to the particle phase from the gas
phase
dy_dt_array[num_species+size_step*num_species:num_species+
num_species*(size_step+1)]=dm_dt[0:num_species]

# Subtract the net condensational mass flux from the gas phase
concentrations
dy_dt_array[0:num_species]=dy_dt_array[0:num_species]-np.sum(
dy_dt_gas_matrix, axis=1)
```

**Listing 4.14** Initializing concentrations of ozone prior to running a sectional volatility basis set model. *Associated code - chap4_alg6_droplet_growth_coupled.py.*

Coupling the gaseous and condensed phases increases the complexity of the differential equations we need to solve. We have also now increased the computational complexity of our code base. Previously, we have noted that a typical gas phase simulation with 800 chemicals has to solve an ODE with 800 states. Through equation REF, a mixed phase simulation with 16 size bins will have 13,600 (800 + 800Œ16) states. In addition the size of the Jacobian matrix, as required by implicit ODE solvers, will increase in a quadratic way from 800*800 to 13,600*13,600. It is true that we have noted the sparsity of the Jacobian gas phase mechanisms. However, increasing the number of size bins and accounting for gas-to-particle partitioning will always increase the size of the Jacobian matrix. For gas phase only simulations, applying simple rules of differentiation to the first two terms of Equation (4.12) allow us to arrive at an analytical Jacobian. Adding the droplet growth equation however does not guarantee we can extend this manually. One might decide that retaining an ideal homogeneous solution lends itself to an analytical Jacobian. However, one must also then decide whether that is an approach that can respond to additional and/or changes in processes captured by the model.

Most ODE solvers do not require the user to provide an analytical Jacobian since a finite difference approximation can be used. Whilst this is applicable to most function definitions, the finite difference approximation can have high-performance costs due to multiple evaluations of the RHS function. Automatic differentiation of code has the convenience of automatically generating a Jacobian matrix from the software itself. Based on the fact that all programs are combination of primitive instructions, an autodifferentiation library could generate the derivative of any program according to the chain rule and predefined derivatives of primitive instructions [12]. This is included in the automatic box model generator JlBox, where the only limitation is that the RHS function must be fully written in the Julia language. One could argue that since that particular model is based in Julia already, this removes some challenges associated with sharing and developing mixed language solutions. On the other hand, if significant effort goes into developing a new process description in another language then the developer must decide what balance is required between implementation of an existing model or translating this into the same language as the host model. In the following section, we provide an example of using JlBox to build a coupled multiphase phase box-model from a chemical mechanism file, using the developments presented in this chapter.

### 4.1.3   An Example Using the Sectional Model Generator JlBox

Existing aerosol model frameworks encompass the procedures outlined in Sections 4.1–4.1.2 and allow us to simulate an evolving gaseous and condensed phase through a chemical mechanism file. Both JLBox and PyBox use the same numerical structure, coupling a fully moving sectional approach to a gas phase model. They do differ in their approaches to construct and run the simulation as dictated by the environment both Julia and Python offer. Figure 4.5 compares a high-level overview of both models, whilst fundamental differences are summarized by Huang and Topping [12] and are summarized later.

Both frameworks use the UManSysProp package to calculate the pure component and mixture properties required to simulate gas to particle partitioning. We refer the reader to Section 4.2 to understand more about chemical identifiers and property prediction suites, including a description of the UManSysProp package.

| | PyBox | JlBox |
|---|---|---|
| Language | Python+Numba or Fortran | Pure Julia |
| Gas kinetics | Static code generation | Sparse matrix manipulation |
| Property calculation | Python code calling UManSysprop | Translated Julia code calling UMan-SysProp |
| Partitioning | Fortran code | Translated Julia code |
| RHS function | Python code calling Fortran/numba | Julia code |
| ODE solver | CVODE_BDF | CVODE_BDF or native solvers |
| Sparse Jacobian | N/A | Support with GMRES linear solver |
| Jacobian matrix | String printed Fortran or Numba code [finite difference differentiated Jacobian for gas+aerosol] | Translated Julia code [automatic differentiated Jacobian for gas+aerosol] |

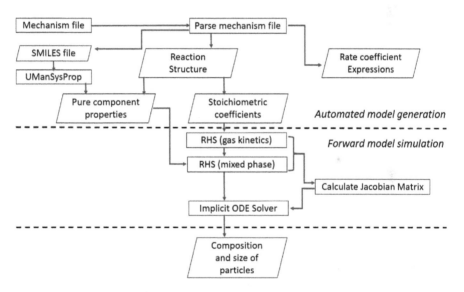

**Figure 4.5** Workflow of parsing an identifier for property prediction technique purposes.

To supplement a previous demonstration of using Julia to explicitly solve the droplet growth equation in Chapter 2, we now use JlBox to simulate a coupled simulation based on a chemical mechanism file that describes the degradation of the VOC $\alpha$-pinene.

To begin, navigate to the JlBox source code location, usually found in the folder .julia/dev/JlBox if setup via the Julia package manager. In the example folder, you will find the file Simulation_aerosol_sparse.jl. Code Listing 4.15, taken from the same file, presents the definition of a function **configure_aerosol()** that we use to specify which mechanism file our simulation is constructed from, along with ambient conditions and any preexisting aerosol. In this example, we run a simulation for 1 h from midday with a pre-existing ammonium sulfate aerosol. We use initial concentrations of 10 and 30 ppb for ozone and alpha-pinene, respectively.

```
function configure_aerosol()
    file="C:/Users/Dave/.julia/dev/JlBox/data/MCM_APINENE.eqn.txt"
    temp=288.15 # Kelvin
    RH=0.5 # RH/100% [0 - 0.99]https://www.overleaf.com/project/5
      e747c4ba312160001090beb
    hour_of_day=12.0 # Define a start time  24 hr format
    start_time=hour_of_day*60*60 # seconds, used as t0 in solver
    simulation_time= 3600.0 # seconds
    batch_step=300.0 # seconds
    temp_celsius=temp-273.15
    Psat_w=610.78*exp((temp_celsius/(temp_celsius+238.3))*17.2694)# Saturation
      VP of water vapour, to get concentration of H20
    Pw=RH*Psat_w
    Wconc=0.002166*(Pw/(temp_celsius+273.16))*1.0e-6 #kg/cm3
    H2O=Wconc*(1.0/(18.0e-3))*6.0221409e+23 #Convert from kg to molecules/cc
    Cfactor= 2.55e+10 #ppb-to-molecules/cc
    reactants_initial_dict=Dict(["O3"=>10.0,"APINENE"=>30.0,"H2O"=>H2O/Cfactor])
    constantdict=Dict([(:temp,temp)])

    num_bins=16
    #Lognormal Distribution
    total_conc=100 #Total particles per cc
    size_std=2.2 #Standard Deviation
    lowersize=0.01 #microns
    uppersize=1.0 #microns
    meansize=0.2 #microns

    # - Specify the core material.
    # This code is currently setup to consider *ammonium sulphate* as the core
    y_core_init=1.0e-3.+zeros(Float64,num_bins) #Will hold concentration of core
      material, only initialise here [molecules/cc]
    core_density_array=1770.0.+zeros(Float64,num_bins) #[kg/m3] - need to make
      sure this matches core definition above
    core_mw=132.14.+zeros(Float64,num_bins) #[g/mol]
    core_dissociation=3.0 #Define this according to choice of core type. Please
      note this value might change

    vp_cutoff=-6.0
    sigma=72.0e-3 # Assume surface tension of water (mN/m)
    property_methods=Dict("bp"=>"joback_and_reid","vp"=>"nannoolal","critical"=>
      "nannoolal","density"=>"girolami")
    diff_method="fine_seeding"
    config=JlBox.AerosolConfig(file,temp,RH,start_time,simulation_time,
      batch_step,
                        H2O,Cfactor,reactants_initial_dict,constantdict,
      num_bins,
                        total_conc,size_std,lowersize,uppersize,meansize,
      y_core_init,
                        core_density_array,core_mw,core_dissociation,
      vp_cutoff,
                        sigma,property_methods,diff_method)
    config
end
```

**Listing 4.15** Specifying initial conditions are model parameter options for a simulation conducted using JlBox.

The string variable `file` points to the location of the chemical mechanism file we wish to use, whilst proceeding variables define the temperature, humidity, and start time of the simulation. One can also see the number of size bins we use in our sectional representation as variable `num_bins`. Notice that we also specify a dictionary `property_methods` that contains the choice of property prediction techniques for both saturation vapor pressures and subcooled liquid density (see Section 4.2).

```
function configure_aerosol_solver_sparse()
    prec = JlBox.default_prec()
    psetup = JlBox.default_psetup("fine_seeding","fine_analytical", 200)
    ndim=5000
    solver=Sundials.CVODE_BDF(linear_solver=:FGMRES,prec=prec,psetup=psetup,
      prec_side=2,krylov_dim=ndim)
    sparse=true
    reltol=1e-6
    abstol=1.0e-3
    dtinit=1e-8
    dtmax=100.0
    positiveness=false
    diff_method="fine_seeding"
    solverconfig=JlBox.SolverConfig(solver,sparse,reltol,abstol,dtinit,dtmax,
      positiveness,diff_method)
    solverconfig
end

config = configure_aerosol()
solverconfig = configure_aerosol_solver_sparse()
@time sol, reactants2ind, param_dict = JlBox.run_simulation(config, solverconfig
    )
df = JlBox.postprocess_gas(sol, reactants2ind, config)
df_SOA = JlBox.postprocess_aerosol(sol, param_dict, config)
```

**Listing 4.16** Specifying the solver to be used for JlBox.

In code Listing 4.16, also taken from the same `Simulation_aerosol_sparse.jl` file, we specify the solver to be used and subsequent tolerances through a call to a function **configure_aerosol_solver_sparse()**, whilst we run the simulation through the function call **JlBox.run_simulation(...)**. We refer the reader to Huang and Topping [12] for more information on the solvers available through JlBox, including the construction of preconditioners and the impact on computational performance. Indeed, we do briefly discuss the range of solver configurations appropriate for the type of systems found in aerosol simulations in Chapter 2. For the remainder of this example, additional function call separate postprocessing procedures on both the gaseous and condensed phase. This allows the user to convert the information on concentrations of every compound in both phases into total mass and size. To run this default simulation, we can open the interaction Julia interactive shell and type the command **include("Simulation_aerosol_sparse.jl")**. You should then see output that resembles information provided in Listing 4.13 and the generated dataframe that stores information on total SOA mass every `batch_step` seconds in Table 4.2:

```
num_eqns: 836, num_reactants: 305
num_reactants_condensed: 156
Dry core mass = 11.269128463820683
=============Aerosol Simulation Config=============
Mechanism file: C:/Users/Dave/.julia/dev/JlBox/data/MCM_APINENE.eqn.
    txt
Start time t0 (s): 43200.0, Simulation time (s): 3600.0, Saving
    interval (s): 300.0
Temperature (K): 288.15, Relative Humidity (%): 0.5
Initial Condition (ppm): Dict("O3" => 10.0,"H2O" =>
    8.374434416160123e6,"APINENE" => 30.0)
Num_bins: 16, Vp_cutoff (log10(Pa)): -6.0
Property methods: Dict("vp" => "nannoolal","critical" => "nannoolal"
    ,"density" => "girolami","bp" => "joback_and_reid")
Jacobian method: fine_seeding
===================Solver Config===================
Using solver: CVODE_BDF{:Newton,:FGMRES,JlBox.var"#103#104",JlBox.
    var"#105#106"{Int64,typeof(JlBox.aerosol_jac_fine_seeding!),
    typeof(JlBox.aerosol_jac_fine_analytical!)}}(0, 0, 5000, false,
    10, 5, 7, 3, 10, JlBox.var"#103#104"(), JlBox.var"#105#106"{
    Int64,typeof(JlBox.aerosol_jac_fine_seeding!),typeof(JlBox.
    aerosol_jac_fine_analytical!)}(20, JlBox.
    aerosol_jac_fine_seeding!, JlBox.aerosol_jac_fine_analytical!),
    2)
Using sparse Jacobian: true
Reltol: 0.0001, Abstol: 0.001
DtInit: 1.0e-6, DtMax: 100.0
Positiveness detection: false
===================================================
Current Iteration: 10, time_step: 1.000000e-06, SOA(ug/m3): 4.314471
    e-20
...
...
```

Table 4.2   Total SOA output from a simulation of α-pinene to a sectional distribution.

**Output from JlBox**

| Output row | Time (s) | SOA ($\mu g \cdot m^{-3}$) |
|---|---|---|
| 1 | 0.0 | 0.0 |
| 2 | 300.0 | 0.00857018 |
| 3 | 600.0 | 0.0700665 |
| 4 | 900.0 | 0.20116 |
| 5 | 1200.0 | 0.402361 |
| 6 | 1500.0 | 0.673837 |
| 7 | 1800.0 | 1.01517 |
| 8 | 2100.0 | 1.42552 |
| 9 | 2400.0 | 1.90382 |
| 10 | 2700.0 | 2.44881 |
| 11 | 3000.0 | 3.05897 |
| 12 | 3300.0 | 3.73238 |
| 13 | 3600.0 | 4.46661 |

### 4.1.4 Modal Model for Condensational Growth

The moment-based modal aerosol approach introduced in Section 1.3.3 provides a simpler representation of the size distribution than sectional approaches and can be leveraged to simulate the growth of particles due to the condensation of vapor-phase constituents. Compared to the examples of sectional models presented thus far, the number of equations required to simulate the gas-particle partitioning of newly formed condensable vapors to the existing aerosol size distribution is drastically reduced in a modal model. As a reminder, the modal approach exploits the fact that size distributions are typically well described by log-normal distributions. Thus, the method that follows involves translating the log-normal aerosol properties (i.e., number, geometric mean diameter, and geometric standard deviation) to the conserved modal moments, predicting the change in each moment due to mass transfer, and then translating the new moments back to log-normal aerosol properties. For condensation (evaporation) processes, the flux of mass added to (subtracted from) the particle population should be equal and opposite to the gas phase source. Further, there should be strictly no change to particle number for this process. Thus, the governing equations for condensational growth of the zeroth ($M_0$), second ($M_2$), and third ($M_3$) moments of a modal aerosol representation is

$$\frac{dM_0}{dt} = \frac{dN}{dt} = 0 \tag{4.16}$$

$$\frac{dM_2}{dt} = \frac{4}{\pi\rho} \sum (C_{g,i} - C_{a,i}) \beta_2 \tag{4.17}$$

$$\frac{dM_3}{dt} = \frac{6}{\pi\rho} \sum (C_{g,i} - C_{a,i}) \beta_3 \tag{4.18}$$

where $\rho$ is the density of the chemical components in the particle phase, $C_{a,i}$ is the aerosol phase concentration of $I$, and $C_{g,i}$ is its gas phase concentration. The beta-condensation sink term accounts for the effect of transition between the free molecular and continuum regime solutions of condensation [13], [12]:

$$\beta_k = \frac{I_{fm,k} I_{ct,k}}{I_{fm,k} + I_{ct,k}} \tag{4.19}$$

$$I_{fm,k} = \frac{\pi\alpha\bar{c}}{4} M_{k-1} \tag{4.20}$$

$$I_{ct,k} = 2\pi D_v M_{k-2} \tag{4.21}$$

where the free molecular term is proportional to the accommodation coefficient ($\alpha$), the particle speed ($\bar{c}$), and second moment ($M_2$), while the continuum regime term is a function of the vapor diffusivity (Dv) and the first moment (M1). This form of the condensation sink approximates the flux at both size limits and within the transition regime using a harmonic mean estimation approach as described in Whitby et al. [12]. Using these formulae and the example code in Section 4.1.2, the change in the aerosol size distribution due to condensation of a nonreacting distribution of organic gases may be calculated with the following code.

```python
sigmag1 = np.log(1.7) # Natural Log of Geometric standard deviation
Dp      = 150 * 1.0E-9 # Geometric Mean particle diameter [150nm]
N       = 100.0 # Total number of particles [per cm-3]

# Initialize all 3 moments of initial distribution
M0 = N * 1.0E6  # [particles per m-3]
M2 = M0 * Dp**2 * np.exp( 2 * sigmag1**2 )   # [m2 m-3]
M3 = M0 * Dp**3 * np.exp( 4.5 * sigmag1**2 ) # [m3 m-3]
# --------------------------------------------------------------

# -------- Initialize Size Distribution of Particle Core abundance
    using the above size distribution -------
density_core = 1400.0  # kg m-3
core_mw      = 200.0   # g mol-1

dry_mass = np.pi/6.0 * density_core * M3 * 1.0E9 # ug m-3
print("Initial dry aerosol mass = ", dry_mass)
core_abundance = dry_mass * 1.0E-6 / core_mw * Avo * 1.0E-6
# [molec cm-3]
print("Initial core abundance = ", core_abundance )

# New define an array that holds the molecular abundance of each gas
    and its concentration
# in each aerosol mode [molecules / cc]
# Add 3 additional terms for each mode at the end representing M0,
    M2, and M3
array = np.zeros(( num_species + num_species + 3 ), dtype=float)
array[ 0:num_species ] = gas_abundance
array[ num_species:2*num_species ] = 1.0e-10 # assuming we start
    with no condensed
                                     # material in the
    aerosol phase
nchem = num_species + num_species
array[nchem      :nchem+1 ] = M0
array[nchem + 1 :nchem+2 ] = M2
array[nchem + 2 :nchem+3 ] = M3

###########################################################
####################################

# RHS function [with no ozone effects]
def dy_dt(array,t):

    # Load gas species concentrations in local array
    Cg_i_m_t = array[0:num_species]

    # We are working with 1 aerosol mode, which has an involatile
    core
    dy_dt_array = np.zeros(( nchem + 3 ), dtype='float64')
    dM0_dt = 0.
    dM2_dt = 0.
    dM3_dt = 0.

    # 1) Calculate total moles and mole fraction of each species in
    this mode,
    #    and assign modal parameters
    temp_array    = array[ num_species:nchem ]        # [molec/cc]
    total_moles   = np.sum(temp_array) + core_abundance # [molec/cc
    ]
    mole_fractions = temp_array / total_moles             # [frac]
```

```
M0 = array[ nchem   ]  # 0th moment = Number
M2 = array[ nchem+1 ]  # 2nd moment prop. to surface area
M3 = array[ nchem+2 ]  # 3rd moment prop. to volume

# 2) Calculate density
density_array = np.zeros((num_species+1), dtype='float64')
density_array[0:num_species] = density_org[0:num_species] # [kg
 m-3]
density_array[num_species]   = density_core                # [kg
 m-3]

mass_array = np.zeros((num_species+1), dtype='float64')
mass_array[0:num_species] = (temp_array/Avo) * mw_array    # [g
 cm-3]
mass_array[num_species] = (core_abundance/Avo)*core_mw     # [g
 cm-3]

total_mass = np.sum(mass_array)                            # [g
 cm-3]
mass_fractions_array = mass_array / total_mass             # [
 frac]
density = 1.0 / (np.sum( mass_fractions_array / density_array ))
  # [kg m-3]

#3) Calculate Diameter and sigma from the three moments above.

sigmag1 = ( 1.0/3.0 * np.log( M0 ) - np.log( M2 ) + 2.0/3.0 * np
.log( M3 ) ) **0.5
Dp = ( M3 / ( M0 * np.exp( 4.5 * sigmag1**2 ) ) ) **(1.0/3.0)  #
 [m]
M1 = M0 * Dp * np.exp( 0.5 * sigmag1**2 )  # [m m-3]

#4) Free Molecular Regime Calculation for the 2nd and 3rd
moments
Ifm2 = np.pi * alpha_d_org * mean_them_vel / 4. * M1 # [ m2 m-3
 s-1]
Ifm3 = np.pi * alpha_d_org * mean_them_vel / 4. * M2 # [ m3 m-3
 s-1]

#5) Continuum Regime Calculation for the 2nd and 3rd Moments
Ict2 = 2.0 * np.pi * DStar_org[:]*1.0e-4 * M0   # [m2 s-1 m-3]
Ict3 = 2.0 * np.pi * DStar_org[:]*1.0e-4 * M1   # [m3 s-1 m-3]

#6) Combine Ifm and Ict for moment- and species-dependent Beta
correction terms
Beta2 = ( Ifm2 * Ict2 ) / ( Ifm2 + Ict2 )  # [m2 m-3 s-1]
Beta3 = ( Ifm3 * Ict3 ) / ( Ifm3 + Ict3 )  # [m3 m-3 s-1]

#7) Modal growth equation
Cstar_i_m_t = ((Cstar*1.0e-6)/(mw_array))*1.0e-6*Avo

dm2_dt = 4.0 / np.pi * np.sum( (Cg_i_m_t - Cstar_i_m_t*
mole_fractions) * 1.0e6 / rho_l * Beta2 )
dm3_dt = 6.0 / np.pi * np.sum( (Cg_i_m_t - Cstar_i_m_t*
mole_fractions) * 1.0e6 / rho_l * Beta3 )
dy_dt_array[ num_species:nchem ]  = ( Cg_i_m_t[:] - Cstar_i_m_t*
mole_fractions) * Beta3
dy_dt_array[ 0:num_species ]  = -1.0 * dy_dt_array[num_species
:2*num_species]
dy_dt_array[ nchem    ] = 0.  # Particle number does not change
. This is included in the ODE
                              # set for demonstration since it
would change for other processes
                              # like coagulation. It should be
removed for operational use if
```

```
                                      # only condensation is being
    considered.
    dy_dt_array[ nchem + 1 ] = dm2_dt
    dy_dt_array[ nchem + 2 ] = dm3_dt

    return dy_dt_array
```

**Listing 4.17** Implementing a modal model in Python to predict condensation of vapors to a particle population using the volatility basis set model without ozone reaction. See Python source code for complete implementation of constants and initial parameters as well as an example of how to integrate this function and plot results. *Associated code - chap4_alg7_droplet_growth_modal.py.*

For a range of initial $\sigma_g$ values, the initial and final size distributions are shown in Figure 4.6. As depicted, this approach results in the expected narrowing of the size distributions with growth of the mode to larger sizes due to condensation. The system can also be expanded to consider multiple coexisting aerosol modes with a fraction of the condensing (or evaporating) vapor going to (coming from) each mode. The diameter and width of each mode will respond based on its own initial concentration and

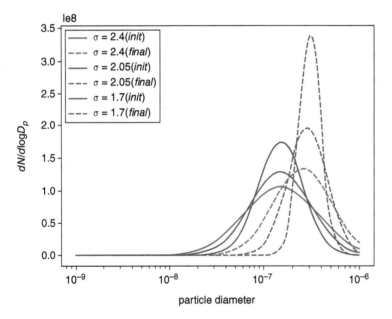

**Figure 4.6** Sensitivity of the initial (init; before condensation) and final (after condensation) modal distributions to the initial modal standard deviation. Condensation is occurring without simultaneous ozone reaction. The initial median diameter is 150 nm and the initial number concentration is 100 particles $\cdot cm^{-3}$.

the concentration of the condensable vapor among the other modes and in the gas phase.

Code snippet 4.18 implements the simultaneous ozonolysis of the initial VBS distribution of organic vapors and condensation of semivolatiles to a particle population. As seen in the sectional approach, the condensed mass increases considerably due to a shift in the volatility distribution. In this case, the system with ozonolysis predicts condensation of 3.30 $\mu \cdot g\,m^{-3}$ while the system without ozonolysis condenses substantially less 1.99 $\mu g \cdot m^{-3}$.

```python
def dy_dt_ozone(array,t):

    # Load gas species concentrations in local array
    Cg_i_m_t = array[0:num_species]

    # We are working with 1 aerosol mode, which has an involatile
    # core
    dy_dt_array = np.zeros(( nchem + 3 ), dtype='float64')
    dM0_dt = 0.
    dM2_dt = 0.
    dM3_dt = 0.

    # Calculate the rate of each reaction explicitly
    # rate_1 = ozone_abundance*array[0]*rate_coefficient
    rate_1 = ozone_abundance*array[1]*rate_coefficient
    rate_2 = ozone_abundance*array[2]*rate_coefficient
    rate_3 = ozone_abundance*array[3]*rate_coefficient
    rate_4 = ozone_abundance*array[4]*rate_coefficient
    rate_5 = ozone_abundance*array[5]*rate_coefficient
    rate_6 = ozone_abundance*array[6]*rate_coefficient
    rate_7 = ozone_abundance*array[7]*rate_coefficient
    rate_8 = ozone_abundance*array[8]*rate_coefficient
    rate_9 = ozone_abundance*array[9]*rate_coefficient

    # Now write out the RHS for each volatity bin according to our
    simple
    # mechanism
    dy_dt_array[0] = rate_1
    dy_dt_array[1] = rate_2 - rate_1
    dy_dt_array[2] = rate_3 - rate_2
    dy_dt_array[3] = rate_4 - rate_3
    dy_dt_array[4] = rate_5 - rate_4
    dy_dt_array[5] = rate_6 - rate_5
    dy_dt_array[6] = rate_7 - rate_6
    dy_dt_array[7] = rate_8 - rate_7
    dy_dt_array[8] = rate_9 - rate_8
    dy_dt_array[9] = -1.0*rate_9

    # 1) Calculate total moles and mole fraction of each species in
    this mode,
    #     and assign modal parameters
    temp_array      = array[ num_species:nchem ]            # [molec/cc
    ]
    total_moles     = np.sum(temp_array) + core_abundance # [molec/cc
    ]
    mole_fractions = temp_array / total_moles              # [frac]

    M0 = array[ nchem   ]  # 0th moment = Number
    M2 = array[ nchem+1 ]  # 2nd moment prop. to surface area
    M3 = array[ nchem+2 ]  # 3rd moment prop. to volume
```

```
# 2) Calculate density
density_array = np.zeros((num_species+1), dtype='float64')
density_array[0:num_species] = density_org[0:num_species] # [kg
m-3]
density_array[num_species]    = density_core                # [kg
m-3]

mass_array = np.zeros((num_species+1), dtype='float64')
mass_array[0:num_species] = (temp_array/Avo) * mw_array    # [g
cm-3]
mass_array[num_species] = (core_abundance/Avo)*core_mw     # [g
cm-3]

total_mass = np.sum(mass_array)                            # [g
cm-3]
mass_fractions_array = mass_array / total_mass            # [
frac]
density = 1.0 / (np.sum( mass_fractions_array / density_array ))
  # [kg m-3]

#3) Calculate Diameter and sigma from the three moments above.
sigmag1 = ( 1.0/3.0 * np.log( M0 ) - np.log( M2 )
+ 2.0/3.0 * np.log( M3 ) ) **0.5
Dp = ( M3 / ( M0 * np.exp( 4.5 * sigmag1**2 ) ) )
 **(1.0/3.0)  # [m]
M1 = M0 * Dp * np.exp( 0.5 * sigmag1**2 )   # [m m-3]

#4) Free Molecular Regime Calculation for the 2nd and 3rd
moments
Ifm2 = np.pi * alpha_d_org * mean_them_vel
/ 4. * M1  # [ m2 m-3 s-1]
Ifm3 = np.pi * alpha_d_org * mean_them_vel
/ 4. * M2  # [ m3 m-3 s-1]

#5) Continuum Regime Calculation for the 2nd and 3rd Moments
Ict2 = 2.0 * np.pi * DStar_org[:]*1.0e-4 * M0   # [m2 s-1 m-3]
Ict3 = 2.0 * np.pi * DStar_org[:]*1.0e-4 * M1   # [m3 s-1 m-3]

#6) Combine Ifm and Ict for moment- and species-dependent Beta
correction terms
Beta2 = ( Ifm2 * Ict2 ) / ( Ifm2 + Ict2 )  # [m2 m-3 s-1]
Beta3 = ( Ifm3 * Ict3 ) / ( Ifm3 + Ict3 )  # [m3 m-3 s-1]

#7) Modal growth equation
Cstar_i_m_t = ((Cstar*1.0e-6)/(mw_array))*1.0e-6*Avo

dm2_dt = 4.0 / np.pi * np.sum( (Cg_i_m_t - Cstar_i_m_t*
mole_fractions) * 1.0e6 / rho_l * Beta2 )
dm3_dt = 6.0 / np.pi * np.sum( (Cg_i_m_t - Cstar_i_m_t*
mole_fractions) * 1.0e6 / rho_l * Beta3 )
dy_dt_array[ num_species:nchem ]  = ( Cg_i_m_t[:] - Cstar_i_m_t*
mole_fractions) * Beta3
dy_dt_array[ 0:num_species ]  = dy_dt_array[0:num_species] -
dy_dt_array[num_species:2*num_species]
dy_dt_array[ nchem      ] = 0.  # Particle number does not change
. This is included in the ODE
                            # set for demonstration since it
would change for other processes
                            # like coagulation. It should be
removed for operational use if
                            # only condensation is being
considered.
dy_dt_array[ nchem + 1 ] = dm2_dt
dy_dt_array[ nchem + 2 ] = dm3_dt
```

```
    return dy_dt_array
```

Listing 4.18 Implementing a modal model in Python to predict condensation of vapors to a particle population using the volatility basis set model with ozone reaction of a distribution of organic vapors. See Python source code for complete implementation of constants and initial parameters as well as an example of how to integrate this function and plot results. *Associated code - chap4_alg7_droplet_growth_modal.py.*

The accuracy of the modal model for condensational growth is limited by a number of factors. First, the harmonic mean approximation for the condensational sink term $\beta$ (Eq 4.19) neglects the Kelvin equation for vapor pressure over the curved particle surface. Second, because the diameter growth rate decreases with increasing particle size for a fixed mass flux, it is likely that smaller modes will grow to comparable sizes as larger ones and a strategy for transferring particles between the modes or combining the modes together is needed. Approaches for this type of mode merging have been documented by Binkowski and Roselle [21] and Whitby et al. [22]. Apparent mode splitting may also occur for very small particle sizes when new particle formation occurs simultaneously with coagulation of slightly larger particles. Sartelet et al. [23] developed and evaluated an approach for mode splitting against results from a sectional aerosol model, concluding that their solution was promising for use in 3D chemical transport models.

## 4.2  Chemical Identifiers and Parsing Structures

Automating the prediction of pure component and mixtures properties requires structural information of individual compounds to be interrogated to identify appropriate features. There are many options available for representing a chemical structure, choice of which is influenced by factors that include human readability, support within existing proprietary and open-source platforms, and also integration with aerosol software platforms. Many thousands of individual aerosol components ensure that explicit manual calculation of these properties is laborious and time-consuming. The emergence of explicit automatic mechanism generation techniques [10] including up to many millions of individual gas phase products as aerosol precursors renders the process impossible and automation is necessary. Furthermore, to identify key aerosols components and thus resolve their environmental impacts we must be able to replicate chemical characteristics measured in real/simulated atmospheres. A comprehensive experimental determination of individual organic components of atmospheric aerosols is not available, leading to more "ensemble" measurements on chemical signatures of mixtures. Thus, individual component complexity is lost in predictive models as "proxy" compounds are used to replicate such characteristics. Through automation of component property estimation, combined with a gas/aerosol transfer model, properties of aerosol distributions as determined by state-of-the-science atmospheric instrumentation can be predicted in a level of detail largely unexplored.

Heller et al. [24] give a succinct overview on the provenance in how we represent and store information on individual compounds for use in databases and simulation software. To begin, a chemical identifier is a text label that denotes a chemical substance. It is of course highly desirable that this identifier is unambiguous. Whilst some systems enable different labels to be assigned to the same compound, each should still only refer to that specific compound. Recent developments of registry-lookup identifiers include the Chemical Abstracts Service (CAS) Registry numbers by the American Chemical Society [https://www.cas.org/cas-data/cas-registry], Compound IDs (CID) assigned by PubChem [https://pubchem.ncbi.nlm.nih.gov/], and identifiers assigned by ChEMBL [http://chembl.blogspot.com/2011/08/chembl-identifiers.html] and ChemSpider [http://www.chemspider.com/]. You may have come across CAS registry numbers before. A CAS number is a unique numerical identifier assigned by the CAS to every chemical substance described in the open-scientific literature. The number itself contains no information about the substance itself but represents the order in which the compound was identified through CAS.

Known problems associated with widespread use of registry-lookup identifiers include the inability to move beyond substances that already exist, such as the ability to generate a new substance visually or through a set of rules. The alternative approaches are structural representations.

Line notations represent structures as a linear string of characters. Common molecular file formats include Wiswesser Line Notation (WLN) [25], ROSDAL [26], and SYBL [27]. IUPAC and NIST recently developed the IUPAC International Chemical Identifier (InChI) [28]. Another linear notation using short ASCII strings is the SMILES format (Simplified Molecular Input Line Entry System)—a simplified chemical notation that allows a user to represent a two-dimensional chemical structure in linear textual form [29]. For example, the smiles notation for carbon dioxide is "$O=C=O$," whereas cyclohexane is represented as "C1CCCCC1." The notation is commonly employed in commercial and public software for prediction of chemical properties. It can be imported by most molecule editors for conversion into 2D/3D models and has a wide base of software support and extensive theoretical backing [www.daylight.com]. Whilst there are usually a large number of valid generic SMILES which represent a given structure, part of the power of SMILES is that a canonicalization algorithm to generate one special generic SMILES among all valid possibilities; this is known as the "unique SMILES". With standard SMILES, the name of a molecule is synonymous with its structure; with unique SMILES, the name is universal. SMILES describing atoms and bonds, but no chiral or isotopic information, are known as generic SMILES. There are usually a large number of valid generic SMILES which represent a given structure. It also allows molecules be "broken up" into important molecular substructure information required for property predictions via a straightforward extension of the same notation.

Constructing SMILES can be done by manually following the set of rules described extensively in the SMILES theory manual provided by Daylight Chemical Information Systems [www.daylight.com]. One could also generate a SMILES string through any number of drawing packages. Figure 4.7 displays the structure for pinonic acid taken from the PubChem database service, along with the generated identifiers in Table 4.3.

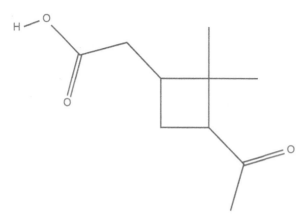

**Figure 4.7** Chemical structure depiction for pinonic acid. *Source:* PubChem URL: https://pubchem.ncbi.nlm.nih.gov. Description: Data deposited in or computed by PubChem.

**Table 4.3** List of chemical identifiers for pinonic acid.

**Multiple identifiers for pinonic acid**

| Identifier | String | Notes |
|---|---|---|
| IUPAC Name | 2-(3-acetyl-2,2-dimethylcyclobutyl)acetic acid | Computed by LexiChem 2.6.6 (PubChem release 2019.06.18) |
| InChI | InChI=1S/C10H16O3/c1-6(11)8-4-7(5-9(12)13)10(8,2)3/h7-8H,4-5H2,1-3H3,(H,12,13) | Computed by InChI 1.0.5 (PubChem release 2019.06.18) |
| Canonical SMILES | CC(=O)C1CC(C1(C)C)CC(=O)O | Computed by OEChem 2.1.5 (PubChem release 2019.06.18) |
| Molecular formula | $C_{10}H_{16}O_3$ | Computed by PubChem 2.1 (PubChem release 2019.06.18) |

To use the SMILES format requires a flexible and efficient facility to extract substructure information from each string that is meaningful to each property predictive technique. To achieve this we can use the OpenBabel chemical toolbox, an open project which has the ability to filter and search molecular files using the SMARTS format, and other methods [30]. As noted in the Daylight manual https://www.daylight. com/dayhtml/doc/theory/theory.smarts.html, SMARTS is a language that allows you to specify substructures using rules that are straightforward extensions of SMILES [30]. Combinations of atomic and bond primitives along with logical operators allows the user to extract specific features. All SMILES expressions are also valid SMARTS expressions. However, the semantics changes because SMILES describes molecules, whereas SMARTS describes patterns. For example, an example given in the official documentation states that the SMILES "C1=CC=CC=C1" (cyclohexatriene) is interpreted as the benzene molecule. This molecule will be matched by the SMARTS c1ccccc1, which is interpreted as the pattern "6 aromatic carbons in a ring." The SMARTS "C1=CC=CC=C1" makes a pattern ("six aliphatic carbons in a ring with alternating

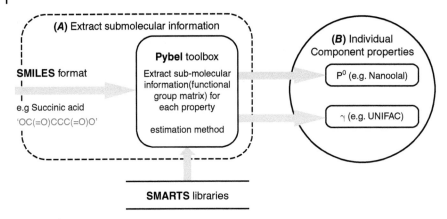

**Figure 4.8** Workflow of parsing an identifier (A) for property prediction technique purposes (B).

single and double bonds") which will not match benzene. It will, however, match the nonaromatic phenylate cation with SMILES C1=CC=CC=[CH+]1. For more information, we encourage the reader to develop their own SMARTS matching exercises to better understand this works. Figure 4.8 displays a hypothetical workflow of passing a SMILES string identifier for succinic acid into the OpenBabel toolbox, via the Python interface Pybel, and then extracting features through a set of SMARTS for use in property predictive techniques.

In Section 4.3, we discuss how we might implement the workflow in Figure 4.8 using example code listings from the UManSysProp package [31].

Why is this of use to us when building property predictive techniques? As we will see shortly, when we use the Python interface for the OpenBabel package we can start to use dictionaries to combine chemical identifiers and properties derived from feature extractions. OpenBabel [30] is a free-to-use toolbox for performing a number of operations around chemical identifiers.

## 4.3 Coding Property Prediction Techniques

We need a mechanism for interrogating SMILES strings based on a collection of features encoded as SMARTS. Code Listing 4.19 presents a simple example of importing the Python interface (**Pybel**) to the **OpenBabel** module and then creating a "Molecule" object, which we have stored as variable `pybel_object`, using the function **readstring** and specifying the SMILES format and passing the SMILES string for pinonic acid. Before we proceed to interrogate this object using a SMARTS string, code Listing 4.20 shows how to access various attributes of our new molecule object.

```
from openbabel import pybel
pybel_object = pybel.readstring('smi','CC(=O)C1CC(C1(C)C)CC(=O)OC')
```

Listing 4.19 Creating an OpenBabel Molecule object.

Listing 4.20 also includes a simple example of searching for certain features in the SMILES string. After we create a Pybel Molecule object, we create a SMARTS object using the **Smarts** function by passing it a SMARTS identifier. In this case we have simply used SMARTS notation that matches any carbon atom. Once we have created the SMARTS object, which we store as `smarts_object`, we then search the previously created molecule object using the command `smarts_objects.findall(pybel_object)`. This returns a list of matches for each carbon atom found along the SMILES string, the length of which equates to the total number of matches found. This is displayed in Listing 4.21 which provides the output when you run the file *chap4_alg8_Openbabel.py*.

```python
# Print attributes of our Molecule object
print("Molecular weight = ", pybel_object.molwt)
print("Molecular formula = ", pybel_object.formula)

# Print the
smarts_object = pybel.Smarts('[#6]')
print("SMARTS matches = ", smarts_object.findall(pybel_object))
print("Number of SMARTS matches = ", len(smarts_object.findall(
    pybel_object)))
```

**Listing 4.20** Access attributes of the OpenBabel Molecule object and also match the location and number of a single SMARTS pattern.

```
Molecular weight =  198.25882000000001
Molecular formula =  C11H18O3
SMARTS matches =  [(1,), (2,), (4,), (5,), (6,), (7,), (8,), (9,),
    (10,), (11,), (14,)]
Number of SMARTS matches =  11
```

**Listing 4.21** Output from example script to create an OpenBabel molecule object and parse this by a single SMARTS identifier.

You could extend these simple examples and develop a script that automatically loops through a series of SMILES strings and extract relevant features. Specifying those features will depend on the specific use case. In Section 4.3.1, we introduce the concept of group contribution techniques and provide an example of methods for predicting vapor pressures and liquid densities. In those examples, we reference and use functionalities included within the UManSysProp suite of tools. In every chapter, we discuss the options available for building numerical models. Likewise, when developing a facility for property estimations one can make a number of decisions around not only the structure of code used in individual files, but also the relative referencing to generic modules and data. It is worthwhile briefly exploring this, using UManSysProp as an example, to facilitate better understanding of the narrative and cross-reference functions in the proceeding sections. However, to repeat, we use UManSysProp as an example of an existing set of tools that you can expand. UManSysProp was designed to act as a standalone package, providing a range of tools that can be integrated into existing software products. It was also designed to be used as part of a web server application and provide predictive capability for a wide range of uses within and beyond atmospheric science. Indeed, the majority of the property predictive techniques provided through UManSysProp were originally designed for industrial and/or wider applications. The choice of structure for the collection of Python scripts, and their format, was thus influenced by multiple factors. In the proceeding diagram, we see the structure of the UManSysProp package through the distributions of folders and a subset of files. In this layout, we find the use of a common file we have not discussed so far; the use of a file named *__init__.py*. Whilst we have imported existing modules as separate *.py* files, embedding a *__init__.py* file in a directory allows that directory to be imported as a regular package. As stated in the Python documentation:

> A regular package is typically implemented as a directory containing an __init__.py file. When a regular package is imported, this __init__.py file is implicitly executed, and the objects it defines are bound to names in the package's namespace. The __init__.py file can contain the same Python code that any other module can contain, and Python will add some additional attributes to the module when it is imported.

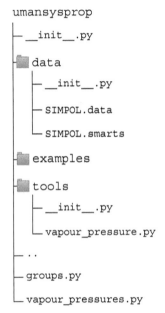

```
umansysprop
├── __init__.py
├── 📁 data
│    ├── __init__.py
│    ├── SIMPOL.data
│    └── SIMPOL.smarts
├── 📁 examples
├── 📁 tools
│    ├── __init__.py
│    └── vapour_pressure.py
├── ..
├── groups.py
└── vapour_pressures.py
```

From the perspective of a general user, the structure is perhaps unimportant. Indeed, in code Listing 4.22, we import the file *vapour_pressures.py* and its internal functions to predict the vapor pressures of a specific SMILES string. However, notice how in this example we also make sure the location of our local copy of UManSysProp is on our path using the **sys** module and then we import the functions **vapour_pressures** and **boiling_points**. How have we been able to import the entire directory of UManSysProp and access a set of functions? Looking back at the structure diagram, the first *__init__.py* file contains information about the version number and contributors, but its presence alone allows the directory to be marked as a Python module by the Python interpreter. Again, to a general user this may be unimportant. However, if someone wishes to expand the UManSysProp functionality, or modify existing functions, one has to understand the relevant linkages between modules. In this instance, the relevant flow is as follows:

- Once a given method in the *vapour_pressures.py* file is called, there are subsequent calls to functions within the file *groups.py*.
- In the *groups.py* file, whilst these initial functions are specific to each predictive technique, they rely on general functions that parse each Pybel molecule object to (1)

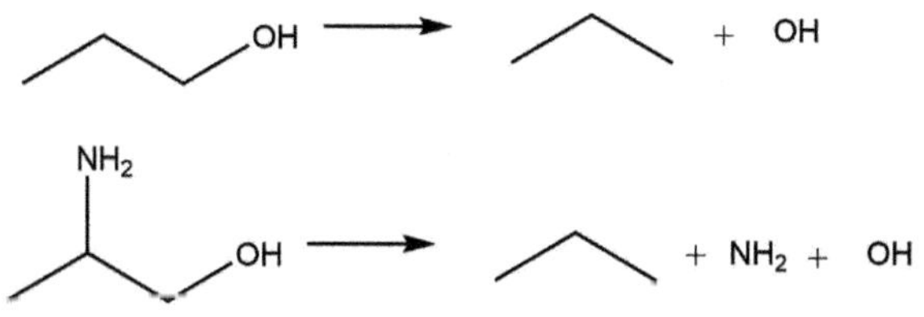

**Figure 4.9** GCMs split a compound into its carbon backbone and its functional groups.

extract and count the relevant structure features and (2) combine these with model parameters to complete the implementation of a given technique.

- The aforementioned generic functions are able to extract the relevant features using SMARTS data stored within functions and files in the **data** folder. These functions are defined in the *__init__.py* file, in the **data** folder, that read relevant files that contain property prediction technique parameters and SMARTS strings.

```
from openbabel import pybel
import sys
# Here you would need to put the relevant path in for your
    UManSysProp distribution.
sys.path.append('...enter relevant path...')
from umansysprop import boiling_points
from umansysprop import vapour_pressures

# Define a temperature [Kelvin]
temperature=298.15

# Create a Pybel Molecule object
pybel_object = pybel.readstring('smi','CC(=O)C1CC(C1(C)C)CC(=O)OC')

# Calculate vapour pressures [log10 (atm)]
vapour_pressure_estimate=vapour_pressures.evaporation(pybel_object,
    temperature)
```

**Listing 4.22** Importing the UManSysProp directory structure as a package, calculating the vapor pressure of an individual component.

In Section 4.3.3, we walk through the steps needed to embed another technique within the UManSysProp structure. Again, you may wish to create your own package, but following our instructions should provide the basis to expand the existing functionality of a local copy of the UManSysProp suite if so desired.

### 4.3.1 Group Contribution Methods

Group contribution methods (GCMs) are designed to capture the functional dependencies on predicting absolute values of physical constants. GCMs break a compound down into a base molecule (typically the carbon backbone) and its functional groups as shown in Figure 4.9. The contribution to the property of interest from a functional group should be the same regardless of what it is added to, so knowing the property of the base molecule and the contribution from the functional group, it is possible to estimate the property of interest.

This concept holds true in many cases, but most of these are for straight chain hydrocarbons or monofunctional compounds. When there are multiple functional groups, this simplistic method begins to break down as it does not account for any effects that interactions between the functional groups within a compound may have on the property of interest. Some GCMs attempt to account for this by using additional terms looking at secondary interactions.

### 4.3.2 Vapor Pressure Prediction Methods

The most common vapor pressure prediction techniques used in atmospheric aerosol studies are GCMs. Within the version of UManSysProp included in this book, there are currently four vapor pressure GCMs: Nanoolal et al. [32], Myrdal and Yalkowsky [33], EVAPORATION [34], and SIMPOL [35].Of these methods, Nanoolal et al. [32] and Myrdal and Yalkowsky [33] are combined methods that also require a boiling point as an input, whereas EVAPORATION [34] and SIMPOL [35] only require the structure of a compound to make a prediction. The boiling point is another property of a compound that is possible to predict using a GCM. UManSysProp currently contains three boiling point GCMs. These are Joback and Reid [36], Stein and Brown [37], and Nannoolal et al. [38]. Joback and Reid has known biases [39] and Stein and Brown [37] is an updated version of Joback and Reid [36]. For this reason, Stein and Brown [37] is largely recommended over Joback and Reid [36]. In an assessment article by O'Meara et al. [40] the use of Stein and Brown [37] and Nannoolal et al. [38] to predict boiling points were used in conjunction with several vapor pressure GCMs. For the GCMs looked at in that study, it was concluded that there was no significant difference between Stein and Brown [37] and Nannoolal et al. [38].

Whilst in Section 4.3 we discussed the general structure of UManSysProp, in the following sections we provide a narrative of how specific prediction techniques are added to the existing functionality. These are designed to provide further training around the use of Python to integrate the open-source informatics functionality provided by the OpenBabel suite of tools. Again, whilst the structures discussed allow you to expand UManSysProp specifically, the general approach is transferable.

### 4.3.3 Example: Adding the SIMPOL Method to UManSysProp

In Section 4.3, we can infer that to add a new technique to UManSysProp, we need to modify a file that acts as the interface to a user (e.g., *vapour_pressures.py*), whilst creating a bridge between *groups.py* and the data held within the data folder. In the following, we demonstrate this with an example.

The SIMPOL group contribution method can be used to predict the sub-cooled liquid vapor pressure of a compound as a function of temperature $(T)$ only requiring the structure of the compound and a temperature as an input. For each compound, $i$, SIMPOL assumes the following relationship between vapor pressure and temperature shown in Equation (4.22),

$$log_{10}p^o_{l,i}(T) = \Sigma_k \nu_{k,i} b_k(T) \tag{4.22}$$

where $p^o_{l,i}$ is the subcooled liquid vapor pressure, $\nu_{k,i}$ is the number of groups of type $k$, and $b_K(T)$ is the contribution to $log_{10} p^o_{l,i}(T)$ by each group of type $k$. The temperature dependence of $b_k(T)$ is defined in Equation (4.23),

$$b_k(T) = \frac{B_{1,k}}{T} + B_{2,k} + B_{3,k}T + B_{4,k}ln(T) \tag{4.23}$$

where the values for $B_{1,k}$ to $B_{4,k}$ for each instance of $k$ are available in [35].

Within SIMPOL, a total of 30 structural groups are considered, as well as a zeroth group used in all calculations. We use the integer variable $k$ as a numeric reference to each of these groups. For example, $k = 1$ corresponds to the total number of carbons within the compound of interest. From Equation (4.23) we can see that, for each $k$, there are four model parameters. Following the convention used within UManSysProp, these model parameters are saved within a file in the **data** folder. This text file, saved under the name **SIMPOL.data** contains five columns; the first row containing columns titles representing the four model parameters $B_{1,k}$, $B_{2,k}$, $B_{3,k}$, and $B_{4,k}$. The specific **.data** file extension used within UManSysProp simply allows us to write generic functions that identify any file that contain parameters central to any given prediction technique.

| # group | Bk1 | Bk2 | Bk3 | Bk4 |
|---|---|---|---|---|
| 0 | -4.26938E+02 | 2.89223E-01 | 4.42057E-03 | 2.92846 |
| | E-01 | | | |
| 1 | -4.11248E+02 | 8.96919E-01 | -2.48607E-03 | 1.40312 |
| | E-01 | | | |
| 2 | -1.46442E+02 | 1.54528E+00 | 1.71021E-03 | -2.78291 |
| | E-01 | | | |
| 3 | 3.50262E+01 | -9.20839E-01 | 2.24399E-03 | -9.36300 |
| | E-02 | | | |

Listing 4.23 Format for SIMPOL.data.

As we have already noted, alongside model parameters for a given technique, it is necessary to identify which groups/functionalities are present within a compound. Continuing our use of SMILES strings, the features that are defined for use within SIMPOL can range from very simple, such as the number of carbon atoms present within a compound to the more complex, such as the presence of C=C–C=O in a *non*-aromatic ring. Therefore, each value for $k$ will also have a single or set of SMARTS strings. Following the convention used within UManSysProp, the SMARTS used to identify these features are stored in a file with the **.smarts** extension. These **.smarts** files are built around a structure where the first column lists the relevant integer values for k and the second column containing the SMARTS that represents k. Code Listing 4.24 displays the first eight lines of the file **SIMPOL.smarts**.

```
1     [#6]
2     ([#6][!$([NX3][CX3]=O),$([CX3](=O)[NX3])])
301   [#6,#7,#8;r3;a]
302   [#6,#7,#8;r4;a]
303   [#6,#7,#8;r5;a]
304   [#6,#7,#8;r6;a]
305   [#6,#7,#8;r7;a]
306   [#6,#7,#8;r8;a]
```

Listing 4.24 SMARTS strings relevant to the SIMPOL method.

On the first line, we see k is set to 1 which corresponds to the SMART string [#6]. The SMART [#6] searches for all instances where the element with atomic number 6 occurs.

[#6] is used in this instance rather than C as in SMARTS C would only identify aliphatic carbons and would therefore miss any aromatic carbons. From line 3 we see a change in how k is defined. Whilst we acknowledge the method is based on 30 features of a molecule, in some instances it is difficult to extract a given feature using one unique SMARTS. For $k = 3$, the goal here was to check whether or not an aromatic ring is present, and if an aromatic ring is present, how many there are. This is more complex to search for than the number of carbons and, rather than defining one set of SMARTS for $k = 3$, we break this down as follows.

There are multiple SMARTS used to identify one feature. This is due to aromatic rings not having a fixed size, and the need to not only identify whether an aromatic ring is present but also the number of rings. For $k = 301$, "#6, #7, #8" searches for either carbon, nitrogen, or oxygen atoms. The term "r3" searches for a ring of size 3 and "a" searches for aromaticity. The character "," acts as an OR operator whilst ";" acts as an AND operator. This means that k entry 301 searches for atoms in a ring of size 3, that is also aromatic, and can be either a carbon, nitrogen, or oxygen. "r4" will search for a ring of size 4, "r5" will search for a ring of size 5 *ect.* If the task is to identify whether an aromatic ring alone was present, this could be solved simply with the SMART string [r;a], with "r" being the SMARTS to search for an atom in a ring, "a" being the SMARTS for an atom that is aromatic, and ";" requiring both "r" and "a" to be true. We define k from 301 to 306 such that they do not provide the total number of aromatic rings present, but instead the total number of atoms in aromatic rings. Whilst this is not the desired information for the SIMPOL method, the information that is provided is used to determine the number of rings in a separate function, once the SMARTS have been extracted from the Pybel molecule object, which will be shown shortly.

Similarly, code Listing 4.25 demonstrates how we have defined SMARTS to capture nitro functionality. The index **k** = 16 is used for the nitro functionality. Whilst the correct SMILES representation of a nitro group is $[N^+](=O)[O^-]$, it is not uncommon for the SMILES to be represented as $N(=O)=O$. The index **k** = 1601 searches for a nitro group where no charges are included in the input, whereas index **k** = 1602 searches for nitro groups where charges are present.

```
1601    [#6][$([NX3](=O)=O)]
1602    [#6][$([NX3+](=O)[O-])]
```

Listing 4.25 SMARTS describing a nitro group.

Now that the **.data** file and **.smarts** file have been constructed, we need a mechanism to provide this information to our property predictive techniques. To recap, the workflow in UManSysProp is a user access a function within a single file in the main directory, which in turn connects with this information in the **data** folder through the *groups.py* file. The first file in this instance is, again, the *vapour_pressures.py* file. Code Listing 4.26 displays the SIMPOL function defined in that file, displays the code equivalent of Equations (4.23) and (4.22).

```
def simpol(compound, temperature):
    m = groups.simpol(compound)

    b1 = groups.aggregate_matches(m, data.SIMPOL_1)
    b2 = groups.aggregate_matches(m, data.SIMPOL_2)
    b3 = groups.aggregate_matches(m, data.SIMPOL_3)
```

```
    b4 = groups.aggregate_matches(m, data.SIMPOL_4)

    bkT = b1 / temperature + b2 + b3 * temperature + b4 * log(
    temperature)
    return bkT
```

**Listing 4.26** Estimating vapor pressure using SIMPOL. Equations (4.23) and (4.22) are combined to calculate $\log_{10} P_{L,i}(T)$.

The input variable `compound` is a Pybel molecule object, whilst variable `temperature` is, as the name suggests, a floating point variable with a value that represents the desired temperature in Kelvin.

On line 2, you can see there is a call to another function called **simpol** imported from the **groups** module (thus *groups.py* file). This function returns a dictionary of SMARTS identifiers, which we have already introduced as a set of integer values $k$ in the text above, and the number of each feature as key:value pairs. Behind the scenes, this is where the Pybel parser interrogates a SMILEs string, converted to an OpenBabel molecule object, with SMARTS. The second call to an internal function **groups.aggregate_matches** combines this dictionary with model parameters and calculates the terms $B_{1,k}$ to $B_{4,k}$ in Equation (4.23). You will also notice we pass in expressions **data.SIMPOL_1** to **data.SIMPOL_4**. Whilst we have presented a subset of the text files that include SMARTS and data required to implement the SIMPOL method, this syntax may seem confusing. Again, this is a result of the design choice and thus structure of UManSysProp. In this instance, the *__init__.py* file in the **data** folder does not simply act to enable the folder to be marked as a Python module by the Python interpreter but also contains a series of functions to convert the information stored in our text files to be used by the Pybel molecule object parser and property estimation techniques alike.

In code Listings 4.27 and 4.28, we present a subset of the *__init__.py* file that reads the model parameters and SMARTS strings through bespoke functions **_read_data** and **_read_smarts**, respectively. These are also defined in the *__init__.py* file and are not presented here, but act to read each text file in turn.

```
SIMPOL_1                              = _read_data('SIMPOL.data',
      value_col=1)
SIMPOL_2                              = _read_data('SIMPOL.data',
      value_col=2)
SIMPOL_3                              = _read_data('SIMPOL.data',
      value_col=3)
SIMPOL_4                              = _read_data('SIMPOL.data',
      value_col=4)
```

**Listing 4.27** Subset of the *__init__.py* file that provides link to extracting model parameters for the SIMPOL method.

```
SIMPOL_SMARTS                         = _read_smarts('SIMPOL.smarts'
      )
```

**Listing 4.28** Subset of the *__init__.py* file that provides link to extracting SMART strings for the SIMPOL method.

Whilst we do not provide the full code listing of all internal functions in this book, they can be accessed through the open-source UManSysProp suite. Nonetheless, it is worth highlighting that features within the **simpol** function held within the *groups.py* file that have been designed to account for specific combinations of functionalities.

In code Listing 4.29, we present a snapshot of that function. If you recall, previously we noted that if we wish to check whether or not an aromatic ring is present, and count how many there are, this is more complex to search for than the number of carbons and required us to break this down into multiple SMARTS definitions. In code Listing 4.29, result['0'] is set to 1 as this is the case for all SIMPOL calculations as defined by Pankow and Asher (2008). The dictionary entry result['1'] captures the feature extracted from the SMILES string described by m['1']. This is described in the SIMPOL.smarts file. Programmatically, each carbon present in the SMILES string will increase m[1] by 1. Therefore, application of this script to ethane (SMILES string = CC) will set result['1'] to 2, and propane (SMILES string = CCC) will set result['1'] to 3.

Dictionary entries result['3'] and result['4'] are used to count the number of rings (aromatic or aliphatic, respectively) within a molecule. The SMARTS that are used to search for the number of rings actually count each atom within a ring. For this reason, if all atoms in a three-membered rings are counted, then to get the number of rings, it is necessary to divide by 3. This works when multiple rings within a molecule do not share the same atoms. However, in the case of fused rings this function would break down. Taking naphthalene as an example, there are 2 six-membered rings, but only 10 carbons making up these 2 rings. This would cause result['4'] to return a noninteger value for the number of rings. To mitigate this, the **ceil** function is used to round up to the next integer value.

```
def simpol(compound):
    m = matches(data.SIMPOL_SMARTS, compound)
    result = {}
    #Zeroth group (constant term used in all calculations)
    result['0'] = 1
    #total number of carbons present
    result['1'] = m[1]
    #number of carbons on acid side of amide bond
    result['2'] = m[2]
    #Aromatic ring present
    #for one ring contribution is one. A ring containing 3 atoms will
      give answer of 3 therefore
    #must be divided by 3 to give the correct contribution
    #m[301] for 3 membered rings
    #m[302] for 4 membered rings
    #up to m[306] for 8 membered rings
    #ceil function used to compensate for fused rings
    result['3'] =    ceil(
                    m[301] / 3 + m[302] / 4 + m[303] / 5
                    + m[304] / 6 + m[305] / 7 + m[306] / 8
                    )
    #Aliphatic ring present
    #same theory as result['3']
    result['4'] =    ceil(
                    m[401] / 3 + m[402] / 4 + m[403] / 5
                    + m[404] / 6 + m[405] / 7 + m[406] / 8
                    )
```

**Listing 4.29** Subset of the **simpol** function held within the *groups.py* file that returns the required number of features to implement the SIMPOL method.

```
#total number of nitroester groups present
#group 3001 is for the nitro part represented as -N(=O)=O
#group 3002 is for the nitro part represented as -[N+](=O)[O-]
result['30'] = m[3001] + m[3002]
if result['30'] > 0:
    result['11'] = 0
```

Listing 4.30 Subset of the **simpol** function held within the *groups.py* file that returns the required number of features to implement the SIMPOL method.

Likewise, in code Listing 4.30, the dictionary entry `result['30']` represents a count of the nitroester functionality. As a nitroester contains an ester group, it will return a positive match for the ester functionality, and whilst it is true that as the ester functionality is present, the proximity of the nitro group changes its chemical behavior significantly and for the purpose of property estimation, a nitro-ester and an ester are distinctly different groups. Therefore, for the cases where `result['30']` is true, `result['11']` (result for an ester group) must be equal to zero.

All of the aforementioned linkages are specific to UManSysProp but illustrate how to structure a set of generic functions in a particular way.

## 4.4 Subcooled Liquid Density

Converting condensed abundance and number concentrations into particle size requires an estimate of individual compound and mixture density. This also requires knowledge, or assumptions, about the phase state of the particle. We have also found in Chapter 1 that density of individual components can also be used to predict the volume of each solute in an ideal solution. As briefly covered in Section 4.3.1, we can combine group contribution techniques with molecular structure parsing methods to automate the prediction of component density. Barley et al. [41] evaluated the performance of seven GCMs for estimating liquid density, concluding that a value of $1350 \text{ kg} \cdot \text{m}^{-3}$ could be used for most estimations of secondary organic aerosol. Given the variable levels of information held in mechanism files, we nonetheless provide an overview of translating theory to code for one common technique here and the use of the UMan-SysProp package should you wish to replicate or build your own property prediction technique.

In the following, we provide a narrative on how the method of Girolami (1994) [42] is translated into code. Whilst this is a relatively simple technique, this allows us to further demonstrate the utility of functions provided by the Pybel package. This method also does not rely on the use of critical property estimations [41].

Equation 4.24 presents the simple formula described by [42], where the key assumption is that only the atomic stoichiometry of the liquid is important and not its molecular structure. Where this assumption breaks down, correction factors account for deviations that arise from hydrogen bonding group:

$$\rho_i = \frac{M_i}{5V_{atomic}} c \tag{4.24}$$

where $M_i$ is the molecular weight of compound $i$ and $V_{atomic}$ is the summed contribution the volumes of its constituent atoms using relative volumes given in Table

**Table 4.4** Copy of elements and relative volumes from Girolami, to be used in Equation (4.24).

**Relative volumes for predicting liquid density**

| Element | Relative volume |
| --- | --- |
| H | 1 |
| First short period, Li–F | 2 |
| Second short period, Na–Cl | 4 |
| First long period, K–Br | 5 |
| Second long period, Rb–I | 7.5 |
| Third long period, Cs–Bi | 9 |

4.4. $c$ is a correction factor to account for additional structural features. Barley et al. [41] summarize these to include 10% correction for each hydrogen bonding functional group present (up to a maximum of 30%) and a 10% correction for each unfused ring and a 7.5% correction for each fused ring (all within a maximum correction of 30%).

Code Listing 4.31 displays the function definition to predict density according to Girolami from the UManSysProp package. The input parameter `compound` is a Pybel object, created in a previous step using the OpenBabel package to read a SMILES string as covered in Section 4.2. The first step is to create a dictionary that stores feature names and the number of those features as key:value pairs. As with vapor pressure estimation techniques, there are multiple functions within the **group** module in UManSysProp, each designed to match a particular library of SMARTS, thus molecular features, to a SMILES string which is stored as an OpenBabel object `compound`. In this instance, once that dictionary has been returned, the **aggregate_matches** function sums the number of matches for each group multiplied by the relative volume for each group. The same **aggregate_matches** function is then used to identify features that require a correction factor to be applied to the previous estimate of liquid density, held in the `result` variable.

```
def girolami(compound):
    comp = groups.composition(compound)
    m = groups.girolami(compound)
    m['H'] = comp['H']
    volumes = {
        'H':              1.0,
        'short_period_1': 2.0,
        'short_period_2': 4.0,
        'long_period_1':  5.0,
        'long_period_2':  7.5,
        'long_period_3':  9.0,
        }
    result = groups.aggregate_matches(
        m, coefficients=volumes, groups=volumes.keys())
    result = comp['mass'] / (5.0 * result)
    # Make adjustments based on the presence of specific groups
    group_multipliers = {
        'hydroxyl_groups':         10.0,
        'hydroperoxide_groups':    10.0,
        'peroxyacid_groups':       10.0,
```

```
                'carboxylic_acid_groups':            10.0,
                'primary_secondary_amino_groups': 10.0,
                'amide_groups':                      10.0,
                'unfused_rings':                     10.0,
                'fused_rings':                        7.5,
                }
        percentage_increase = min(30.0, groups_aggregate_matches(
            m, coefficients=group_multipliers,groups=group_multipliers.
        keys()
            ))
        result += result * percentage_increase / 100.0
        return result
```

**Listing 4.31** UManSysProp function that takes a Pybel object and extracts the atomic contributions to calculate liquid density according to Girolami.

Code Listing 4.32 shows the **girolami** function included within the **groups** module.

```
def girolami(compound):
    m = matches(data.GIROLAMI_SMARTS, compound)
    result = {}
    result['short_period_1'] = m[2]
    result['short_period_2'] = m[3]
    result['long_period_1'] = m[4]
    result['long_period_2'] = m[5]
    result['long_period_3'] = m[6]
    result['hydroxyl_groups'] = m[7]
    result['hydroperoxide_groups'] = m[701]
    result['carboxylic_acid_groups'] = m[8]
    result['peroxyacid_groups'] = m[801]
    result['primary_secondary_amino_groups'] = m[9]
    result['amide_groups'] = m[10]
    all_rings = sum((
        m[1101] / 3,
        m[1102] / 4,
        m[1103] / 5,
        m[1104] / 6,
        m[1105] / 7,
        m[1106] / 8,
        ))
    all_rings = 0 if all_rings < 1 else ceil(all_rings)
    result['fused_rings'] = m[1201] + m[1202] + m[1203]
    result['unfused_rings'] = all_rings - result['fused_rings']
    return result
```

**Listing 4.32** UManSysProp function to parse a Pybel object into atomic and structural features.

# Bibliography

[1] C. L. Heald and J. H. Kroll. The fuel of atmospheric chemistry: Toward a complete description of reactive organic carbon. *Science Advances*, 1–8, 2, 2020. https://www.science.org/doi/10.1126/sciadv.aay8967

[2] Charles E. Kolb and Douglas R. Worsnop. Chemistry and composition of atmospheric aerosol particles. *Annual Review of Physical Chemistry*, 63:471–491, 5, 2012.

**[3]** M. E. Jenkin, J. C. Young, and A. R. Rickard. The MCM v3.3.1 degradation scheme for isoprene. *Atmospheric Chemistry and Physics*, 15:11433–11459, 10, 2015.

**[4]** M. E. Jenkin, K. P. Wyche, C. J. Evans, T. Carr, P. S. Monks, M. R. Alfarra et al. Development and chamber evaluation of the MCM v3.2 degradation scheme for $\beta$–caryophyllene. *Atmospheric Chemistry and Physics*, 12:5275–5308, 2012.

**[5]** C. Bloss, V. Wagner, M. E. Jenkin, R. Volkamer, W. J. Bloss, J. D. Lee et al. Development of a detailed chemical mechanism (MCMv3.1) for the atmospheric oxidation of aromatic hydrocarbons. *Atmospheric Chemistry and Physics*, 5:641–664, 2005.

**[6]** M. E. Jenkin, S. M. Saunders, V. Wagner, and M. J. Pilling. Protocol for the development of the master chemical mechanism, MCM v3 (part b): Tropospheric degradation of aromatic volatile organic compounds. *Atmospheric Chemistry and Physics*, 3:181–193, 2003.

**[7]** S. M. Saunders, M. E. Jenkin, R. G. Derwent, and M. J. Pilling. Protocol for the development of the master chemical mechanism, MCM v3 (part a): Tropospheric degradation of non-aromatic volatile organic compounds. *Atmospheric Chemistry and Physics*, 3:161–180, 2003.

**[8]** Michael E. Jenkin, Sandra M. Saunders, and Michael J. Pilling. The tropospheric degradation of volatile organic compounds: A protocol for mechanism development. *Atmospheric Environment*, 31:81–104, 1 1997.

**[9]** Isaac Kwadjo Afreh, Bernard Aumont, Marie Camredon, and Kelley Claire Barsanti. Using gecko-a to derive mechanistic understanding of SOA formation from the ubiquitous but understudied camphene. *Atmospheric Chemistry and Physics Discussions*, pages 1–27, 2020.

**[10]** B. Aumont, S. Szopa, and S. Madronich. Modelling the evolution of organic carbon during its gas-phase tropospheric oxidation: Development of an explicit model based on a self generating approach. *Atmospheric Chemistry and Physics*, 5:2497–2517, 2005.

**[11]** David Topping, Paul Connolly, and Jonathan Reid. Pybox: An automated box-model generator for atmospheric chemistry and aerosol simulations. *Journal of Open Source Software*, 3(28):755, 2018.

**[12]** Langwen Huang and David Topping. Jlbox v1.1: A julia-based multi-phase atmospheric chemistry box model. *Geoscientific Model Development*, 14:2187–2203, 4, 2021.

**[13]** J. H. Seinfeld and S. N. Pandis. *Atmospheric Chemistry and Physics: From Air Pollution to Climate Change*. J. Wiley & Sons, New York, USA, 1998. https://www.wiley.com/en-us/Atmospheric+Chemistry+and+Physics:+From+Air+Pollution+to+Climate+Change,+3rd+Edition-p-9781118947401

**[14]** Roberto Sommariva, Sam Cox, Chris Martin, Kasia Boroska, Jenny Young, Peter K. Jimack et al. Atchem (version 1), an open-source box model for the master chemical mechanism. *Geoscientific Model Development*, 13:169–183, 1, 2020.

**[15]** Python 3.9.6 documentation. https://docs.python.org/3/

**[16]** Valeriu Damian, Adrian Sandu, Mirela Damian, Florian Potra, and Gregory R. Carmichael. The kinetic preprocessor KPP—a software environment for solving chemical kinetics. *Computers and Chemical Engineering*, 26:1567–1579, 11, 2002.

[17] Siu Kwan Lam, Antoine Pitrou, and Stanley Seibert. Numba: A LLVM-Based Python JIT Compiler. Association for Computing Machinery (ACM), New York, USA 7:1–6, 2015. https://dl.acm.org/doi/abs/10.1145/2833157.2833162

[18] Pearu Peterson. F2py: A tool for connecting Fortran and python programs. *International Journal of Computational Science and Engineering*, 4:296–305, 2009.

[19] Christian Andersson, Claus Führer, and Johan Åkesson. Assimulo: A unified framework for ode solvers. *Mathematics and Computers in Simulation*, 116:26–43, 6, 2015.

[20] F. Binkowski and U. Shankar. The regional particulate matter model 1. model description and preliminary results. *Journal of Geophysical Research*, 100:26191–26209, 1995.

[21] Francis S. Binkowski and Shawn J. Roselle. Models-3 community multiscale air quality (cmaq) model aerosol component 1. model description. *Journal of Geophysical Research: Atmospheres*, 108(D6), 2003.

[22] Evan Whitby, Frank Stratmann, and Martin Wilck. Merging and remapping modes in modal aerosol dynamics models: a "dynamic mode manager". *Journal of Aerosol Science*, 33(4):623–645, 2002.

[23] K. N. Sartelet, H. Hayami, B. Albriet, and B. Sportisse. Development and preliminary validation of a modal aerosol model for tropospheric chemistry: Mam. *Aerosol Science and Technology*, 40(2):118–127, 2006.

[24] Stephen R. Heller, Alan McNaught, Igor Pletnev, Stephen Stein, and Dmitrii Tchekhovskoi. Inchi, the IUPAC international chemical identifier. *Journal of Cheminformatics*, 7:23, 5, 2015.

[25] William J. Wiswesser. The wiswesser line formula notation. *Chemical & Engineering News Archive*, 30(34):3523–3526, 1952.

[26] H.-G. Rohbeck. Representation of structure description arranged linearly. In Jürgen Gmehling, editor, *Software Development in Chemistry 5*, pages 49–58, Berlin, Heidelberg, 1991. Springer Berlin Heidelberg.

[27] Sheila Ash, Malcolm A. Cline, R. Webster Homer, Tad Hurst, and Gregory B. Smith. Sybyl line notation (sln): A versatile language for chemical structure representation. *Journal of Chemical Information and Computer Sciences*, 37:71–79, 1997.

[28] Wendy A. Warr. Representation of chemical structures. *WIRES computational molecular science*, 1:557–579, 4, 2011. https://doi.org/10.1002/wcms.36

[29] David Weininger, Arthur Weininger, and Joseph L. Weininger. Smiles. 2. algorithm for generation of unique smiles notation. *Journal of Chemical Information and Computer Sciences*, 29:97–101, 1989.

[30] Noel M. O'Boyle, Michael Banck, Craig A. James, Chris Morley, Tim Vandermeersch, and Geoffrey R. Hutchison. Open babel: An open chemical toolbox. *Journal of Cheminformatics*, 3, 10, 2011.

[31] David Topping, Mark Barley, Michael K. Bane, Nicholas Higham, Bernard Aumont, Nicholas Dingle et al. Umansysprop v1.0: An online and open-source facility for molecular property prediction and atmospheric aerosol calculations. *Geoscientific Model Development*, 9:899–914, 3, 2016.

[32] Yash Nannoolal, JUrgen Rarey, and Deresh Ramjugernath. Fluid Phase Equilibria Estimation of pure component properties Part 3. Estimation of the vapor pressure of non-electrolyte organic compounds via group contributions and group interactions. *Fluid Phase Equilibria*, 269:117–133, 2008.

[33] Paul B Myrdal and Samuel H Yalkowsky. Estimating Pure Component Vapor Pressures of Complex Organic Molecules. *Industrial & Engineering Chemistry Research*, 36(6):2494–2499, 1997.

[34] S. Compernolle, K. Ceulemans, and J. F. Müller. Evaporation: A new vapour pressure estimation method for organic molecules including non-additivity and intramolecular interactions. *Atmospheric Chemistry and Physics*, 11(18):9431–9450, 2011.

[35] J. F. Pankow and W. E. Asher. SIMPOL.1: A simple group contribution method for predicting vapor pressures and enthalpies of vaporization of multifunctional organic compounds. *Atmospheric Chemistry and Physics*, 8(10):2773–2796, 2008.

[36] K. G. Joback, R. C. Reid, and C. Reid. Estimation of Pure-Component Properties From Group-Contributions. *Chemical Engineering Communications*, 157:233–243, 1987.

[37] S. E. Stein and R. L. Brown. Estimation of Normal Boiling Points from Group Contributions. *Journal of Chemical Information and Computer Sciences*, 34:581–587, 1994.

[38] Yash Nannoolal, Jürgen Rarey, Deresh Ramjugernath, and Wilfried Cordes. Estimation of pure component properties Part 1. Estimation of the normal boiling point of non-electrolyte organic compounds via group contributions and group interactions. *Fluid Phase Equilibria*, 226:45–63, 2004.

[39] M. H. Barley and G. McFiggans. The critical assessment of vapour pressure estimation methods for use in modelling the formation of atmospheric organic aerosol. *Atmospheric Chemistry and Physics*, 10(2):749–767, January, 2010.

[40] Simon O'Meara, Alastair Murray Booth, Mark Howard Barley, David Topping, and Gordon Mcfiggans. An assessment of vapour pressure estimation methods. *Physical Chemistry Chemical Physics*, 16(16):19453–19469, 2014.

[41] Mark H. Barley, David O. Topping, and Gordon McFiggans. Critical assessment of liquid density estimation methods for multifunctional organic compounds and their use in atmospheric science. *Journal of Physical Chemistry A*, 117:3428–3441, 4, 2013.

[42] Gregory S. Girolami. A simple "back of the envelope" method for estimating the densities and molecular volumes of liquids and solids. *Journal of Chemical Education*, 71:962–964, 1994.

# 5

# Coagulation

*Coagulation* is the process of two particles colliding with each other, forming a new particle of a larger size. Coagulation causes the total number concentration of an aerosol population to decrease and the size distribution to shift toward larger sizes. The total mass of the population is conserved under coagulation. If the particles are both solid, this process is also called *aggregation*, leading to non-spherical particle aggregates. If the particles are liquid, i.e., if they are solution droplets, then they tend to coalesce after collision, forming one larger solution droplet. If one is solid, and the other is liquid, the solid particle may partially or completely dissolve in the liquid one, or the liquid particle may partially or completely coat the solid particle.

Coagulation is caused by particles having different velocity vectors, because two particles moving in the same direction with the same speed cannot collide. There are multiple mechanisms that can introduce relative velocities between particles, including Brownian motion, gravitational settling, laminar shear flow, and turbulent flow. Interparticle forces can modify these relative velocities induced by external forces, including van der Waals forces, Coulomb forces, and hydrodynamic forces. In the real atmosphere, several coagulation mechanisms can act simultaneously, but their relative importance is strongly dependent on the particle sizes.

## 5.1   Coagulation Probabilities and Rates

At the particle level, coagulation can be modeled as a stochastic process. We consider a parcel of air with volume $V$, containing a population of aerosol particles. The probability of coagulation between any two particles is quantified by the *coagulation kernel*, $K(m_1, m_2)$, which depends on the masses $m_1$ and $m_2$ of the particles. The probability that the two particles coagulate within a time interval $\Delta t$ is then given by $p = K \Delta t / V$. Because the probability $p$ is dimensionless, we see that $K$ must have units of $m^3 s^{-1}$. In this model we are treating coagulation as a stochastic process, and more precisely a Markov process [1], assuming that the coagulation events at time $t$ only depend on the state of the system at time $t$. In the deterministic limit, the coagulation kernel represents the mean rate coefficient of coagulation. The functional form of the coagulation kernel is determined by the underlying

*Introduction to Aerosol Modelling: From Theory to Code.*
First Edition. Edited by David Topping and Michael Bane.
© 2022 John Wiley & Sons Ltd. Published 2022 by John Wiley & Sons Ltd.

physics of the coagulation process. Here, we use as example the *Brownian coagulation kernel*.

The Brownian kernel captures the effect of relative velocities that arise due to Brownian motion of the particles as they are constantly bombarded by the surrounding gas (air) molecules. Small particles respond to this bombardment readily, while larger particles are much more sluggish in their response. In addition, large particles having a large surface area constitute large targets for the fast-moving small particles. Therefore, coagulation of particles with different sizes are more likely than of similar-sized particles.

The Brownian coagulation kernel used in many applications is given by Fuchs [2], here expressed as a function of particle diameters $d_{p1}$ and $d_{p2}$, assuming spherical particles:

$$K(d_{p1}, d_{p2}) = 2\pi(D_1 + D_2)(d_{p1} + d_{p2})$$

$$\times \left( \frac{d_{p1} + d_{p2}}{d_{p1} + d_{p2} + 2(g_1^2 + g_2^2)^{1/2}} + \frac{8(D_1 + D_2)}{(\bar{c}_1^2 + \bar{c}_2^2)^{1/2}(d_{p1} + d_{p2})} \right)^{-1}, \quad (5.1a)$$

$$\bar{c}_i = \left( \frac{8k_B T}{\pi m_i} \right)^{1/2}, \quad (5.1b)$$

$$\ell_i = \frac{8D_i}{\pi \bar{c}_i}, \quad (5.1c)$$

$$g_i = \frac{1}{3d_{pi}\ell_i} \left( (d_{pi} + \ell_i)^3 - (d_{pi}^2 + \ell_i^2)^{3/2} - d_{pi} \right), \quad (5.1d)$$

$$D_i = \frac{k_B T C_c}{3\pi \mu d_{pi}}, \quad (5.1e)$$

where $\bar{c}_i$ is the mean thermal speed of particle $i$, $m_i$ is the mass of particle $i$, $\ell_i$ is the mean free path of particle $i$, $g_i$ is a variable of convenience, $D_i$ is the Brownian diffusivity of particle $i$, $k_B$ is the Boltzmann constant, $T$ is the temperature, $\mu$ is the viscosity of air, and $C_c$ is the Cunningham correction factor.

For a derivation of this kernel function, see Seinfeld and Pandis [3]. A visual representation of this kernel is presented in Figure 5.1. It is easy to see that coagulation events between differently-sized particles are many orders of magnitude more likely than events between particles of similar sizes.

Now that we have established the coagulation kernel, we will move on to formulating how an aerosol population evolves due to coagulation. We will describe this first from a stochastic viewpoint, and then move to the deterministic large-number-limit.

We will first simulate coagulation as a stochastic process using Gillespie's method [4, 5], considering all possible interactions among a population of particles. We will quickly discover that this method is too slow to apply for atmospherically relevant systems that typically contain a large number of particles, but it is conceptually simple and instructive to follow. We will then illustrate three methods to speed up the simulation, which eventually lead to the well-known population balance equation, or Smoluchowski equation [6], in the large-number-limit.

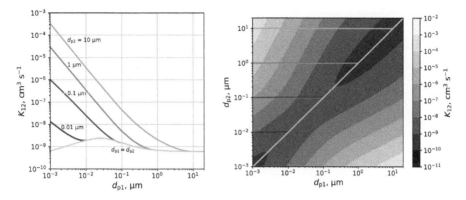

**Figure 5.1** (Left) Brownian coagulation kernel as a function of particle diameter $d_{p1}$ and selected particle diameters $d_{p2}$ for coagulation at 25°C. (Right) Brownian coagulation kernel as a function of particle diameters $d_{p1}$ and $d_{p2}$.

## 5.2 Stochastic Coagulation with Discrete Particle Masses

### 5.2.1 Gillespie's Basic Algorithm for Discrete Number Concentrations

Let us define a set of particles by an array $Q = [Q_1, \dots, Q_N]$, where each entry represents the scale factor of a base mass $m_b$ of a particle, for example $Q = [3, 1, 2, 2, 3, 1, 1, 3]$. This array has eight particles, where three particles have mass $1 \times m_b$, two particles have mass $2 \times m_b$, and three particles have mass $3 \times m_b$. We write $Q_i$ for the scale factor of the $i$-th particle. For now, we will assume that all particles consist of the same chemical compound, and we will return to the more realistic case of a coagulating multi-component aerosol in Section 5.4.1.

The algorithm for Gillespie's method is described in Algorithm 5.1 and implemented in *chap5_alg1_basic_method.py*. In time $\tau$, the probability that particle $i$ coagulates with particle $j$ is $K_{ij}\tau/V$, where $V$ is the volume that the particles reside in and $K_{ij} = K(Q_i m_b, Q_j m_b)$ is the coagulation kernel between the two particles. We use $\tau$ rather than $\Delta t$ here because it is traditional when describing stochastic simulation algorithms. Gillespie's method starts at $t = 0$ and then repeats a set of actions for each step. We consider all possible events, which for us are coagulations between each pair $(i, j)$ of particles. For each possible event, we compute a random time for when this event would happen if it was the only possible event. This random time is $\tau_{ij} = -\log(u_{ij})/\lambda_{ij}$, where $u_{ij} \sim \text{Unif}(0, 1)$ is a random number between 0 and 1, and $\lambda_{ij} = K_{ij}/V$ is the *probability rate* of events between particles $i$ and $j$. This expression comes from the fact that the time to the next event in a Markov process is exponentially distributed with parameter $\lambda_{ij}$.

The sampling of a random number to calculate $\tau_{ij}$ (Line 9 in Algorithm 5.1) is implemented as in code listing 5.1:

```
K12 = kernel_helper.GetKernel(Q[i]*mb, Q[j]*mb)
lamb = K12/V
u = np.random.rand()
tau[i,j] = -np.log(u)/lamb
```

**Listing 5.1** Sampling of a random number to calculate $\tau_{ij}$.

---

**Algorithm 5.1:** Gillespie's basic method for stochastic coagulation.

1  **Data:** $Q[i]$ is the mass scale of particle $i = 1, \dots, N$ relative to base mass $m_b$.

2  $t \leftarrow 0$

3  **while** $t < t_f inal$ **do**

4      **for** $i \leftarrow 1, \dots, N$ **do**

5          **for** $j \leftarrow 1, \dots, (i-1)$ **do**

6              $K \leftarrow \texttt{Kernel}(Q[i]m_b, Q[j]m_b)$

7              $\lambda \leftarrow K/V$

8              $u \sim \text{Unif}(0, 1)$

9              $\tau[i, j] \leftarrow -\log(u)/\lambda$

10         **end**

11     **end**

12     $(i, j) \leftarrow \text{argmin}_{(i', j')} \tau[i', j']$

13     append $(Q, Q[i] + Q[j])$

14     delete $Q[i]$

15     delete $Q[j]$

16     $N \leftarrow N - 1$

17     $t \leftarrow t + \tau[i, j]$

18 **end**

---

Having computed $\tau_{ij}$ for each particle pair, we then choose the smallest of these (which corresponds to the first event) and this is the one that really happens. We coagulate the $(i, j)$ pair by removing particles $i$ and $j$ from the array and adding a new particle with mass $(Q_i + Q_j)m_b$. For example, if we coagulated the first two particles in our example array (particles of scale factor 1 and 3) then we would remove these from the array and add a new particle of scale factor 4 $(= 1+3)$ to the end of the array. At the same time, we advance the current simulation time by $t \leftarrow t + \tau_{ij}$. This is implemented in Python using np.argmin (returning the index of the minimum entry in the *flattened* tau array), np.unravel_index (which finds the index tuple from a flat index), and the append and delete functions to modify the $Q$ array. Code listing 5.2 provides the code that corresponds to Lines 12–15 in Algorithm 5.1:

```
index_array = np.argmin(tau)
(p1,p2) = np.unravel_index(index_array, shape=tau.shape)
Q.append(Q[p1]+Q[p2]) # Append new particle
Q.delete(p1) # Remove particle 1
Q.delete(p2) # Remove particle 2
```

Listing 5.2 Implementing lines 12 to 15 in Algorithm 5.1.

A Python command not specifically related to Algorithm 5.1 is time.time(), which we frequently use for timing purposes and will come in handy for Exercise (f). For example, to measure an elapsed time we can use code listing 5.3:

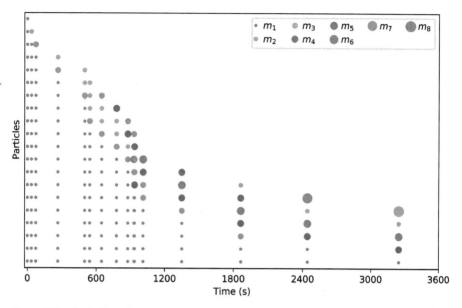

**Figure 5.2** Evolution of a set of particles using Algorithm 5.1 (Gillespie's basic method for stochastic coagulation) with an initial population consisting of 20 particles of mass $m_b$. Each particle is represented by a dot, where the color and the size of the dot represents the particle mass.

```
t1 = time.time()
# do some calculation here
t2 = time.time()
print(t2-t1)
```

**Listing 5.3** Timing a function in Python.

Figure 5.2 visualizes the application of Algorithm 5.1. Each particle is represented by a dot, where the size of the dot represents the particle mass. The population is plotted after each coagulation event, updating the number of partices and their sizes as appropriate. The initial population consisted of 20 particles of size $m_b$. After 3600 s, the particle number was reduced to five particles. The figure also shows that the time interval, $\tau_{ij}$, between two consecutive coagulation events varied and tended to become larger as the number of particles decreased.

If you run the Python script *chap5_alg1_basic_method.py* repeatedly, the result will look slightly different each time, which is expected for a stochastic process. This is accomplished since, by default, Python uses a different initialization of the random number generator each time that it starts, which results in a different sequence of random numbers every time the code is executed. To force Python to use the same sequence of random numbers, the so-called random seed needs to be specified at the beginning. Note that while setting the seed gives the same random numbers on one machine (which can be advantageous for debugging), they can differ for different machines.

### 5.2.2 First Speedup: Binning Particles

Exercise (d) teaches us a problem with Algorithm 5.1: It requires us to check all the particle pairs, which means that the cost of each step scales with $O(N^2)$ for $N$ particles. As the particle number increases, it takes much longer for the simulation to complete. To perform a simulation for realistic atmospheric conditions, the computational cost and the memory demands become prohibitive.

There are several ways in which we can speed up our code. The first step is to observe that all the particles of the same mass act the same as each other. This means we do not need to check them all individually, but can group them according to their mass or size. Grouping, or binning, the particles in this way will also save memory. The resulting algorithm is shown in Algorithm 5.2.

---

**Algorithm 5.2:** Gillespie's method for binned stochastic coagulation.

---

1  **Data:** $N[k]$ is the number of particles in bin $k = 1, \dots, N_{\text{bin}}$
2  $t \leftarrow 0$
3  **while** $t < t_{\text{final}}$ **do**
4     **for** $k \leftarrow 1, \dots, N_{\text{bin}}$ **do**
5         **for** $\ell \leftarrow 1, \dots, k$ **do**
6             $K \leftarrow \texttt{Kernel}(km_{\text{b}}, \ell m_{\text{b}})$
7             **if** $k \neq \ell$ **then**
8                 $N_{\text{pairs}} \leftarrow N[k]N[\ell]$
9             **else**
10                 $N_{\text{pairs}} \leftarrow \frac{1}{2}N[k](N[\ell] - 1)$
11             **end**
12             $\lambda \leftarrow N_{\text{pairs}}K/V$
13             $u \sim \text{Unif}(0, 1)$
14             $\tau[k, \ell] \leftarrow -\log(u)/\lambda$
15         **end**
16     **end**
17     $(k, \ell) \leftarrow \text{argmin}_{(k', \ell')} \tau[k', \ell']$
18     $N[k] \leftarrow N[k] - 1$
19     $N[\ell] \leftarrow N[\ell] - 1$
20     $N[k + \ell] \leftarrow N[k + \ell] + 1$
21     $t \leftarrow t + \tau[k, \ell]$
22 **end**

---

This means we need to change how we store particles. Rather than storing all their masses individually, we will store the *number of particles of each mass*. The particle mass is still a multiple of the base mass $m_{\text{b}}$. For example, rather than storing $[1, 1, 1, 1, 2, 2, 2, 2, 2, 2, 3, 3, 4, 4, \dots]$, we will store $[4, 6, 2, 2, \dots]$. This new array contains the variables $[N_1, N_2, N_3, N_4, \dots]$, where $N_k = N[k]$ is the number of particles of mass $k \times m_{\text{b}}$. We usually call the different size categories "size bins" or simply "bins." Here, all particles in one bin have the same size. The particle "size" can be expressed in terms of particle volume, particle mass, or particle diameter. In this chapter we choose particle mass.

Rather than thinking about reaction events occurring between pairs of particles, instead we can now think about events between *pairs of bins*, which reduces the problem size substantially. For two bins containing $N_1$ and $N_2$ particles, there are $N_1 \times N_2$ possible particle interactions. The event rate is thus $N_1 \times N_2 \times K_{12}$. We can now use Gillespie's method to simulate events between pairs of bins. Each time a bin pair experiences an event, we coagulate one particle from each bin, and place it in its destination bin. Since the particle masses are still multiples of $m_b$, each newly created particle will fit exactly in a destination bin.

An important special case is to compute the rate of events of particles *within* a bin. If the bin contains $N$ particles, then the number of pairs is $\binom{N}{2} = \frac{1}{2}N(N-1)$. Note that the factor of $\frac{1}{2}$ is not needed for bin pairs $(k, \ell)$ where $k \neq \ell$ because we only include bin pairs with $\ell \leq k$ to avoid double counting particle pairs.

Listing 5.4 corresponds to Lines 3–14 in Algorithm 5.2 and shows the modified loops over the number of bins rather than individual particles.

```
...
while t < t_final and np.sum(N) > 1:
    tau = np.full((N_bin,N_bin),np.Inf, dtype=float)
    for ind_k, k in enumerate(bin_size):
        for ind_l, l in enumerate(bin_size[0:ind_k+1]):
            K12 = kernel_helper.GetKernel(k*mb,l*mb)
            if (k == l):
                N_pairs = (N[ind_k]*(N[ind_l]-1)) / 2
            else:
                N_pairs = N[ind_k]*N[ind_l]
            lamb = N_pairs*K12/V
            u = np.random.rand()
            if (lamb > 0):
                tau[ind_k,ind_l] = -np.log(u)/lamb
```

Listing 5.4 Looping through particle size bins to implement Algorithm 5.2.

Observe that in the code above we use `lamb` for $\lambda$, because `lambda` is a reserved keyword in Python. Another way in which the code differs from the description in Algorithm 5.2 is that the `for` loops each use two different variables, for example `ind_k` and `k`, whereas Algorithm 5.2 simply used $k$ for the loop. This is because Python uses 0-indexed arrays, where the first entry in an array in indexed by `N[0]` and the second by `N[1]`. This means we need an "indexing" loop variable like `ind_k` to take values $0, 1, ...$, while the actual size variable `k` takes values $1, 2, ...$. In the code this is achieved using `enumerate()` on an explicit `bin_size` array that contains the values $1, 2, ..., (k + 1)$.

Another interesting feature in the code above is that we take care to only compute `tau` if `lamb > 0`, because otherwise we would have a division-by-zero error. If we have `lamb == 0` then `tau` will retain its initial value of `Inf`, which means we will never select this event that has zero rate.

Figure 5.3 (left panel) visualizes the result of Algorithm 5.2, starting with a particle population that consists of 100 particles of mass $m_b$ and showing the results of one simulation. The lines represent the time evolution of particle number $N_k$ for $k = 1, 2, ..., 10$. The number of particles $N_1$ of mass $1 \times m_b$ decreases as a result of coagulation, which

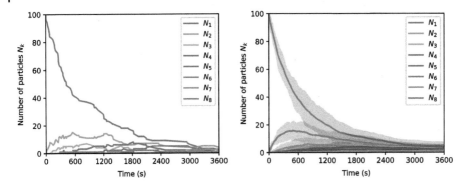

**Figure 5.3** Left: One simulation of the evolution of the number of particles having different masses using Algorithm 5.2 (Gillespie's method for binned stochastic coagulation). The number of particles of mass $N_1 = 1 \times m_b$ decreases as a result of coagulation, which produces particles with masses $N_2 = 2 \times m_b$, $N_3 = 3 \times m_b$, and so on, with their number in turn decreasing as coagulation proceeds. Right: The solid line is the average of 20 simulations, and the shaded band indicates the 95% confidence interval.

produces particles with masses $2 \times m_b$, $3 \times m_b$, and so on, with their numbers in turn decreasing as coagulation proceeds. Observe that the results look somewhat "noisy," which is caused by the stochastic nature of the algorithm. To reduce the noise, we repeated the simulation several times, producing an ensemble of simulations, and then calculate the ensemble average, which approaches the expected value. The result is shown in the right panel of Figure 5.3, where the solid line represents the average of 20 individual simulations and the shaded band indicates the 95% confidence interval.

### 5.2.3  Second Speedup: Discretize Time and Use Tau-leaping

The first speedup described in section 5.2.2 was helpful because it enabled us to treat all particles that looked the same (i.e., had the same mass) at once. However, a fundamental problem remains in that we still have to perform each coagulation event one by one.

To further speed up the calculations, we will need to start grouping coagulation events together. That is, we will now consider an interval of time of length $\Delta t$ and resolve all the equivalent events in this timestep in one go. This algorithm is called *tau-leaping* [7].

Algorithm 5.2 needs to be modified as follows, which results in Algorithm 5.3. Let us consider two bins that contain $N_1$ and $N_2$ particles, respectively. The rate of coagulation events for particles in these two bins is $\lambda = K_{12}/V$. In an interval of length $\Delta t$ the number of coagulation events for this bin pair is a random variable $r$ that has a Poisson probability distribution with parameter $\lambda \Delta t$. This can be justified by the fact that for uncorrelated random events with mean rate $\lambda$, the probability distribution of the number of the events in a time interval $\Delta t$ is described by Poisson($\lambda \Delta t$). This is exactly what we are concerned with in the case of coagulation.

With discretizing the time in this way, we make an important approximation. We take $\lambda = N_1 N_2 K/V$ at the beginning of the timestep and assume that $\Delta t$ is small enough so that $\lambda$ does not change during the timestep. In reality, $\lambda$ will change somewhat during the timestep because $N_1$ and $N_2$ change due to coagulations, but we ignore this, which introduces an error.

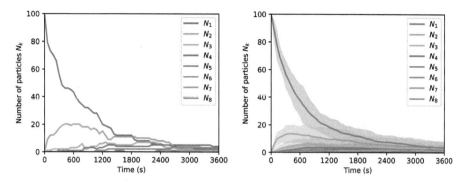

**Figure 5.4** Left: One simulation of the evolution of the particle number of different masses using Algorithm 5.3 (Tau-leaping method for binned stochastic coagulation using discretized time). This algorithm uses a fixed timestep $\Delta t$, set here to 6 s. Right: The solid line is the average of 20 simulations, and the shaded band indicates the 95% confidence interval.

---

**Algorithm 5.3:** Tau-leaping method for binned stochastic coagulation using discretized time.

---

1   **Data:** $N[k]$ is the number of particles in bin $k = 1, \dots, N_{\text{bin}}$
2   $t \leftarrow 0$
3   **while** $t < t_{\text{final}}$ **do**
4     **for** $k \leftarrow 1, \dots, N_{\text{bin}}$ **do**
5       **for** $\ell \leftarrow 1, \dots, k$ **do**
6         $K \leftarrow \texttt{Kernel}(k m_{\text{b}}, \ell m_{\text{b}})$
7         **if** $k \neq \ell$ **then**
8           $N_{\text{pairs}} \leftarrow N[k]N[\ell]$
9         **else**
10          $N_{\text{pairs}} \leftarrow \frac{1}{2}N[k](N[\ell] - 1)$
11         **end**
12         $\lambda \leftarrow N_{\text{pairs}}K/V$
13         $N_{\text{event}} \sim \text{Poisson}(\lambda \Delta t)$
14         $N[k] \leftarrow N[k] - N_{\text{event}}$
15         $N[\ell] \leftarrow N[\ell] - N_{\text{event}}$
16         $N[k + \ell] \leftarrow N[k + \ell] + N_{\text{event}}$
17       **end**
18     **end**
19     $t \leftarrow t + \Delta t$
20 **end**

---

Figure 5.4 (left panel) shows the result of using Algorithm 5.3. Qualitatively, this figure looks just like Figure 5.3 for Algorithm 5.2, but we note a few important differences. In Figure 5.3, the time series for the particle number for the individual simulation appears more noisy for simulation times less than 600 s. This is because the timestep $\tau$ in Algorithm 5.3 was much smaller during this first phase than the constant timestep $\Delta t$ chosen for Algorithm 5.3.

Algorithm 5.2 can be viewed as the benchmark, since each coagulation event is individually resolved. Compared to this Algorithm 5.3 incurs an error due to the time discretizations. If the timestep $\Delta t$ is small enough, this error will be negligible, but it will become apparent when $\Delta t$ is chosen too large—something to explore in Question (j).

A Python implementation of the core of Algorithm 5.3 is provided in code listing 5.5. Just like for Algorithm 5.2, we loop over the number of bins. But rather than sampling the time $\tau$ between events, we now sample for the number of events, using the np.random.poisson() function.

```
while t < t_final and np.sum(N) > 1:
    for ind_k, k in enumerate(bin_size):
        for ind_l, l in enumerate(bin_size[0:ind_k+1]):
            K12 = kernel_helper.GetKernel((k)*mb,(l)*mb)
            if (k == l):
                N_pairs = (N[ind_k]*(N[ind_l]-1)) / 2
            else:
                N_pairs = N[ind_k]*N[ind_l]
            if (N_pairs < 0):
                print(N_pairs,k,l, N[ind_k],N[ind_l])
            lamb = N_pairs*K12/V
            N_event = np.random.poisson(lamb*dt)
            N[ind_k] -= N_event
            N[ind_l] -= N_event
            if (N[ind_k] < 0):
                N[ind_k] = 0
            if (N[ind_l] < 0):
                N[ind_l] = 0
            dest = min(k + 1, N_bin)
            N[dest-1] += N_event
    hist[index,:] = N
    index += 1
    t = times[index]
```

Listing 5.5 Python implementation of Algorithm 5.3.

One consequence of the approximations in Algorithm 5.3 is that N_event might be larger than the number of particles we have available, in which case an entry in the N array would become negative. To avoid this physically impossible situation, we check for it and explicitly set N[ind_k] = 0 in this case.

The time step is fixed in Algorithm 5.3 and to control the time stepping we use the following code.

```
t_final = 3600.0
nt = 60
times = np.linspace(0,t_final,nt+1)
dt = t_final / nt
hist = np.zeros((len(times),N_part))
```

Observe that above we set the final time, t_final, and number of time steps, nt, and from this we calculate the time step, dt. This guarantees a whole number of time steps in the interval, so that we don't finish on a half step. By using an integer index we also avoid the problem with $t = t + dt$ that we might just miss the expected final time due to round-off error and have one too many or too few time steps.

The compute time and the memory usage for Algorithms 5.1, 5.2, and 5.3 scale differently with the number of computational particles, which is illustrated in Figure 5.5.

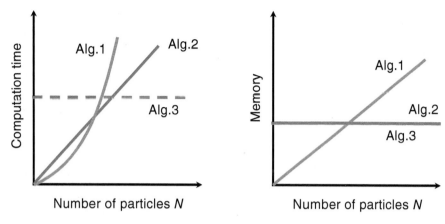

**Figure 5.5**  Scaling of computation time and memory usage with the number of computational particles for Algorithms 5.1, 5.2, and 5.3.

The computational cost for Algorithm 5.1 scales as $O(N^3)$, so, for example, doubling the number of computational particles results in a simulation that takes eight times as long. This is because the cost per coagulation event scales with $O(N^2)$ and the number of coagulation events in a given time interval always scales with $O(N)$, assuming the volume $V$ scales with $N$ to keep a constant number concentration. The memory usage scales linearly with the number of particles.

In contrast, the computational cost for Algorithm 5.2 scales linearly with the number of computational particles because the cost per coagulation event does not depend on the number of computational particles, so we only have the linear scaling with the total number of coagulation events, which is itself linear in $N$. Not shown on this graph is the fact that the computational cost also depends on the number of bins $N_{bin}$, namely the cost per event is $O(N_{bin}^2)$. Therefore the total cost for this algorithm scales as $O(N{\times}N_{bin}^2)$. The memory usages does not depend on the number of particles, but it scales linearly with $N_{bin}$.

Lastly, the cost of Algorithm 5.3 does not depend on the number of computational particles. However, it does depend on the timestep $\Delta t$ and the number of bins $N_{bin}$ as $O(N_{bin}^2/\Delta t)$. While Algorithm 5.3 is clearly much faster than the earlier algorithms, we must remember that this is achieved through an approximation, which also introduces some error. In practice, we are almost always happy to accept some small error for a large gain in speed.

### 5.2.4  Third Speedup: Large-number-limit Using Continuous Number Concentration

For the third speedup, we will exploit the fact that we consider a large volume $V$ and a small timestep $\Delta t$. That is, we are interested in the limits $V \to \infty$ and $\Delta t \to 0$. Keeping the number concentration constant, this implies that we consider a very large number of particles. One can show that for the large-number limit, the variance of number concentration goes to zero, and we therefore only need to track the mean. This removes the stochasticity, and rather than using Markov jump processes (a stochastic treatment), the mathematical model will now be a set of ordinary differential equations (a deterministic treatment). For a more sophisticated mathematical derivation see [8].

We start out by formulating a population balance equation for the number of particles in bin $k$ at time $t + \Delta t$, $N_k(t + \Delta t)$, assuming that the number of particles in bin $k$ at time $t$, $N_k(t)$, is known.

The number of particles in bin $k$ at time $t + \Delta t$ is given by

$$N_k(t + \Delta t) = N_k(t) - \sum_{\ell=1}^{k} N_{\text{event},k,\ell} - \sum_{\ell=k}^{\infty} N_{\text{event},k,\ell} + \sum_{\ell=1}^{k-1} \frac{1}{2} N_{\text{event},\ell,(k-\ell)}, \tag{5.2}$$

where the three summation terms represent the loss of particles from bin $k$ due to coagulation events with smaller particles, the loss of particles from bin $k$ due to coagulation with larger particles, and the gain of particles in bin $k$ due to coagulation of two smaller particles.

Rearranging this equation, and remembering that the number of coagulation events, $N_{\text{event}}$, is Poisson-distributed with the expected value of

$$E[N_{\text{event},k,\ell}] = \lambda_{k,\ell} \Delta t = N_k N_\ell K / V, \tag{5.3}$$

we can formulate the expected value of the change in the number of particles in bin $k$ from time $t$ to time $t + \Delta t$ as

$$E[N_k(t + \Delta t) - N_k(t)] = -\sum_{\ell=1}^{k} \Delta t \lambda_{k,\ell} - \sum_{\ell=k}^{\infty} \Delta t \lambda_{k,\ell} + \sum_{\ell=1}^{k-1} \frac{1}{2} \Delta t \lambda_{\ell,k-\ell}. \tag{5.4}$$

Dividing this equation by $\Delta t$ and $V$, we obtain

$$\frac{1}{\Delta t} E\left[\frac{N_k(t + \Delta t)}{V} - \frac{N_k(t)}{V}\right] = -\sum_{\ell=1}^{k-1} \frac{\lambda_{k,\ell}}{V} - 2\frac{\lambda_{k,k}}{V} - \sum_{\ell=k+1}^{\infty} \frac{\lambda_{k,\ell}}{V}$$

$$+ \sum_{\ell=1}^{k-1} \frac{1}{2} \frac{\lambda_{\ell,k-\ell}}{V}. \tag{5.5}$$

Observe that the $2\lambda_{k,k}/V$ term arises from the first and last terms, respectively, in the first two sums of Equation 5.4, which now exclude the $\ell = k$ index.

Let us define the number concentration in bin $k$ as $C_k = N_k/V$. The rate of coagulation per volume $V$ then becomes, for large volumes $V$:

$$\text{for} \quad k \neq \ell : \qquad \lim_{V \to \infty} \frac{\lambda_{k,\ell}}{V} = \frac{N_k N_\ell K_{k\ell}}{V^2} = C_k C_\ell K_{k\ell}, \tag{5.6}$$

$$\text{for} \quad k = \ell : \qquad \lim_{V \to \infty} \frac{\lambda_{k,\ell}}{V} = \frac{1}{2} \frac{N_k(N_k - 1)K_{k\ell}}{V^2} = \frac{1}{2} C_k C_\ell K_{k\ell}. \tag{5.7}$$

Applying $V \to \infty$ to Equation 5.5 and using Equations 5.6 and 5.7, it becomes

$$\frac{1}{\Delta t}(C_k(t + \Delta t) - C_k(t)) = -\sum_{\ell=1}^{k-1} C_k C_\ell K_{k,\ell} - C_k C_k K_{k,k} - \sum_{\ell=k+1}^{\infty} C_k C_\ell K_{k,\ell}$$

$$+ \sum_{l=1}^{k-1} \frac{1}{2} C_\ell C_{k-\ell} K_{\ell,k-\ell} \tag{5.8}$$

$$= -\sum_{\ell=1}^{\infty} C_k C_\ell K_{k,\ell} + \sum_{\ell=1}^{k-1} \frac{1}{2} C_\ell C_{k-\ell} K_{\ell,k-\ell}. \tag{5.9}$$

Finally, letting $\Delta t \to 0$ we obtain the following ordinary differential equation for the particle number concentration $C_k$ in size bin $k$:

$$\frac{dC_k}{dt} = -\sum_{\ell=1}^{\infty} C_k C_\ell K_{k,\ell} + \sum_{\ell=1}^{k-1} \frac{1}{2} C_\ell C_{k-\ell} K_{\ell,k-\ell}. \tag{5.10}$$

This equation is known as the *discrete population balance equation*. Note that this equation still assumes that particles have masses that are multiples of a base mass. We will relax this assumption in Section 5.3. The number concentrations $C_k$, however, are not integer counts anymore, but real numbers.

Algorithms 5.4 and 5.5 illustrate how Equation 5.10 can be solved numerically. In practice, one would use a more sophisticated time-stepping algorithm than the simple Euler Forward algorithm shown in Algorithm 5.5, which is shown here only as an illustration.

---

**Algorithm 5.4:** Coagulation rates for the discrete Smoluchowski coagulation equation (5.10).

---

1  **Function** CoagulationRates $(C)$
2     **Data:** $C[k]$ is the number concentration of particles in bin $k = 1, \dots, N_{\text{bin}}$
3     **Result:** $r[k]$ will be the coagulation rate $dC[k]/dt$ for bin $k = 1, \dots, N_{\text{bin}}$
4     **for** $k \leftarrow 1, \dots, N_{\text{bin}}$ **do**
5        $r[k] \leftarrow 0$
6        **for** $\ell \leftarrow 1, \dots, N_{\text{bin}}$ **do**
7           $K \leftarrow \text{Kernel}(k m_{\text{b}}, \ell m_{\text{b}})$
8           $r[k] \leftarrow r[k] - C[k]C[\ell]K$
9        **end**
10       **for** $\ell \leftarrow 1, \dots, (k-1)$ **do**
11          $K \leftarrow \text{Kernel}(\ell m_{\text{b}}, (k-\ell) m_{\text{b}})$
12          $r[k] \leftarrow r[k] + \frac{1}{2} C[\ell]C[k-\ell]K$
13       **end**
14    **end**
15    **return** $r$

---

---

**Algorithm 5.5:** Simulation algorithm for the discrete Smoluchowski coagulation equation (5.10).

---

1  **Data:** $C[k]$ is the number concentration of particles in bin $k = 1, \dots, N_{\text{bin}}$
2  $t \leftarrow 0$
3  **while** $t < t_{\text{final}}$ **do**
4     $r \leftarrow \text{CoagulationRates}(C)$
5     **for** $k \leftarrow 1, \dots, N_{\text{bin}}$ **do**
6        $C[k] \leftarrow C[k] + \Delta t \, r[k]$
7     **end**
8     $t \leftarrow t + \Delta t$
9  **end**

---

In Python, the function to calculate the coagulation rates can be written as per the code provided in listing 5.6. Observe once again the use of two variables `ind_k` and `k` together with `enumerate()` to accommodate the 0-indexed arrays in Python.

```python
def CoagulationRates(C):
    N_bin = len(C)
    r = np.zeros(N_bin)
    bin_masses = np.arange(1,N_bin+1)
    for ind_k,k in enumerate(bin_sizes):
        # Destruction
        for ind_l,l in enumerate(bin_masses):
            K = kernel_helper.GetKernel((k)*mb,(l)*mb)
            r[ind_k] -= C[ind_k]*C[ind_l]*K
        # Production
        for ind_l,l in enumerate(bin_masses[0:(ind_k)]):
            K = kernel_helper.GetKernel(l*mb,(k-l)*mb)
            r[ind_k] += .5*K*C[ind_l]*C[ind_k-ind_l-1]

    return(r)
```

**Listing 5.6** Function to calculate coagulation rates.

Note that in the "Production" loop above the code evaluates all index pairs twice. For example, if `k` is 4 then we evaluate both the case (`l = 1, k-l = 3`) and the case (`l = 3, k-l = 1`), which are identical. This loop could be thus rewritten to save approximately half of the computational cost, although it will then be somewhat more complex.

Figure 5.6 shows a simulation of Equation (5.10) using Algorithms 5.4 and 5.5. Because the stochastic nature of the solution has been eliminated, we are not sampling any random numbers, and the curves for the number concentration in each size bin, $C_k$, look completely smooth. This simulation looks very similar to the average curves in the right panels of Figures 5.3 and 5.4, but it is not exactly the same as either of them. This is because the deterministic limit obtained by taking infinitely many

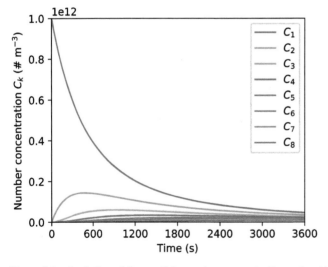

**Figure 5.6** Evolution of the particle number concentration using Algorithm 5.5 (Simulation algorithm for the discrete Smoluchowski coagulation equation). In the large-number limit, the solution is not stochastic anymore.

particles ($V \rightarrow \infty$) and $\Delta t \rightarrow 0$ is not exactly the same as the average behavior of the stochastic system. However, for most applications related to atmospheric aerosols the deterministic limit is a reasonable approximation. If the number concentration is very small, however, then stochastic fluctuations can be important drivers of the system evolution. An example of this is in the coalescence of water droplets to form rain drops [9].

We have now seen three different representations of aerosol populations: the $Q$ array from Algorithm 5.1 that stores individual particle sizes, the $N$ array from Algorithms 5.2 and 5.3 that stores the number of particles of each size, and the $C$ array from Algorithms 5.4 and 5.5 that stores number concentrations. These different data structures are key to the computational and memory improvements made in the corresponding algorithms. Figure 5.7 shows a schematic visualization of these different data structures.

**Figure 5.7** The progression of data structures for storing particle representations used in Algorithms 5.1–5.5. For Algorithm 5.1, particles are represented as a simple array, where each entry represents the scale factor of a base mass. For Algorithms 5.2 and 5.3, rather than storing all particle masses individually, we store the number of particles of each mass. For Algorithms 5.4–5.5, we introduced the continuous number concentration as a result of taking the large-number limit. In all cases, we still assume that particles have masses that are multiples of a base mass. We will introduce continuous particle masses in Section 5.3.

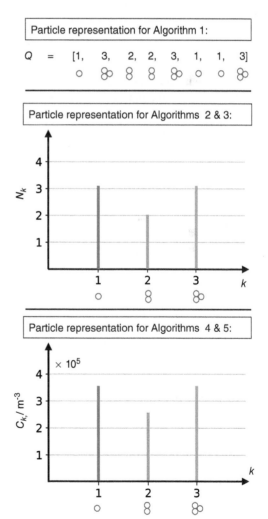

## 5.3 Coagulation with Continuous Particle Masses

Although particles are in fact composed of atoms, it is not practical to treat them as multiples of some base-size mass. In this section, we will therefore move to a version of the coagulation equation that uses *real numbers for particle masses*. That is, particle mass will now be regarded as a *continuous* variable, rather than as a *discrete* variable as we used in Section 5.2.

The population balance equation then becomes continuous. This is the basis for the codes used in most aerosol modeling applications. Formally, starting from Equation 5.10, we replace $n_k$ by the continuous number distribution density function $n(m, t)$, where $n(m, t)dm$ is the number concentration of particles at time $t$ with mass between $m$ and $m + dm$. This yields

$$\frac{\partial n(m, t)}{\partial t} = -\int_0^\infty K(m', m)n(m', t)n(m, t)\, dm'$$

$$+ \frac{1}{2}\int_0^m K(m - m', m')n(m - m', t)n(m', t)\, dm'. \tag{5.11}$$

This equation is called the *Smoluchowski coagulation equation*.

In the literature, the number distribution density function, $n$, is frequently written using independent variables other than the particle mass $m$, with popular choices being the particle volume $v$, the particle diameter $d_p$, or the logarithm of the particle diameter $\log_{10} d_p$. Two number distribution density functions, $n(a)$ and $\tilde{n}(b)$, using two different independent variables, $a$ and $b$, are related to each other via

$$dN = n(a)da = \tilde{n}(b)db, \tag{5.12}$$

$$n(a) = \tilde{n}(b)\frac{db}{da}. \tag{5.13}$$

For example, if $a = m$ and $b = d_p$, then $n(m) = \frac{2}{\pi \rho_p d_p^2}\tilde{n}(d_p)$, assuming that $m(d_p) = \rho_p \frac{\pi}{6}d_p^3$, with $\rho_p$ being the particle density.

The solution strategy of Equation (5.11) depends on the chosen aerosol representation (e.g., particle-resolved, sectional, or modal). In the following section, we will demonstrate how Equation (5.11) can be solved using a particle-resolved representation (Section 5.3.1), which builds on the material presented in Section 5.2. We will then compare and contrast this with using a sectional aerosol representation and a modal aerosol representation in Sections 5.3.2 and 5.3.3.

### 5.3.1 Particle-resolved Approach for Coagulation

Using the continuous version of the population balance equation (5.11) introduces a new problem, namely, if particle masses are not multiples of each other, we cannot avoid storing all the individual particles. Specifically, we cannot apply the memory-saving method from Section 5.2.2, where we stored the number of particles of the same mass. We could still apply Gillespie's method described in Section 5.2.1, but this has the problem that the cost scales with the square of the number of particles, and hence becomes extremely expensive when large numbers of particles need to be tracked, as is the case for most applications of atmospheric aerosols.

The solution to this is that, first, storing particles in bins is still more efficient, but now the particles will not have identical sizes within one bin. Therefore, they need to be stored as a list rather than just by count. Second, tau-leaping and discretized time (Section 5.2.3) can still be used to simulate in time, but now we need to use an accept-reject algorithm to determine which coagulation events to perform.

Algorithm 5.6 shows the resulting algorithm. This algorithm is the basis of the stochastic particle-resolved aerosol model PartMC [10], which will be discussed in detail in Section 5.5.

Algorithm 5.6 is based on Algorithm 5.3, except that each bin now contains a set of similar-sized but not identically-sized particles. This raises the question of how to correctly calculate the number of coagulation events, given that the coagulation rate will differ slightly between every pair of particles from two bins. The answer is to use an *accept-reject* sampling method. For a given pair of bins $(k, \ell)$, we first compute an upper bound $K_{max}$ for the coagulation kernel between particles from these two bins (Line 7 in Algorithm 5.6) and use this to find the number of candidate coagulation events $N_{event}$ (Line 14).

Now, we don't simply assume that all of these $N_{event}$ candidate coagulation events actually occur. Instead, in the second step we choose $N_{event}$ uniformly-sampled particle pairs from the two bins and for each pair we compute the true kernel $K$ between them (Line 18) and *accept* the coagulation event with probability $K/K_{max}$ (Line 20). If the event is accepted then we actually coagulate the two particles, and otherwise we *reject* the event and do nothing.

This two-step procedure samples coagulation events from the exact kernel rate distribution, with no approximation error from the use of bins. The bins are only used so that we can construct the upper bound $K_{max}$. Figure 5.8 shows an illustration of the accept-reject idea in a simplified setting where one particle is held fixed and its coagulation partner is sampled. In the figure, the stair-step lines show the upper bound $K_{max}$ while the continuous curve shows the true kernel $K$. The initial sample of potential coagulation partners in a certain bin is represented by the rectangular bar with top at $K_{max}$, while the area below the curve in each bin is the set of particles which will be accepted. The shaded blue regions thus represent the rejected particles.

Figure 5.8 makes it clear that using a higher value of $K_{max}$ will not change the correctness of the algorithm or the statistics of the sampled particles in any way, but it will simply mean that we are somewhat more inefficient because we need to sample more particles initially, only to reject more of them. From the figure we also see that using narrower bins will reduce the number of rejected particles, but we must also recognize that this must be balanced against the cost of iterating over the bins, so the optimally-efficient number of bins will be some intermediate value. In practice, about 10 bins per decade of particle diameters is appropriate for most atmospherically relevant coagulation kernels.

Most of the computation in Algorithm 5.6 only deals with information about each bin, irrespective of how many particles are in a bin. However, ultimately, we will need to coagulate individual particles, which means the computational cost will be proportional to the number of actual coagulation events (which is order $O(N)$), not the number of particle pairs (order $O(N^2)$), which makes this method feasible.

Algorithm 5.6 requires calculating an estimate of the maximum kernel per bin pair. The Python implementation for this is provided in code listing 5.7. We sample each

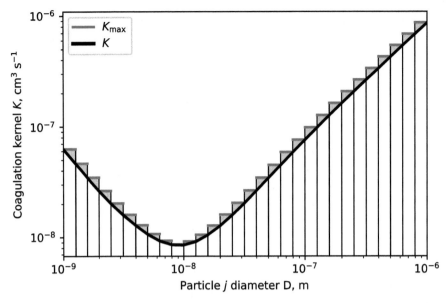

**Figure 5.8** Accept-reject for a particle of size 10 nm. The blue-shaded regions indicate the particles that are expected to be sampled in the first step as candidate coagulation pairs but which will then be rejected in the second step.

bin pair three times and pick the largest value for $K_{max}$. Of course, this method is not perfectly accurate, but a good compromise between efficiency and accuracy.

```python
MaxKernel = np.zeros((N_bin,N_bin))
nsample = 3
for i in range(N_bin):
    for j in range(N_bin):
        k_max = 0.0
        i_sub = np.linspace(vol_edges[i],vol_edges[i+1],nsample)
        j_sub = np.linspace(vol_edges[j],vol_edges[j+1],nsample)
        for ii in range(nsample):
            for jj in range(nsample):
                B_sub = kernel_helper.GetKernel(i_sub[ii],j_sub[jj])
                k_max = max(k_max, B_sub)
        MaxKernel[i,j] = k_max
```

**Listing 5.7** Maximum kernel per bin pair for Algorithm 5.6.

Another interesting issue arises because performing coagulation events will inevitably reduce the number of particles, decreasing our ability to accurately resolve the particle population. A remedy for this is to simply double the particles if the number decreases below a threshold, and simultaneously double the computational volume (because we do not want to change the number concentration due to this operation). This is done in code listing 5.8:

```python
if (np.sum(N) < N_part /2):
    for i_bin in range(N_bin):
        M[i_bin] = M[i_bin] + M[i_bin]
        N[i_bin] = 2* N[i_bin]
    V *= 2
```

**Listing 5.8** Checking if particle number has dropped below a user-defined threshold.

---

**Algorithm 5.6:** Algorithm that shows the particle-resolved approach for solving the population balance equation, Equation (5.11).

---

1   **Data:** $N[k]$ is the number of particles in bin $k = 1, \dots, N_{\text{bin}}$
2   **Data:** $m[k][i]$ is the mass of the $i$th particle in bin $k = 1, \dots, N_{\text{bin}}$, for
       $i = 1, \dots, N[k]$
3   $t \leftarrow 0$
4   **while** $t < t_{\text{final}}$ **do**
5     **for** $k \leftarrow 1, \dots, N_{\text{bin}}$ **do**
6        **for** $\ell \leftarrow 1, \dots, k$ **do**
7           $K_{\text{max}} \leftarrow \texttt{MaxKernel}(k, \ell)$
8           **if** $k \neq \ell$ **then**
9              $N_{\text{pairs}} \leftarrow N[k]N[\ell]$
10           **else**
11              $N_{\text{pairs}} \leftarrow \frac{1}{2}N[k](N[\ell] - 1)$
12           **end**
13           $\lambda \leftarrow N_{\text{pairs}}K_{\text{max}}/V$
14           $N_{\text{event}} \sim \text{Poisson}(\lambda \Delta t)$
15           **for** $q \leftarrow 1, \dots, N_{\text{event}}$ **do**
16              $i \sim \text{Uniform}(\{1, \dots, N[k]\})$
17              $j \sim \text{Uniform}(\{1, \dots, N[\ell]\})$
18              $K \leftarrow \texttt{Kernel}(m[k][i], m[\ell][j])$
19              $r \leftarrow \text{Uniform}([0, 1])$
20              **if** $r < K/K_{\text{max}}$ **then**
21                 $m_{\text{new}} = m[k][i] + m[\ell][j]$
22                 find the bin index $k_{\text{new}}$ for the new particle mass $m_{\text{new}}$
23                 $N[k] \leftarrow N[k] - 1$
24                 $N[\ell] \leftarrow N[\ell] - 1$
25                 $N[k_{\text{new}}] \leftarrow N[k_{\text{new}}] + 1$
26                 delete the $i$th entry in $m[k]$
27                 delete the $j$th entry in $m[\ell]$
28                 append $m_{\text{new}}$ to $m[k_{\text{new}}]$
29              **end**
30           **end**
31        **end**
32     **end**
33     $t \leftarrow t + \Delta t$
34 **end**

---

In a complete aerosol simulation that includes particle sources, it is also advantageous to include the reverse process whenever the number of particles exceed 2 * N_part. That is, when this occurs we discard a randomly chosen half of the particles and correspondingly halve the computational volume $V$.

### 5.3.2 Sectional Approach for Coagulation

Recall from Section 1.3.2 that the sectional aerosol representation is distribution-based and breaks up the size distribution into discrete size bins along the size axis. To illustrate how coagulation can be simulated using a sectional aerosol representation, we use the method by Jacobson [11]. This semi-implicit method divides the particle volume coordinate into a number of discrete bins. It has the desirable properties of conserving total particle volume, requiring no iterations for advancing by a single timestep, and being numerically stable regardless of timestep size.

We consider bins $i$ and $j$, which have bin center volumes of $v_i$ and $v_j$. If we consider all particles in each bin to have volume equal to the bin center, then coagulation of particles in these two bins will result in particles of volume $V_{i,j} = v_i + v_j$. We then need to distribute these coagulated particles between two neighboring model bins $k$ and $k + 1$ because they aren't of exactly the correct size for any single bin. This distribution is shown schematically in Figure 5.9, where the resulting particle concentration due to $i, j$ interactions is fractionally split between bins $k$ and $k + 1$.

The distribution between the two neighboring bins is specified by a partitioning factor $f_{i,j,k}$, which gives the volume fraction of particles from bin pair $(i, j)$ coagulations that go into bin $k$. This factor is given by

$$f_{i,j,k} = \begin{cases} \left( \dfrac{v_{k+1} - V_{i,j}}{v_{k+1} - v_k} \right) \dfrac{v_k}{V_{i,j}}, & \text{if } v_k \le V_{i,j} < v_{k+1} \text{ and } k < N_{\text{bin}}, \\ 1 - f_{i,j,k-1}, & \text{if } v_{k-1} \le V_{i,j} < v_k \text{ and } k > 1, \\ 1, & \text{if } V_{i,j} > v_k \text{ and } k = N_{\text{bin}}, \\ 0, & \text{otherwise,} \end{cases} \tag{5.14}$$

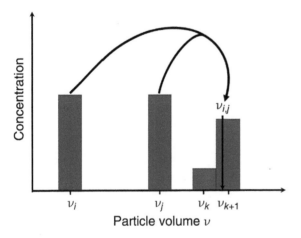

**Figure 5.9** Illustration of the partitioning of the resulting particle volume to destination bins $k$ and $k + 1$ after coagulation between two bins $i$ and $j$.

where $N_{\text{bin}}$ is the number of bins. Denoting the number concentration at time step $t$ in bin $k$ by $n_k^t$, the Jacobson algorithm computes the updated number concentrations $n_k^{t+1}$ by

$$v_k n_k^{t+1} = \frac{v_k n_k^t + \Delta t \sum_{j=1}^{k} \sum_{i=1}^{k-1} f_{i,j,k} K_{i,j} v_i n_i^{t+1} n_j^t}{1 + \Delta t \sum_{j=1}^{N_{\text{bin}}} (1 - f_{k,j,k}) K_{k,j} n_j^t}, \tag{5.15}$$

where $K_{i,j}$ is the coagulation kernel for particles in bins $i$ and $j$. This equation requires no iterations as all the terms on the right-hand side of the equation are known. Note that this even applies to the number concentration term $n_i^{t+1}$ of the summation term in the numerator of Equation (5.15), since this is the final concentration calculated for a previous bin.

The time evolution specified by Equation (5.15) can be written algorithmically as shown in Algorithms 5.7 and 5.8.

---

**Algorithm 5.7:** Semi-implicit method for coagulation for sectional aerosol representation according to Jacobson [11].

---

1 **Function** SectionalUpdate($n$)
2    Compute $f_{i,j,k}$ from Equation (5.14) for all $i, j, k = 1, \dots, N_{\text{bin}}$
3    **for** $k \leftarrow 1, \dots, N_{\text{bin}}$ **do**
4       $P \leftarrow 0$
5       $D \leftarrow 0$
6       **for** $j \leftarrow 1, \dots, k$ **do**
7          **for** $i \leftarrow 1, \dots, k-1$ **do**
8             $P \leftarrow P + f_{i,j,k} K_{i,j} v_i n_i n_j$
9          **end**
10       **end**
11       **for** $j \leftarrow 1, \dots, N_{\text{bin}}$ **do**
12          $D \leftarrow D + (1 - f_{k,j,k}) K_{k,j} n_j$
13       **end**
14       $n_k \leftarrow \frac{1}{v_k} \frac{v_k n_k + \Delta t P}{1 + \Delta t D}$
15    **end**

---

---

**Algorithm 5.8:** Simple time-stepping algorithm for Jacobson [11] method.

---

1 **Data:** $n[k]$ is the number concentration of particles in bin $k = 1, \dots, N_{\text{bin}}$
2 $t \leftarrow 0$
3 **while** $t < t_{\text{final}}$ **do**
4    $n \leftarrow$ SectionalUpdate($n$)
5    $t \leftarrow t + \Delta t$
6 **end**

---

### 5.3.3 Modal Approach for Coagulation

Similar to the sectional aerosol representation (and in contrast to the particle-resolved approach), the modal representation is also distribution-based and uses overlapping log-normal functions (see Section 1.3.3 for a reminder). Using a moment-based modal aerosol representation is computationally efficient and therefore a popular choice for many regional and global chemical transport models. In this method, the number of parameters needed to describe the vector of single particles or discrete size bins is reduced to the number of moments selected to describe a small number of representative modes; see section 1.3.3 for an overview.

A full discussion of modal approaches for coagulation and a derivation of methods to solve the resulting system of equations is presented in Whitby [12], while Binkowski [13] summarizes these methods and demonstrates how the approach is implemented in a large-scale atmospheric model. In this section, we repeat their findings for a simplified application that only considers the effects of coagulation, and provide modern numerical code to solve the problem.

For our example, we represent a hypothetical size distribution of fine particles with two modes (number distributions) $n_i$ and $n_j$, corresponding to an Aitken mode ($i$) at smaller diameters and an Accumulation mode ($j$) at larger diameters. Whitby [12] derive the differential equations describing the change due to coagulation of particles in modes $i$ and $j$ as:

$$\frac{\partial M_{k,i}}{\partial t} = \frac{1}{2} \int_0^\infty \int_0^\infty (d_{p1}^3 + d_{p2}^3)^{\frac{k}{3}} K(d_{p1}, d_{p2}) n_i(d_{p1}) n_i(d_{p2}) \, dd_{p1} \, dd_{p2} \qquad (5.16a)$$

$$- \int_0^\infty \int_0^\infty d_{p1}^k K(d_{p1}, d_{p2}) n_i(d_{p1}) n_i(d_{p2}) \, dd_{p1} \, dd_{p2} \qquad (5.16b)$$

$$- \int_0^\infty \int_0^\infty d_{p1}^k K(d_{p1}, d_{p2}) n_i(d_{p1}) n_j(d_{p2}) \, dd_{p1} \, dd_{p2}, \qquad (5.16c)$$

$$\frac{\partial M_{k,j}}{\partial t} = \frac{1}{2} \int_0^\infty \int_0^\infty (d_{p1}^3 + d_{p2}^3)^{\frac{k}{3}} K(d_{p1}, d_{p2}) n_j(d_{p1}) n_j(d_{p2}) \, dd_{p1} \, dd_{p2} \qquad (5.16d)$$

$$- \int_0^\infty \int_0^\infty d_{p1}^k K(d_{p1}, d_{p2}) n_j(d_{p1}) n_j(d_{p2}) \, dd_{p1} \, dd_{p2} \qquad (5.16e)$$

$$+ \int_0^\infty \int_0^\infty (d_{p1}^3 + d_{p2}^3)^{\frac{k}{3}} K(d_{p1}, d_{p2}) n_i(d_{p1}) n_j(d_{p2}) \, dd_{p1} \, dd_{p2} \qquad (5.16f)$$

$$- \int_0^\infty \int_0^\infty d_{p2}^k K(d_{p1}, d_{p2}) n_i(d_{p1}) n_j(d_{p2}) \, dd_{p1} \, dd_{p2}, \qquad (5.16g)$$

where $M_{k,i}$ and $M_{k,j}$ are the $k$th moments for the $i$ and $j$ modes, respectively. Terms (5.16a)–(5.16b) and (5.16d)–(5.16e) correspond to intramodal (particles within the same mode) coagulation and terms (5.16c) and (5.16f)–(5.16g) account for intermodal (particles from two different modes) coagulation.

Modal models typically assume that intramodal coagulation results in particles that stay in the same mode, while intermodal coagulation results in particles in the larger mode, leading to term (5.16f). Another common assumption evident in Equations (5.16) is that the volume of particles is assumed to be conserved, that is,

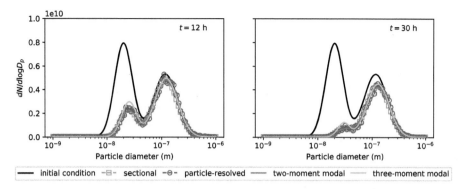

Figure 5.10 Comparison of size distributions for particle-resolved, sectional, and modal methods after 12 (left) and 30 (right) hours of simulation.

the volume of the new particle is equal to the sum of the volumes of the colliding particles.

In practice, one also assumes that the number size distributions $n_i(d_{p1})$ and $n_j(d_{p1})$ are log-normal as discussed in Chapter 1, which determines the strategy for solving Equations (5.16). The log-normal function for each mode is determined by three parameters, namely geometric mean diameter ($d_g$), geometric standard deviation ($\sigma_g$), and total number concentration ($N_t$). The aim of the model simulation is then to predict the time evolution of these three unknown parameters. Rather than formulating differential equations for these parameters directly, one solves Equations (5.16) for three moments per mode and then applies Equations (1.25), (1.30) and (1.31) to deduce $d_g$, $\sigma_g$, $N_t$.

To simplify further, for the application in regional or global models, the geometric standard deviation is usually assumed to be constant, which means that only two moments need to be tracked, usually $M_0$ (i.e., total number concentration) and $M_3$ (proportional to the total volume concentration). Relying on a combination of asymptotic forms of the coagulation kernel $K$, Whitby [12] demonstrated an analytical approach that approximates the exact solution of the right-hand side of Equations (5.16) for the zeroth, third, and sixth moments, which results in the set of equations that we use for our example below.

We will demonstrate here the solution strategy for a two-moment approach, using the zeroth and third moments. As mentioned above, this implies that the standard deviation of the distributions is assumed to be constant. In Figure 5.10, we also show the solution for a three-moment approach using $M_0$, $M_3$, and $M_6$. For more details on the equations for the rate of change of $M_6$, refer to [12].

The rate of change of the zeroth and third moments in modes $i$ and $j$ may be formally written as:

$$\frac{\partial}{\partial t} M_{0,i} = -\text{Ca}_{0,ii} - \text{Ca}_{0,ij}, \tag{5.17a}$$

$$\frac{\partial}{\partial t} M_{0,j} = -\text{Ca}_{0,jj}, \tag{5.17b}$$

$$\frac{\partial}{\partial t} M_{3,i} = -\mathrm{Ca}_{3,ij}, \tag{5.17c}$$

$$\frac{\partial}{\partial t} M_{3,j} = +\mathrm{Ca}_{3,ij}, \tag{5.17d}$$

where the right-hand side is obtained by analytically solving the double integrals in Equations (5.16). The right-hand-side variables will be defined below.

Recalling that the zeroth moments $M_{0,i}$ and $M_{0,j}$ are the total number concentrations in modes $i$ and $j$, we see that the number concentrations in both modes decrease as a result of intramodal coagulation, whereas only the number concentration in mode $i$ decreases due to intermodal coagulation. This reflects the convention that intermodal coagulation results in particles assigned to the larger mode ($j$). The third moments $M_{3,i}$ and $M_{3,j}$, which are proportional to the total volumes in modes $i$ and $j$, respectively, are not changed by intramodal coagulation, but change due to intermodal coagulation, where the loss of mode $i$ is the gain of mode $j$.

The analytical solution of the right-hand sides of Equations (5.16) is challenged by the size dependence of the coagulation kernel ($K$), which takes a different form in the free-molecular (fm) regime and near-continuum (nc) regimes. To handle this, Whitby [12] average the free-molecular and continuum regime solutions (see Appendix H of Whitby [12]) via harmonic mean. This results in a form that reduces to the limiting cases at small or large particle sizes, and approximates the transition between the two limits at intermediate sizes:

$$\mathrm{Ca}_{0,ii} = \frac{\mathrm{Ca}_{0,ii}^{\mathrm{nc}} \cdot \mathrm{Ca}_{0,ii}^{\mathrm{fm}}}{\mathrm{Ca}_{0,ii}^{\mathrm{nc}} + \mathrm{Ca}_{0,ii}^{\mathrm{fm}}} \tag{5.18a}$$

$$\mathrm{Ca}_{0,jj} = \frac{\mathrm{Ca}_{0,jj}^{\mathrm{nc}} \cdot \mathrm{Ca}_{0,jj}^{\mathrm{fm}}}{\mathrm{Ca}_{0,jj}^{\mathrm{nc}} + \mathrm{Ca}_{0,jj}^{\mathrm{fm}}} \tag{5.18b}$$

$$\mathrm{Ca}_{0,ij} = \frac{\mathrm{Ca}_{0,ij}^{\mathrm{nc}} \cdot \mathrm{Ca}_{0,ij}^{\mathrm{fm}}}{\mathrm{Ca}_{0,ij}^{\mathrm{nc}} + \mathrm{Ca}_{0,ij}^{\mathrm{fm}}} \tag{5.18c}$$

$$\mathrm{Ca}_{3,ij} = \frac{\mathrm{Ca}_{3,ij}^{\mathrm{nc}} \cdot \mathrm{Ca}_{3,ij}^{\mathrm{fm}}}{\mathrm{Ca}_{3,ij}^{\mathrm{nc}} + \mathrm{Ca}_{3,ij}^{\mathrm{fm}}}. \tag{5.18d}$$

The near-continuum intermodal zeroth-moment term is given by

$$\mathrm{Ca}_{0,ij}^{\mathrm{nc}} = \int_0^\infty \int_0^\infty K_{\mathrm{nc}}\left(d_{\mathrm{p1}}, d_{\mathrm{p2}}\right) n_i\left(d_{\mathrm{p1}}\right) n_j\left(d_{\mathrm{p2}}\right) \mathrm{d}d_{\mathrm{p1}}\, \mathrm{d}d_{\mathrm{p2}} \tag{5.19a}$$

$$= M_{0,i} M_{0,j} \hat{K}_{\mathrm{nc}} \left[ 2 + A_i \mathrm{Kn}_{g_i} \left( e^{\frac{4}{8} \ln^2(\sigma_{gi})} + \frac{d_{gj}}{d_{gi}} e^{\frac{16}{8} \ln^2(\sigma_{gi})} e^{\frac{4}{8} \ln^2(\sigma_{gj})} \right) \right.$$

$$+ A_j \mathrm{Kn}_{g_j} \left( e^{\frac{4}{8} \ln^2(\sigma_{gj})} + \frac{d_{gi}}{d_{gj}} e^{\frac{16}{8} \ln^2(\sigma_{gj})} e^{\frac{4}{8} \ln^2(\sigma_{gi})} \right)$$

$$\left. + \left( \frac{d_{gi}}{d_{gj}} + \frac{d_{gj}}{d_{gi}} \right) \left( e^{\frac{4}{8} \ln^2(\sigma_{gj})} \right) \left( e^{\frac{4}{8} \ln^2(\sigma_{gi})} \right) \right] \tag{5.19b}$$

$$:= M_{0,i} M_{0,j} \cdot R_{0,ij}^{\mathrm{nc}}\left(d_{gi}, d_{gj}, \sigma_{gi}, \sigma_{gj}\right), \tag{5.19c}$$

with

$$\hat{K}_{nc} = \frac{2k_B T}{3\mu}, \tag{5.20}$$

$$A_i = 1.392 \left(\text{Kn}_{g_i}\right)^{0.0783}, \tag{5.21}$$

$$\text{Kn}_{g_i} = \frac{2\lambda_{air}}{d_{gi}}, \tag{5.22}$$

where $\lambda_{air}$ is the free mean path of air.

The near-continuum intermodal third-moment term is:

$$\text{Ca}_{3,ij}^{nc} = \int_0^\infty \int_0^\infty (d_{p1})^3 K_{nc}(d_{p1}, d_{p2}) n_i(d_{p1}) n_j(d_{p2}) \, dd_{p1} \, dd_{p2} \tag{5.23a}$$

$$= M_{0,i} M_{0,j} \hat{K}_{nc} (d_{gi})^3$$

$$\left[ 2e^{\frac{36}{8}\ln^2(\sigma_{gi})} A_i \, \text{Kn}_{gi} \left( e^{\frac{16}{8}\ln^2(\sigma_{gi})} + \frac{d_{gj}}{d_{gi}} e^{\frac{4}{8}\ln^2(\sigma_{gi})} e^{\frac{4}{8}\ln^2(\sigma_{gj})} \right) \right.$$

$$+ A_j \text{Kn}_{gj} \left( e^{\frac{36}{8}\ln^2(\sigma_{gi})} e^{\frac{4}{8}\ln^2(\sigma_{gj})} + \frac{d_{gi}}{d_{gj}} e^{\frac{64}{8}\ln^2(\sigma_{gi})} e^{\frac{16}{8}\ln^2(\sigma_{gj})} \right)$$

$$\left. + \frac{d_{gj}}{d_{gi}} e^{\frac{16}{8}\ln^2(\sigma_{gi})} e^{\frac{4}{8}\ln^2 \sigma_{gj}} + \frac{d_{gi}}{d_{gj}} e^{\frac{64}{8}\ln^2(\sigma_{gi})} e^{\frac{4}{8}\ln^2(\sigma_{gj})} \right] \tag{5.23b}$$

$$:= M_{0,i} M_{0,j} \cdot R_{3,ij}^{nc}\left(d_{gi}, d_{gj}, \sigma_{gi}, \sigma_{gj}\right). \tag{5.23c}$$

Turning to the free-molecular regime, the intermodal zeroth-moment term is

$$\text{Ca}_{0,ij}^{fm} = \int_0^\infty \int_0^\infty K_{fm}(d_{p1}, d_{p2}) n_i(d_{p1}) n_j(d_{p2}) \, dd_{p1} \, dd_{p2} \tag{5.24a}$$

$$= M_{0,i} M_{0,j} \hat{K}_{fm} b_0^{(1)} \sqrt{d_{gi}}$$

$$\left[ e^{\frac{1}{8}\ln^2(\sigma_{gi})} + \sqrt{\frac{d_{gj}}{d_{gi}}} e^{\frac{1}{8}\ln^2(\sigma_{gj})} \right.$$

$$+ 2\frac{d_{gj}}{d_{gi}} e^{\frac{1}{8}\ln^2(\sigma_{gi})} e^{\frac{4}{8}\ln^2(\sigma_{gj})} + \frac{d_{gj}^2}{d_{gi}^2} e^{\frac{9}{8}\ln^2(\sigma_{gi})} e^{\frac{16}{8}\ln^2(\sigma_{gj})}$$

$$+ \left( \sqrt{\frac{d_{gi}}{d_{gj}}} \right)^3 e^{\frac{16}{8}\ln^2(\sigma_{gi})} e^{\frac{9}{8}\ln^2(\sigma_{gj})}$$

$$\left. + 2\sqrt{\frac{d_{gi}}{d_{gj}}} e^{\frac{4}{8}\ln^2(\sigma_{gi})} e^{\frac{1}{8}\ln^2(\sigma_{gj})} \right] \tag{5.24b}$$

$$:= M_{0,i} M_{0,j} \cdot R_{0,ij}^{fm}\left(d_{gi}, d_{gj}, \sigma_{gi}, \sigma_{gj}\right), \tag{5.24c}$$

with

$$\hat{K}_{fm} = \sqrt{\frac{3k_B T}{\rho_p}}, \tag{5.25}$$

where $\rho_p$ is the particle density.

Finally, the free-molecular intermodal third-moment term has the expression

$$\mathrm{Ca}_{3,ij}^{fm} = \int_0^\infty \int_0^\infty \left(d_{p1}\right)^3 K_{fm}\left(d_{p1}, d_{p2}\right) n_i\left(d_{p1}\right) n_j\left(d_{p2}\right) \mathrm{d}d_{p1}\, \mathrm{d}d_{p2} \qquad (5.26a)$$

$$= M_{0,i} M_{0,j} \hat{K}_{fm} b_3^{(1)} \left(d_{gi}\right)^{\frac{7}{2}}$$

$$\left[ e^{\frac{49}{8}\ln^2(\sigma_{gi})} + \sqrt{\frac{d_{gj}}{d_{gi}}} e^{\frac{36}{8}\ln^2(\sigma_{gi})} e^{\frac{1}{8}\ln^2(\sigma_{gj})} \right.$$

$$+ 2\frac{d_{gj}}{d_{gi}} e^{\frac{25}{8}\ln^2(\sigma_{gi})} e^{\frac{4}{8}\ln^2(\sigma_{gj})} + \frac{d_{gj}^2}{d_{gi}^2} e^{\frac{9}{8}\ln^2(\sigma_{gi})} e^{\frac{16}{8}\ln^2(\sigma_{gj})}$$

$$\left. + \left(\sqrt{\frac{d_{gi}}{d_{gj}}}\right)^3 e^{\frac{100}{8}\ln^2(\sigma_{gi})} e^{\frac{9}{8}\ln^2(\sigma_{gj})} + 2\sqrt{\frac{d_{gi}}{d_{gj}}} e^{\frac{64}{8}\ln^2(\sigma_{gi})} e^{\frac{1}{8}\ln^2(\sigma_{gj})} \right] (5.26b)$$

$$:= M_{0,i} M_{0,j} \cdot R_{3,ij}^{fm}\left(d_{gi}, d_{gj}, \sigma_{gi}, \sigma_{gj}\right). \qquad (5.26c)$$

The terms for the intramodal terms $\mathrm{Ca}_{0,ii}^{nc}$, $\mathrm{Ca}_{0,jj}^{nc}$, $\mathrm{Ca}_{0,ii}^{fm}$, $\mathrm{Ca}_{0,jj}^{fm}$ are constructed analoguously.

The analytical solutions derived in [12] for the continuum-regime intra- and intermodal coagulation rates have negligible deviation from the exact solution and so do not require correction factors to be applied. For the free-molecular regime, the values of the correction factors may be found by approximating the solution to the free-molecular coagulation rates with an accurate integration technique (e.g., Gauss-Hermite quadrature) and normalizing by the analytical approximation for log-normal populations with the same parameters. The values for correction factor in Equations (5.24b) and (5.26b) are reported by [12] as functions of the ratio of the geometric mean diameters of the coagulating modes. We approximate them here as

$$b_0^{(1)} = 0.8, \qquad (5.27a)$$

$$b_3^{(1)} = 0.9. \qquad (5.27b)$$

More complete lookup tables are provided in [12] and are included in the example codes.

This approximate analytical approach introduces at most approximately 20% error in the coagulation rate compared to full numerical integration and use of the full coagulation kernel from Equation (5.1). For situations where participating aerosol modes have $\sigma_g > 1.5$, the maximum error reduces to 10% [12].

Figure 5.10 compares the size distributions after 12 hours and 30 hours of simulation for the particle-resolved (Section 5.3.1), sectional (Section 5.3.2) and modal (Section 5.3.3) methods. For the modal method, we show a two-moment implementation using $M_0$ and $M_3$ and keeping the geometric standard deviations constant, as well as a three-moment implementation using $M_0$, $M_3$, and $M_6$. All simulations started with the same initial condition, consisting of an Aitken mode and an accumulation mode, shown in black. The modal model runs were directly initialized with two modes. These were mapped onto 79 sections in the range from 1 nm to 1 $\mu$m for the sectional model run and 10 000 computational particles for the particle-resolved model run.

All three model simulations show that, after 12 hours of simulation, the total number concentration decreased, to about 56% of the initial total number concentration. This was mostly due to the depletion of Aitken mode particles, and to a lesser extent due to the depletion of the smaller-sized particles in the accumulation mode. The sectional model and the particle-resolved model give very similar results, and the modal model results also follow closely. However, we caution that the case presented here starts out with initial conditions that are captured perfectly by modal models, and no other processes (such as nucelation or condensation) are included, so this is a scenario for which the modal approach is ideally suited. In more realistic simulations we would not always expect a modal model to be so accurate. It is interesting to note that the two-moment implementation of the modal model performs better for this case than the three-moment implementation. That is, tracking more moments does not necessarily lead to more accurate results.

## 5.4 Advanced Coagulation Topics

### 5.4.1 Coagulation for a Multi-dimensional Composition Space

So far in this chapter, we have always assumed that all particles undergoing coagulation consist only of one chemical species. Referring back to Chapter 1, this means we assumed a one-dimensional composition space and ignored the fact that an aerosol is generally a system with many chemical species that can be arranged in many different mixing states, and hence need to be described by a multi-dimensional composition space.

#### Particle-resolved approach
It is straightforward to resolve this multi-dimensional composition space with particle-resolved methods and simulate coagulation as described in Section 5.3.1. In this case, one can think of each particle as an $A$-dimensional vector $\vec{\mu}^i \in \mathbb{R}^A$ with components $(\mu_1^i, \mu_2^i, \mu_3^i, \ldots, \mu_A^i)$, where $\mu_a^i$ is the mass of species $a$ in particle $i$, for $a = 1, \ldots, A$ and $i = 1, \ldots, N_p$. Figure 5.11(a) shows this schematically for a three-dimensional composition space. Once a coagulation event occurs, the composition vector of the resulting particle is simply the sum of the vectors of the coagulating particles.

#### Univariate sectional approach
A univariate sectional model is one where we have only a one-dimensional independent variable, such as particle volume as used in Section 5.3.2. If particles consist of multiple species, the most common univariate approach is to store within each bin the total volume concentration of each species. Each size bin thus represents the average composition for particles of that size—we do not explicitly track how the different chemical species are distributed among the particles within the bin. When coagulation of particles in two bins is simulated, we transfer each species separately using Equation (5.15).

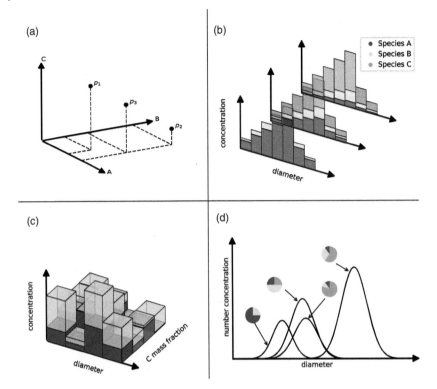

**Figure 5.11** (a) Particle-resolved approach to resolve a multi-dimensional composition space. Shown here is the example of a three-dimensional composition space, made up by three chemical species A, B, and C. Three particles ($p_1$, $p_2$, $p_3$) are shown, each of which has different amounts of *A,B,C* in it. (b) Univariate sectional approach with multiple distributions. Each distribution can contain a different set of chemical species. (c) Multivariate sectional approach with a two-dimensional bin structure (Species C mass fraction and size). (d) Multi-modal approach with several overlapping modes. Each mode can contain a different set of chemical species.

### Univariate sectional approach with multiple distributions

Working with a sectional model that consists of one univariate distribution allows the tracking of size-resolved composition, but composition variation among particles *within* a size bin cannot be resolved. This approach can be refined by introducing multiple one-dimensional distributions (shown in Figure 5.11(b)) that can potentially interact by coagulation. Choices have to be made about which distribution the resulting aerosol should be assigned to, which can increase the number of distributions quickly. For example, Jacobson [14] uses 18 different distributions that differ in the chemical species combination (e.g., three black carbon (BC)-containing distributions with different amounts of non-BC components). Eleven of the 18 distributions arise because of coagulation interactions.

### Multivariate sectional approach

Another possibility is to use a multivariate bin structure [15–19]. A common feature among several of these models is that they use a two-dimensional sectional framework to represent BC-containing particles, with one dimension being dry diameter and the other dimension being BC mass fraction as shown in Figure 5.11(c). Compared to the univariate approach with multiple distributions, this approach can resolve intermediate states of composition in more detail, including the ones that arise due to coagulation. Building on previous two-dimensional sectional frameworks, the MOSAIC-MIX model [20] adds an additional dimension to represent hygroscopicity and shows that this optimizes the calculations of CCN concentrations and aerosol optical properties. Multivariate sectional approaches become computationally expensive very quickly as the number of dimensions increases.

### Multi-modal approach

Modal models inherently assume that if a mode contains more than one species, those species are internally mixed within the mode (i.e., within one mode there is no variation of composition with size). If different modes contain different sets of species, the interaction of coagulation leads to new mixtures, which require the introduction of new modes, similar to the univariate sectional approach with multiple distributions. The Multiconfiguration Aerosol Tracker of Mixing State model (MATRIX) [21] with 16 modes is an example of this approach. The approach is schematically shown in Figure 5.11(d).

## 5.4.2 Other Considerations

Section 5.1 walked us through the process of coagulation by introducing coagulation as a Markov process that in the large-number-limit leads to the deterministic approach (Equation 5.11). We presented this in the context of Brownian coagulation, which is extremely relevant for many applications in aerosol science. However, several issues deserve additional discussion. Here we will briefly introduce the role of external force fields, coagulation mechanisms other than Brownian motion, and the issue of non-spherical particles, and provide references to more-in-depth literature.

### 5.4.2.1 Coagulation Under an External Force Field

#### Interparticle forces

Interparticle forces, such as van der Waals forces or Coulomb forces, can modify the probability of two particles colliding. This can be taken into account by applying a correction factor to the coagulation kernel, which depends on the potential of the force under consideration [3, Chapter 13.A.4]. The impact of van der Waals forces has been investigated for example by Jacobson [22] who showed that these forces can be important to explain the rapid observed evolution of vehicle-exhaust size distributions measured near their points of emission. Charan et al. [23] quantified the role of Coulomb forces for particle dynamics in a Teflon chamber and showed that they have negligible impacts for characteristic chamber ion concentrations. Mahfouz et al. [24] on the other hand showed that the survival of charged nanoparticles (formed by ion-induced new-particle formation) in the atmosphere may be greatly reduced due to enhanced coagulation with charged background particles.

#### Hydrodynamic forces

In our discussion of Brownian coagulation, we assumed that a particle's motion is not impacted by the particle that it is approaching during a coagulation event. However, it turns out that a particle moving in its carrier fluid (air) produces velocity gradients in the fluid, which in turn impact the movement of other approaching particles. The overall result of these fluid mechanical interactions is a decrease of the coagulation rate. Formally this can be taken into account by correcting the common diffusivity $D_1 + D_2$ in Equation 5.1 [3, 25, 26].

### Phoretic forces

Phoretic forces, including thermophoresis and diffusiophoresis, are considered in the context of collisions of cloud droplets with aerosols [27, 28]. Thermophoresis refers to a net transport of particles due to a temperature gradient in the carrier gas, while diffusiophoresis is caused by concentration gradients of one of the carrier gas's constituents (e.g., water vapor). This leads to a scavenging of the aerosol, for mixed-phase clouds potentially leading to freezing of the droplet (i.e., contact nucleation). Young [27] showed that phoretic coagulation rates are less size-dependent than Brownian coagulation rates and tend to dominate for large aerosols under typical atmospheric conditions. Thermophoresis dominates over diffusiophoresis except for aerosols larger than 1 $\mu m$ and both are relatively more important at higher altitudes.

#### 5.4.2.2 Other Coagulation Mechanisms

Section 5.1 focused on Brownian coagulation. However, several other mechanisms exist that can introduce relative velocities among particles, which result in particle collisions and coagulation events. The formalism of all coagulation models remains the same, but the expression for the coagulation kernel $K$ needs to be adjusted to reflect the underlying physics.

### Gravitational collection

This mechanism relies on the fact that heavier particles settle faster than lighter particles, resulting in collisions of heavier and lighter particles. The coagulation kernel for gravitational collection in laminar flow takes the form

$$K(d_{p1}, d_{p2}) = E\left(d_{p1}, d_{p2}\right) \frac{\pi}{4}(d_{p1} + d_{p2})^2 \left(v_{t1}(d_{p1}) - v_{t2}(d_{p2})\right), \qquad (5.28)$$

where $E\left(d_{p1}, d_{p2}\right)$ is the collision efficiency of particles with diameters $d_{p1}$ and $d_{p2}$, and $v_{t1}$ and $v_{t2}$ are the terminal velocities of the particles. A collision efficiency of 1 means that all particles within the cylinder volume below the larger particle are collected by the larger particle. However, in practice, the collision efficiency is usually lower than 1 due to hydrodynamic interactions between the two particles.

It is important to note that Brownian coagulation and coagulation due to gravitational settling dominate in different size regimes. Brownian coagulation is more important for submicron aerosol particles, while gravitational settling becomes more relevant when the settling velocities become appreciable, i.e., for supermicron particles and cloud droplets.

### Turbulent flow

The impact of turbulence on particle collisions is more relevant for the evolution of cloud droplet size distributions than for aerosol size distributions. Arenberg [29] realized that clouds are an inherently turbulent medium where turbulence could influence the droplet coagulation process with the consequence that Equation (5.28) may not capture the relevant physics. More specifically, turbulence increases the collision rate by three mechanisms. First, particle inertia leads to increased relative velocities and less-correlated velocity directions (acceleration effect). Second, the wind field shear produces collisions between particles even with the same inertia (shear effect; [30]). Third, coagulation rates are enhanced because of local concentration increases for particle response times on the order of the Kolmorgorov scale. For this phenomenon the terms "preferential concentration" or "accumulation effect" have been coined. Furthermore, turbulence can also alter the collision efficiency because of local aerodynamic droplet-droplet interactions. Many studies have explored the potential impacts of turbulence on the rain formation process [31]. It is difficult to quantify the coagulation kernel experimentally, and therefore efforts have been made to derive the kernel from direct numerical simulations of particle-laden flow. The challenge here consists in the fact that it is difficult to simulate the conditions of high-Reynolds number flow that is representative of the atmosphere, although much progress has been made over the last two decades [32].

### 5.4.2.3 Non-spherical Particles

Our discussion so far assumed that all particles are spherical. This is a good assumption for deliquesced or water-containing aerosol particles, but may not hold for dry particles, which often start out as non-spherical and do not coalesce after collision but rather form aggregates. Common examples of this in the real atmosphere include mineral dust, soot, dry sea salt aerosol, and pollen. Non-sphericity can impact coagulation rates significantly and has been studied for example by Lee and Shaw [33] for fibrous-type particles, by Rogak and Flagan[34] for fractal-like agglomeration of primary particles, and by Naumann [35] for the coagulation of particles that have fractal shape.

## 5.5 Introduction to Particle-resolved Monte Carlo (PartMC)

PartMC is a box (or 0-D) model that uses the particle-resolved approach to simulate the evolution of aerosol populations. PartMC considers a parcel of air containing $N_{part}$ computational particles where each particle is an $A$-dimensional vector of species masses, where $A$ is the number of aerosol species. The number of computational particles $N_{part}$ varies over time as particles are added and removed, either by dilution, emission, or coagulation events among particles. These particle processes are all represented by stochastically sampling with appropriate probabilities; see [10] for details. The principle of the PartMC coagulation algorithm has been described in Section 5.3.1. The source code can be accessed at https://github.com/compdyn/partmc.

To run PartMC (the program compiled from source code), users are required to set up "input files" and a "spec" file. All PartMC input files are in text format, with the suffix .dat. The "spec" file (with the suffix .spec) specifies the input files and several other settings for PartMC simulations. The PartMC raw output files are in NetCDF format (with a .nc extension), containing the full aerosol state at each output timestep. The NetCDF format is a self-describing, machine-independent data format that supports array-oriented scientific data. A postprocessing script is usually needed to convert the single-particle data to the particle population information for further analysis (e.g., total number concentrations, bulk aerosol mass concentrations, and mixing state information).

In brief, a typical workflow consists of four steps: (1) input file preparation, (2) spec file preparation, (3) PartMC execution, and (4) postprocessing. It is recommended to include the PartMC execution and postprocessing in a Bash script rather than typing the commands manually (interactively). The following sections introduce the elements of the files (input files and the spec file) and the workflow.

### 5.5.1 PartMC Input (.dat) Files Preparation

There are three types of PartMC input files: (1) aerosol input files, (2) environmental input files, and (3) gas input files. The aerosol input files are illustrated in Figure 5.12.

The **aerosol input files** include aerosol material data (aero_data.dat), aerosol background data (aero_back.dat, aero_back_dist.dat, and aero_back_comp.dat), aerosol emission data (aero_emit.dat, aero_emit_dist.dat, and aero_emit_comp.dat), and aerosol initial state data (aero_init_dist.dat and aero_init_comp.dat).

The aerosol material data specifies the aerosol species, as well as their physical properties. The aerosol background data describes the background aerosol information. The PartMC setup assumes that background air is continuously entrained into the PartMC volume at a prescribed entrainment rate. Users need to define the background aerosol distribution file (aero_back_dist.dat) in the aerosol background file (aero_back.dat), and the background composition proportions of species (aero_back_comp.dat) in the aero_back_dist.dat. Similarly, to describe the emission sources, users need to define the aerosol emission distribution (aero_emit_dist.dat) in the file aero_emit.dat, and the emission composition proportions of species (aero_emit_comp.dat) in the aero_emit_dist.dat. For the aerosol initial state, users need to define the file aero_init_comp.dat and aero_init_dist.dat.

Figure 5.13 shows an example how the files aero_init_dist.dat and aero_init_comp.dat are constructed. Each mode is specified by one block of text,

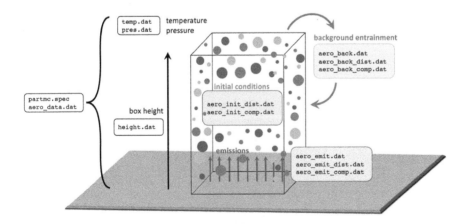

**Figure 5.12** Setup for a PartMC model simulation.

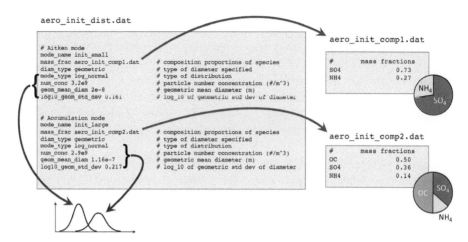

**Figure 5.13** Example of the input files that specify the aerosol initial condition for a PartMC simulation. The mass fractions are dimensionless.

and for each mode a unique composition profile can be assigned. In our example, we specified an Aitken mode with composition as shown in `aero_init_comp1.dat` and an accumulation mode, with composition as shown in `aero_init_comp2.dat`.

The **environmental input files** include the time-varying temperature profile (`temp.dat`), time-varying pressure profile (`pres.dat`), and time-varying mixing height profile (`height.dat`). These files allow users to study the impact of environmental conditions on particle simulations.

The **gas input files** include gas data (`gas_data.dat`), gas background data (`gas_back.dat`), gas emission data (`gas_emit.dat`), and gas initial state data (`gas_init.dat`). The gas input files are only relevant when the dynamic gas-particle mass transfer is included, which requires linking the Model for Simulating Aerosol Interactions & Chemistry (MOSAIC). For the purpose of this chapter, gas-particle partitioning is not relevant, but the files need to exist when running PartMC. Therefore we defined an empty state for these files, consisting of `gas_emit.dat`, `gas_back.dat`, and `gas_init.dat` under `scenarios/5_coag_brownian`.

### 5.5.2 Spec File Preparation

The spec file (e.g., `run_brown_part.spec` under the `scenarios/5_coag_brownian` folder) is the connection between the PartMC program and all other input files. It configures a scenario and sets up the input files for PartMC program. Users need to specify or edit the simulation parameters (e.g., the total ideal number of computational particles, simulation time, frequency of output), the prepared input files, environmental variables (e.g., initial relative humidity, location, and time), and optional model components. For a detailed description we refer the reader to the documentation at http://lagrange.mechse.illinois.edu/partmc/partmc-2.6.1/doc/html/index.html.

### 5.5.3  PartMC Execution and Postprocessing

To execute the PartMC program, users need to specify the path of the PartMC program (build/partmc) and filename of the spec file. PartMC output consists of snapshots of the complete particle-, gas- and environmental state at the chosen output frequency, and requires postprocessing to derive the particle population information of interest. For example, to obtain the total aerosol mass we need to sum up the masses of all particles, while to plot a size distribution we need to sort the particles into prescribed size bins.

Users are encouraged to make use of the existing postprocessing program, or develop their own postprocessing program to extract the desired information for further analysis. We have included several Python scripts that illustrate some basic postprocessing capabilities.

## Bibliography

[1] D. T. Gillespie. *Markov Processes: An Introduction for Physical Scientists*. Academic Press, San Diego, CA, 1992.

[2] N. A. Fuchs, *The Mechanics of Aerosols*, Pergamon Press, New York, 1964.

[3] J. H. Seinfeld and S. Pandis. *Atmospheric Chemistry and Physics*. Wiley, Hoboken, NJ, 2016.

[4] D. T. Gillespie. The stochastic coalescence model for cloud droplet growth. *Journal of the Atmospheric Sciences*, 29:1496–1510, 1972.

[5] D. T. Gillespie. An exact method for numerically simulating the stochastic coalescence process in a cloud. *Journal of the Atmospheric Sciences*, 32:1977–1989, 1975.

[6] M. von Smoluchowski. Versuch einer mathematischen Theorie der Koagulationskinetik kolloider Lösungen. *Zeitschrift für Physikalische Chemie*, 92:129–168, 1916.

[7] D. T. Gillespie. Approximate accelerated stochastic simulation of chemically reacting systems. *Journal of Chemical Physics*, 115(4):1716–1733, 2001.

[8] David F Anderson and Thomas G Kurtz. *Stochastic Analysis of Biochemical Systems*, volume 674. Springer, 2015.

[9] Alexander B. Kostinski and Raymond A. Shaw. Fluctuations and luck in droplet growth by coalescence. *Bulletin of the American Meteorological Society*, 86(2):235–244, 2005.

[10] N. Riemer, M. West, R. A. Zaveri, and R. C. Easter. Simulating the evolution of soot mixing state with a particle-resolved aerosol model. *Journal of Geophysical Research*, 114(D09202), 2009.

[11] Mark Z. Jacobson, Richard P. Turco, Eric J. Jensen, and Owen B. Toon. Modeling coagulation among particles of different composition and size. *Atmospheric Environment*, 28(7):1327–1338, 1994.

[12] E.R. Whitby, P.H. McMurray, U. Shankar und F.S. Binkowski. Modal Aerosol Dynamics Modeling, Technical Report 600/3-91/020, (NTIS PB91-161729/AS National Technical Information Service Springfield, VA), Atmospheric Research and Exposure Assessment Laboratory U.S. Environmental Protection Agency, Research Triangle Park, NC, 1991.

[13] F. Binkowski and U. Shankar. The regional particulate matter model 1. model description and preliminary results. *Journal of Geophysical Research*, 100:26191–26209, 1995.

[14] Mark Z. Jacobson. Analysis of aerosol interactions with numerical techniques for solving coagulation, nucleation, condensation, dissolution, and reversible chemistry among multiple size distributions. *Journal of Geophysical Research*, 107(D19), 2002.

[15] J. Lu and F. M. Bowman. A detailed aerosol mixing state model for investigating interactions between mixing state, semivolatile partitioning, and coagulation. *Atmospheric Chemistry and Physics*, 10(8):4033–4046, 2010.

[16] H. Matsui, M. Koike, Y. Kondo, N. Moteki, J. D. Fast, and R. A. Zaveri. Development and validation of a black carbon mixing state resolved three-dimensional model: Aging processes and radiative impact. *Journal of Geophysical Research*, 118: 2304–2326, 2013.

[17] H. Matsui, Makoto Koike, Yutaka Kondo, Jerome D. Fast, and M. Takigawa. Development of an aerosol microphysical module: Aerosol Two-dimensional bin module for foRmation and Aging Simulation (ATRAS). *Atmospheric Chemistry and Physics*, 14(18):10315–10331, 2014.

[18] N. Oshima, M. Koike, Y. Zhang, and Y. Kondo. Aging of black carbon in outflow from anthropogenic sources using a mixing state resolved model: 2. Aerosol optical properties and cloud condensation nuclei activities. *Journal of Geophysical Research*, 114(D18), 2009.

[19] Shupeng Zhu, Karine N. Sartelet, and Christian Seigneur. A size-composition resolved aerosol model for simulating the dynamics of externally mixed particles: SCRAM (v 1.0). *Geoscientific Model Development*, 8(6):1595, 2015.

[20] Joseph Ching, Rahul A. Zaveri, Richard C. Easter, Nicole Riemer, and Jerome D. Fast. A three-dimensional sectional representation of aerosol mixing state for simulating optical properties and cloud condensation nuclei. *Journal of Geophysical Research*, 121(10):5912–5929, 2016.

[21] S. E. Bauer, D. L. Wright, D. Koch, E. R. Lewis, R. McGraw, L-S. Chang et al. MATRIX (Multiconfiguration Aerosol TRacker of mIXing state): an aerosol microphysical module for global atmospheric models. *Atmospheric Chemistry and Physics*, 8(20):6003–6035, 2008.

[22] Mark Z. Jacobson and John H. Seinfeld. Evolution of nanoparticle size and mixing state near the point of emission. *Atmospheric Environment*, 38(13):1839–1850, 2004.

[23] Sophia M. Charan, Weimeng Kong, Richard C. Flagan, and John H. Seinfeld. Effect of particle charge on aerosol dynamics in Teflon environmental chambers. *Aerosol Science and Technology*, 52(8):854–871, 2018.

[24] Naser G. A. Mahfouz and Neil M. Donahue. Atmospheric nanoparticle survivability reduction due to charge-induced coagulation scavenging enhancement. *Geophysical Research Letters*, 48(8):e2021GL092758, 2021.

[25] Lloyd A. Spielman. Viscous interactions in brownian coagulation. *Journal of Colloid and Interface Science*, 33(4):562–571, 1970.

[26] M. Khairul Alam. The effect of van der waals and viscous forces on aerosol coagulation. *Aerosol Science and Technology*, 6(1):41–52, 1987.

[27] Kenneth C. Young. The role of contact nucleation in ice phase initiation in clouds. *Journal of Atmospheric Sciences*, 31(3):768–776, 1974.

**[28]** Ann M. Fridlind, A. S. Ackerman, Greg McFarquhar, G Zhang, MR Poellot, PJ DeMott et al. Ice properties of single-layer stratocumulus during the mixed-phase arctic cloud experiment: 2. model results. *Journal of Geophysical Research: Atmospheres*, 112(D24), 2007.

**[29]** David Arenberg. Turbulence as the major factor in the growth of cloud drops. *Bulletin of the American Meteorological Society*, 20(10):444–448, 1939.

**[30]** P. G. F. Saffman and J. S. Turner. On the collision of drops in turbulent clouds. *Journal of Fluid Mechanics*, 1(1):16–30, 1956.

**[31]** Wojciech W. Grabowski and Lian-Ping Wang. Growth of cloud droplets in a turbulent environment. *Annual Review of Fluid Mechanics*, 45:293–324, 2013.

**[32]** Sisi Chen, Man-Kong Yau, Peter Bartello, and Lulin Xue. Bridging the condensation–collision size gap: a direct numerical simulation of continuous droplet growth in turbulent clouds. *Atmospheric Chemistry and Physics*, 18(10):7251–7262, 2018.

**[33]** Paul S. Lee and David T. Shaw. Dynamics of fibrous-type particles: Brownian coagulation and the charge effect. *Aerosol Science and Technology*, 3(1):9–16, 1984.

**[34]** Steven N. Rogak and Richard C. Flagan. Coagulation of aerosol agglomerates in the transition regime. *Journal of Colloid and Interface Science*, 151(1):203–224, 1992.

**[35]** Karl-Heinz Naumann. COSIMA-a computer program simulating the dynamics of fractal aerosols. *Journal of Aerosol Science*, 34(10):1371–1397, 2003.

6

# Nucleation: Formation of New Particles from Gases by Molecular Clustering

In atmospheric sciences, the term *nucleation* generally refers to the formation of new secondary aerosol nanoparticles from condensable gases. This gas-to-particle conversion directly from vapors without pre-existing particle surfaces, more specifically referred to as homogeneous nucleation, is the focus of this chapter. On a molecular level, nucleation proceeds through collisions and subsequent clustering of individual gas molecules. The formed molecular clusters normally can also re-evaporate, with the evaporation rate generally decreasing with increasing cluster size. Nucleation is considered to have occurred when the rate at which gas molecules collide and stick to the clusters exceeds the cluster evaporation rate, and the clusters can thus be considered stable. The size of stable atmospheric clusters is typically ~1–2 nm. After the initial formation, the new nanoparticles can grow further by collisions and attachments of gas molecules, and also by colliding with other clusters if the cluster concentrations are sufficiently high.

Strictly speaking, the term *nucleation* is not always completely correctly used in the context of particle formation: generally, nucleation is a phase transition in which a new stable phase is formed in a supersaturated environment. Such phase transition involves overcoming a thermodynamic barrier to reach the thermodynamically favorable phase. For atmospheric new-particle formation, the favorable phase is the particle phase (solid or liquid) and the barrier is manifested as high evaporation rates of the smallest molecular clusters. However, particle formation may also be kinetically controlled, which may occur, for example, at very high gas concentrations. In this case the process does not involve thermodynamic barriers, corresponding to a negligible role of cluster evaporation with respect to the collision rate with gas molecules. In order to keep the terminology correct on a general level, the initial particle formation process is in this chapter referred to as *molecular clustering* or *initial particle formation* instead of nucleation.

The chemical species that participate in atmospheric particle formation in most environments include sulfuric acid ($H_2SO_4$) together with base compounds such as ammonia ($NH_3$) and amines, and potentially also other acids and highly oxidized organic molecules. The new-particle formation processes make a substantial contribution to atmospheric aerosol number concentrations and affect aerosol size

*Introduction to Aerosol Modelling: From Theory to Code.*
First Edition. Edited by David Topping and Michael Bane.
© 2022 John Wiley & Sons Ltd. Published 2022 by John Wiley & Sons Ltd.

distributions. Formation and further growth of airborne molecular clusters is a significant source of ultrafine ($\leq$100 nm) aerosol particles, which typically dominate ambient particle numbers. Through impacts on particle number and size distribution, new-particle formation affects cloud activation (Chapter 2.5) and aerosol radiative forcing.

## 6.1 Modelling Particle Formation: From Atoms to Molecular Cluster Populations

The principal quantity describing the initial nanoparticle formation in aerosol models is the formation rate of new particles per unit volume and unit time, commonly referred to as the nucleation rate or the particle formation rate. This formation rate corresponds to newly-formed stable particles that are not expected to evaporate back to smaller sizes. Within a sectional aerosol population model framework, the new particles are inserted in the smallest aerosol bin (Chapters 1 and 7). In general, the formation rate is obtained by modeling a population of interacting molecules and molecular clusters, and extracting the flux of clusters growing over the size beyond which the clusters are stable against evaporation. The most realistic method for this would be molecular dynamics (MD) simulations of a set of molecules. MD simulations include a description of interactions between the molecules, and follow the trajectory of each individual molecule, eventually counting the number of formed clusters above a threshold size after which evaporation of clusters is negligible. Unfortunately, simulating realistic atmospheric systems long enough to observe particle formation is computationally prohibitively expensive. Moreover, simple classical force fields are not able to capture the essence of molecular interactions for typical atmospheric compounds, but instead quantum mechanics is needed.

Thus, in practice the formation rate for atmospheric systems is predicted by models based on the discrete general dynamic equation (GDE) that simulate the number concentrations of different cluster compositions, assuming effective average collision and evaporation rates for each molecular composition. The true discrete GDE, in which each simulated size is defined by the *exact number of molecules* in a cluster, should not be confused with the aerosol general dynamic equation (Chapter 1) in which each size class is defined by the *volume or mass limits* and thus covers particles of different molecule numbers. The formation rate is given as the total flux of clusters growing to stable sizes.

GDE models are in practice used together with cluster evaporation rates obtained by quantum chemistry, which is nowadays a standard tool for studying the properties of atmospheric clusters. More precisely, quantum chemistry gives the cluster formation free energies, which are converted to evaporation rates by statistical mechanics (Section 6.1.3). Previously, the main approach for formation free energies was the liquid droplet model, based on classical thermodynamics. However, the classical model is not capable of correctly describing the molecular interactions in such small clusters, which causes errors of up to several orders of magnitude in modeled formation rates.

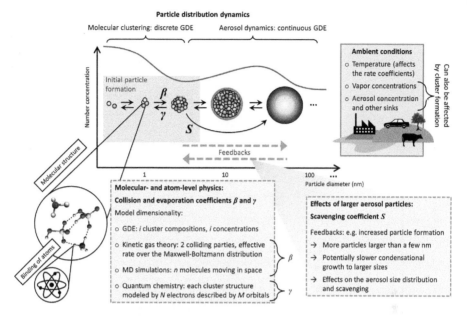

**Figure 6.1** Schematic representation of the initial nanoparticle formation process, the methods and data needed to model the process, the environmental factors affecting it, and its connection to the modeling of larger aerosol particles. Variables *i*, *n*, *N*, and *M* depict different dimensionalities that are related to the modeling of molecular clusters and their configurational space.

Collision rates are normally calculated by the kinetic gas theory (Section 6.1.3). Molecular dynamics methods can be used for limited studies, such as assessing the effects of long-range electrostatic interactions between collision parties on the collision rate constants. Such results can be applied in GDE-based simulations through the input collision rates. Figure 6.1 illustrates the common tools for modeling atmospheric cluster formation, including the discrete GDE framework and the input data from higher-level fundamental physics and chemistry models that can be embedded in it.

When applying the GDE and effective cluster properties and rate constants, the molecular cluster population is modeled largely similarly to an aerosol population (Chapter 1). However, while aerosol dynamics models are based on the continuous form of the GDE—even if the aerosol distribution is split into discrete size bins—molecular cluster dynamics is modeled by the discrete, molecular-resolution GDE. In other words, the clustering process is modeled molecule-by-molecule. Instead of particle size bins, the number concentration as a function of time is solved for each cluster composition, defined by the numbers of different molecules in the cluster. This is illustrated in Figure 6.2, which shows a set of clusters consisting of $H_2SO_4$ and $NH_3$ molecules.

Box 6.1.1 summarizes the essential steps for applying the discrete-GDE-based approach to simulate a selected cluster set. The details of modeling atmospheric clustering by solving the GDE, depicted by Figure 6.1 and listed in Box 6.1.1, are discussed in Sections 6.1.1, 6.1.3, and 6.1.4 below.

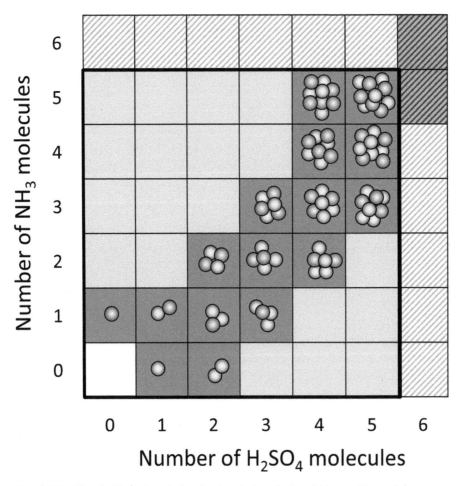

**Figure 6.2** Visualization of a set of molecular clusters that could be used to model $H_2SO_4$–$NH_3$ clustering. The included clusters contain at most five acid and five ammonia molecules, and thus fall within the red rectangle (both light and dark red squares) that limits this composition range on the $H_2SO_4$–$NH_3$ "matrix." The cluster compositions are given by the numbers of $H_2SO_4$ and $NH_3$ molecules (depicted by blue and orange spheres, respectively) in the cluster. The dark red squares that contain a cluster correspond to stable enough compositions that are included in the simulation, and the light red squares are compositions that are not included, as they are expected to be too unstable to exist. The gray squares with raster pattern are compositions outside of the simulation system, with the light and dark gray squares depicting unstable and stable clusters, respectively. The dark gray compositions are allowed to grow out of the simulation system, thus contributing to new particle formation (Sections 6.3 and 6.4.1.1).

> **Info box 6.1.1: GDE-based approach for modeling particle formation in a nutshell**
>
> 1. Define which cluster compositions your system includes. In a simple case of only one molecule type $X$, you might have clusters $X_1$ (the single vapor molecule, i.e. monomer), $X_2$, ..., $X_8$. The largest clusters should be large enough to be stable, i.e. their evaporation should be negligible compared to growth by collisions. (Section 6.1)
> 2. Define what constitutes new particle formation. In the above case, we assume that cluster $X_9$ is stable enough to grow without significant evaporation. Therefore, we define that any collisions that result in cluster $X_{i \geq 9}$ form a stable new particle that is no more part of the simulation. The flux into $X_{i \geq 9}$ thus gives the new particle formation rate. (Sections 6.1.1 and 6.3)
> 3. Define how the collision rate constants between all clusters are calculated. At simplest, hard-sphere collision rates can be used. Effects of attractive forces in collisions involving charged species—and possibly also between two electrically neutral, polar species—need to be taken into account. (Section 6.1.3)
> 4. Define the evaporation rate constants, generally calculated from the formation-free energies $\Delta G$ of the clusters. Even small uncertainties in $\Delta G$ may have a large effect on the results, and therefore it is advisable to use the highest-level quantum chemical method for $\Delta G$ that is computationally feasible. Typically, single-molecule evaporations are the dominant decay pathway. (Section 6.1.3)
> 5. Define the ambient conditions, including the concentration or source rate of the vapor, i.e. the monomers, and the possible loss terms of molecules and clusters, such as adsorption onto the walls of a measurement chamber, and/or coagulation onto larger particles. (Section 6.1.3 and 6.1.4)
>
> Based on the cluster set and the conditions, write the time derivatives of the concentrations of the simulated clusters, known as the discrete general dynamic equation (GDE) or cluster birth–death equation. Solve the GDE to yield the cluster concentrations and the formation rate. Often we are interested in steady-state formation rates, and thus the simulation needs to be run until a steady state is obtained. (Section 6.1.1)

## 6.1.1 The Discrete General Dynamic Equation

The discrete GDE, also called the cluster birth–death equation, gives the time derivative of the concentration $C_i$ of each molecule or cluster composition $i$ as

$$
\frac{\mathrm{d}C_i}{\mathrm{d}t} = \frac{1}{2} \sum_{j<i} \left( \beta_{j,(i-j)} C_j C_{(i-j)} - \gamma_{i \to j,(i-j)} C_i \right)
$$
$$
- \sum_j \left( \beta_{i,j} C_i C_j - \gamma_{(i+j) \to i,j} C_{(i+j)} \right) \tag{6.1}
$$
$$
+ Q_i - S_i C_i,
$$

where $\beta_{i,j}$ is the collision rate constant between clusters $i$ and $j$, and $\gamma_{(i+j) \to i,j}$ is the evaporation rate constant of cluster $(i + j)$ back to $i$ and $j$. The first sum on the right-hand side gives all possible collisions of smaller clusters $j$ and $(i-j)$ that create cluster

$i$, and the reverse evaporation or fragmentation processes of $i$ back to clusters $j$ and $(i-j)$. The factor of $1/2$ prevents counting the formation or decay of $i$ twice as processes involving $j + (i-j)$ and $(i-j) + j$, as these are one and the same process. The second sum gives the collisions of $i$ with all other molecules and clusters, and the evaporations of the larger clusters back to $i$. Typically only single-molecule evaporations are relevant, as the rates of cluster fissions are often insignificant. The terms $Q_i$ and $S_iC_i$ correspond to external sources and sinks of cluster $i$, respectively. Source terms $Q_i$ are normally relevant only for single gas molecules produced by gas phase chemistry, and the sinks correspond to cluster scavenging by larger aerosol particles, the walls of an experimental chamber, or other removal processes.

Clusters can also exist in different charging states, as they can be ionized and neutralized by small atmospheric ions. Ionization and recombination processes can be taken into account in Eq. (6.1) as additional terms similar to other collisions. For this, generic negative and positive ionizing species need to be included in the set of simulated clusters and allowed to collide with the molecules and clusters. For example, the charging of an electrically neutral cluster $i$ by a collision with a negatively charged ionizing species is described by the term $\beta_{i,\text{neg.ion}}C_iC_{\text{neg.ion}}$, which is a loss term for the neutral cluster $i$ and a source term for the corresponding charged cluster $j$. The reverse recombination process is similarly described by a collision of $j$ with an ionizing species of the opposite polarity, given by the term $\beta_{j,\text{pos.ion}}C_jC_{\text{pos.ion}}$ which is a loss term for $j$ and a source term for $i$.

Equation (6.1) is solved numerically to yield the concentrations $C_i$ of all simulated cluster compositions (Sections 6.2 and 6.3). The formation rate of any cluster(s) $k$ of given size(s) and composition(s) can then be obtained directly from the concentrations $C_i$ as the flux of clusters per unit volume and unit time:

$$J_{\{k \mid \text{condition}\}} = \sum_k \frac{1}{2} \sum_{j<k} \left( \beta_{j,(k-j)}C_jC_{(k-j)} - \gamma_{k\to j,(k-j)}C_k \right), \tag{6.2}$$

where $\{k \mid \text{condition}\}$ denotes that the clusters $k$ that contribute to the formation rate must satisfy the given conditions. In practice, when determining the particle formation rate that can be implemented in an aerosol dynamics model, the condition is that the evaporation rates of clusters $k$ are negligible compared to their growth rates.

For a molecular cluster set of a finite size, such as that depicted in Figure 6.2, this means that clusters that are allowed to grow out of the set must be both large enough and have a favorable chemical composition so that their evaporation can be omitted. The formation rate can then be determined as the rate of stable clusters growing out of the simulation system, and Eq. (6.2) reduces to

$$J_{\{k_{\text{out}} \mid \text{condition}\}} = \sum_{k_{\text{out}}} \frac{1}{2} \sum_{j<k_{\text{out}}} \left( \beta_{j,(k_{\text{out}}-j)}C_jC_{(k_{\text{out}}-j)} \right). \tag{6.3}$$

In Eq. (6.3), the "condition" gives the chemical compositions outside of the simulation system that are expected to be stable. The choice of these conditions and the treatment of clusters outside of the system are discussed in detail in Sections 6.3 and 6.4.1.1.

The traditional classical nucleation theory (CNT), which has been widely used in atmospheric particle formation studies, is also based on the discrete GDE, with liquid droplet model used for the cluster properties. The dynamics assumed by CNT are a special case of the cluster birth–death equations, in which only collisions with and evaporations of single molecules are included, and other dynamic processes such as coagulation and scavenging are omitted. In the CNT framework, the discrete GDE is solved analytically by assuming that there is a single thermodynamic barrier. While this results in a relatively simple formulation for the formation rate, the simplified dynamics are a significant source of error at most atmospheric conditions.

As a final remark, it is good to note that the formation rate is not unambiguous as it depends on cluster size. In principle, the formation rate can be determined for stable cluster compositions of any size, but at realistic atmospheric conditions with scavenging sinks and possible cluster coagulation the formation rate is size-dependent. Therefore, the modeled formation rate must always be given together with information on the corresponding particle size.

### 6.1.2 The Discrete Cluster GDE vs. the Continuous Aerosol GDE

It is important to understand the differences between the discrete and the continuous GDE, and the fact that the discrete GDE must be used to model the initial molecular cluster formation. The continuous GDE is a so-called "deterministic" approach: it assumes that all particles that have the same size and composition grow at the same rate when they uptake vapors [1]. This is not the case for small clusters of molecular dimensions, and especially for unstable clusters for which the evaporation frequency $\gamma$ exceeds the collision frequency $\beta_{gas}C_{gas}$ with gas molecules. For such unstable clusters, the continuous GDE would only predict shrinkage and no growth. However, these clusters can indeed grow since collisions with gas molecules are of stochastic nature. Physically this means that while the average evaporation frequency is higher than the average collision frequency for a given cluster composition, some clusters can overcome this obstacle by experiencing several fortuitous consecutive collisions before an evaporation occurs. In terms of fluxes of clusters between different sizes, this can be understood as follows: If cluster $i$ is unstable, the collision flux of clusters from cluster $i$ to a larger size $(i + j)$ is lower than the evaporation flux from $i$ to a smaller size $(i - j)$:

$$\beta_{i,j}C_iC_j < \gamma_{i\to j,(i-j)}C_i. \tag{6.4}$$

However, there is still a positive net flux of clusters from $i$ to $(i+j)$ if the backward evaporation flux from $(i + j)$ to $i$, which depends on the evaporation rate and concentration of $C_{(i+j)}$ (not $C_i$!), is lower than the forward collision flux:

$$\beta_{i,j}C_iC_j > \gamma_{(i+j)\to i,j}C_{(i+j)}. \tag{6.5}$$

This kind of stochastic kinetics, illustrated by Figure 6.3, is described by the discrete GDE but not by the continuous GDE. In "true" nucleation, which involves thermodynamic barriers, the smallest clusters are always unstable, and thus cannot be described by a continuous model. Moreover, also in barrierless clustering the formation of the first clusters involves discrete effects [2, 3]. Therefore, whether true nucleation or not, the initial clustering process must be modeled molecule-by-molecule by the discrete GDE.

### 6.1.3 Rate Constants of the Cluster Dynamics Processes

The quantitative results given by the cluster dynamics equations (Eq. (6.1)) are as good as the input data used in the model, namely the rate constants of the dynamic processes

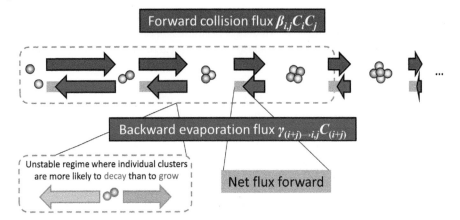

**Figure 6.3** Schematic presentation of fluxes of molecular clusters between consecutive cluster sizes during new particle formation. Note that the figure is simplified, and the arrow lengths should be interpreted qualitatively.

described by the equations. In atmospheric clustering studies, the key parameters are the evaporation rate constants $\gamma$, which are very challenging to determine accurately and have high uncertainties compared to the collision and sink rate constants. The rate constants in the discrete GDE are generally determined by theoretical means, as their extraction from cluster measurements is very challenging and still involves relatively high uncertainties, even though sophisticated Markov Chain Monte Carlo (MCMC) methods for this are under development [4].

The collision rate constants $\beta$ between two electrically neutral parties are most often determined as hard-sphere collision rates

$$
\beta_{i,j} = \left(\frac{3}{4\pi}\right)^{1/6} \left[6k_{\mathrm{B}}T\left(\frac{1}{m_i} + \frac{1}{m_j}\right)\right]^{1/2} \left(V_i^{1/3} + V_j^{1/3}\right)^2 , \tag{6.6}
$$

where $m_i$ and $V_i$ are the mass and volume of cluster $i$, respectively, $T$ is the temperature, and $k_{\mathrm{B}}$ is the Boltzmann constant. While this is considered a reasonable approximation, the sources of uncertainty include possible reduction or enhancement in the collision rates due to collision geometries, steric hindrance, kinetic barriers, and long-range interactions. As mentioned in Section 6.1, more accurate collision rate constants can be calculated by MD simulations or other advanced theoretical approaches. For instance, MD studies have suggested enhancement factors of approximately two compared to the hard-sphere rate for a collision of electrically neutral sulfuric acid molecules due to dipole-dipole interactions [5].

For collisions between charged and neutral species, parameterizations that consider the electrostatic effects should be used (Section 6.4.1.1), as the rates of such collisions can be enhanced by a factor of up to ~10 compared to the hard-sphere rate [see e.g. [6]]. Collision trajectories between small oppositely charged molecules and/or clusters are governed by Coulomb interactions, and thus a single recombination rate coefficient is commonly used instead of rate constants that depend on the cluster identities. The recombination rate constant is approximately $10^3$–$10^4$ times higher than collision rate constants between electrically neutral species.

The uncertainties in the collision rate constants are relatively small compared to those in the evaporation rate constants. In general, the evaporation rate of a molecular complex is deduced from the formation thermodynamics of the complex. Briefly, the rate constant for the evaporation of cluster $(i + j)$ into clusters or molecules $i$ and $j$ is derived using the detailed balance approach [for more details, see e.g. 7] as

$$\gamma_{(i+j)\to i,j} = \beta_{i,j} \frac{P_{\text{ref}}}{k_B T} \exp\left(\frac{\Delta G_{(i+j)} - \Delta G_i - \Delta G_j}{k_B T}\right),\qquad(6.7)$$

where $\beta_{i,j}$ is the rate constant of the reverse process, that is, the collision between clusters $i$ and $j$ (e.g. Eq. (6.6)) and $\Delta G$s are the Gibbs free energies of formation of the clusters, computed at a reference pressure $P_{\text{ref}}$. The primary method to assess the evaporation rate constants of small atmospheric molecular clusters is quantum chemistry, which gives the formation free energy $\Delta G$. Since $\gamma$ is proportional to the exponent of $\Delta G$, even relatively minor uncertainties in $\Delta G$ can propagate to significant uncertainties in $\gamma$, and consequently in the simulated cluster concentrations and formation rates (Eqs. (6.1)–(6.3)). These effects can also be very non-linear, and thus uncertainties in $\Delta G$ cannot be straightforwardly translated to uncertainties in e.g. the formation rate.

The rate constants $S_i$ for cluster removal by external sinks are commonly determined based on hard-sphere rates for scavenging by larger particles, or measurements and semi-empirical parameterizations for deposition onto other surfaces such as laboratory chamber walls, respectively. It is often convenient to determine the loss rate constant for a common gas species, typically $H_2SO_4$, and scale this rate for clusters of different sizes assuming that the rate is a function of cluster diameter and possibly also mass [8, 9].

Describing the cluster population dynamics through effective rate constants is an averaging approach in the sense that it assumes the same properties for all clusters with the same molecular composition. Here, it is assumed that the molecules are always arranged in the same representative configuration, which corresponds to the minimum formation free energy $\Delta G$ of the composition. This optimal global-minimum-energy structure is found by sampling the configuration space [10]. In reality, at atmospheric temperatures, clusters can exist in different local-minimum-energy configurations, corresponding to different electronic structures and hence to different collision and evaporation rates. Especially larger clusters may have several comparable local free energy minima. This can be taken into account by efficient sampling and averaging of the cluster properties over the minima [11]. However, finding a reliable set of local minima is more challenging—and uncertain—than finding the global minimum.

The assumption of average cluster properties naturally applies also to the other rate constants in Eq. (6.1). For collision rate constants, not only the geometry of the clusters, but also that of the collisions plays a role, and the rate constants correspond to effective rates over all possible collision geometries. For many atmospheric clustering species such as $H_2SO_4$ and $NH_3$ that can be expected to form approximately spherical clusters, and that are able to easily attach and form chemical bonds, the effect of different collision geometries can be assumed to be negligible. However, for instance for species that cannot form bonds equally well on all sides, such as in a collision of a water molecule and an organic molecule with a hydrophilic and a hydrophobic end, all collisions may not result in attachment. In such a case, the collision rate constant—or rather the attachment rate constant—should be averaged over a representative set of collision geometries.

The reduction of the full configuration space and the averaged incorporation of molecular-level physics in the discrete GDE is illustrated by the lower left-hand side sub-diagram and text box in Figure 6.1. Finally, similar to aerosol dynamics box models, the discrete GDE approach obviously also assumes a well-mixed air parcel with a constant temperature and other properties, with no significant local fluctuations in the cluster number concentrations.

### 6.1.4 Cluster Formation in Different Atmospheric Environments

Atmospheric new-particle formation rates and clustering mechanisms are affected not only by the properties of the participating molecules, but also by the ambient conditions, including vapor concentrations or source rates, temperature, source rate of atmospheric ions, and concentrations of larger aerosol particles acting as a scavenging sink. These conditions may vary significantly, for instance, from continental to marine environments, and from boundary layer to upper troposphere. Examples of potential effects of the environment on the formation processes are given below.

First, the role of a specific chemical species depends on the ambient conditions: for instance, while $NH_3$ is usually always present and thus likely to contribute to $H_2SO_4$-driven clustering, its role can become minor in the presence of a stronger base species, such as an amine. Similarly, $H_2SO_4$–$NH_3$ clustering may only be a weak source of new particles at warm conditions, but can become a significant clustering mechanism at lower temperatures. Second, the details of the clustering mechanisms may also vary: for example, at low precursor gas concentrations atmospheric ionizing species often play a significant role through the formation of charged molecules which efficiently stabilize molecular clusters. However, the contribution of ions is limited by the atmospheric ion production rate which is independent of the trace gas concentrations. Therefore, ion-mediated processes become insignificant at high gas concentrations as the stability (in terms of the relative collision and evaporation rates) of electrically neutral clusters is increased, and the neutral formation rate exceeds the maximum rate of ion-driven cluster formation. For strongly clustering species, such as a $H_2SO_4$–amine system, ions may be insignificant even at very low gas concentrations.

In general, the factors that limit particle formation processes are different for different atmospheric environments. In remote clean environments with low precursor gas concentrations, the main limitation is the cluster stability (the evaporation vs. the molecular collision terms in Eqs. (6.1) and (6.2)). In polluted environments with high trace gas concentrations, also the concentrations of larger aerosols are typically high. In this case clusters are more stable and grow faster, but cluster concentrations and formation rates are limited by cluster scavenging onto the larger particles. The effects of the identities of the clustering species and the ambient conditions on the particle formation rate and the formation mechanisms are illustrated in Figure 6.4. The figure demonstrates how the dominating species and the role of ions may vary in different environments.

There are also couplings and feedback mechanisms between initial particle formation and further growth of the formed particles. Higher formation rate does not necessarily result in more aerosol particles that are large enough to, for example, act as cloud

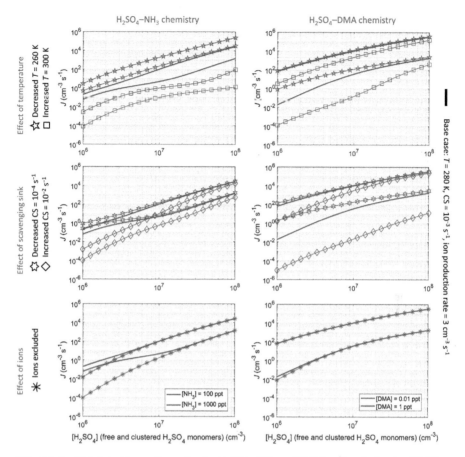

**Figure 6.4** Simulated formation rates for $H_2SO_4$–$NH_3$ and $H_2SO_4$–dimethylamine (DMA) systems for different vapor concentrations, temperatures and scavenging rate constants, and the effect of ions at a typical atmospheric ion production rate (as compared to excluding ions by assuming a zero ion production rate). Note that the purpose of the figure is to demonstrate the various effects that the ambient conditions may have on the formation rates, and the quantitative rates depend on the cluster thermodynamics, used as input in the simulation.

condensation nuclei. This is because the formed particles must grow by vapor condensation to sizes of at least ~50–100 nm, and individual particles grow less if the available condensable gases are distributed onto a larger total number of particles. Formed particles may also scavenge smaller clusters, thus reducing further particle formation.

## 6.2 Coding the Discrete GDE: The Straight-forward Case of a One-component System

We will now construct the discrete GDE for a simple one-component molecular system with no charged clusters using Matlab. The concentrations $C_i$ and their time derivatives $dC_i/dt$ for the different cluster compositions containing $1, 2, 3, \ldots, i_{max}$ molecules are conveniently given in vectors of $i_{max}$ elements. We loop over all compositions $i$, filling in the derivative vector:

```matlab
function dcdt = dgde_1comp(t,c,K,E,S,Q)

% This function contains the time derivatives dc/dt of cluster
%   concentrations c
% Input:
%    t = time (can be a "dummy" variable)
%    c = cluster number concentration vector (clusters m^-3)
%    K = collision rate coefficient matrix (m^3 s^-1)
%    E = evaporation rate coefficient matrix (s^-1)
%    S = cluster scavenging coefficient vector (s^-1)
%    Q = cluster source rate vector (clusters m^-3 s^-1)

dcdt = zeros(size(c));
nmax = length(c);

for i = 1:nmax

    for j = i:nmax

        % The collision removes the smaller clusters...
        K_term = K(j,i)*c(i)*c(j);
        dcdt(i) = dcdt(i) - K_term;
        dcdt(j) = dcdt(j) - K_term;

        if i+j <= nmax

            % ...and creates a larger cluster.
            dcdt(i+j) = dcdt(i+j) + K_term;

            % The corresponding evaporation creates the smaller
        clusters...
            E_term = E(j,i)*c(i+j);
            dcdt(i) = dcdt(i) + E_term;
            dcdt(j) = dcdt(j) + E_term;
            % ...and removes the larger cluster.
            dcdt(i+j) = dcdt(i+j) - E_term;

        end

    end
    % Add the source
    dcdt(i) = dcdt(i) + Q(i);
    % Subtract the loss
    dcdt(i) = dcdt(i) - S(i)*c(i);

end

end
```

**Listing 6.1** The discrete GDE (Eq. (6.1)) for a simple one-component system.

In listing 6.1, all rate constants ("K, E, S, Q") are given as input parameters to the function (see listings 6.2 and 6.3 below). Matrices "K" and "E" contain the collision and evaporation rate constants for each pair of clusters that can collide ($i, j \rightarrow (i + j)$) and evaporate (($i + j) \rightarrow i, j$). For each cluster $i$, the inner loop goes through the collisions with other clusters $j$. In each collision, the two colliding clusters are lost, and a larger cluster ($i + j$) is formed. The corresponding backward evaporation process is conveniently added in the same loop. Finally, possible source and loss terms "Q" and "S" for cluster $i$ are added.

Note that the $j$ loop begins from $i$ in order to account each collision–evaporation process only once, and that for $j = i$, the collision and evaporation terms are added twice. In this way, each possible pair of clusters is correctly taken into account (this must be in line with how the rate constants are determined; see listing 6.3). Note also the order of the $i$ and $j$ indices in the "K" and "E" matrices, set according to the storage order: Matlab, like e.g. Fortran, is column-major, and thus selecting elements column-by-column is faster. "K" and "E" are precalculated and saved to avoid unnecessary calculations at every subroutine call.

Function "dgde_1comp" gives the time derivatives of the cluster concentrations when called with the rate constants ("K, E, S, Q") and current state ("t, c") given as input. First, the rate constants must be determined. In the example below, the collision rate constants "K" are calculated as hard-sphere rates using kinetic gas theory, and the evaporation rate constants "E" are set to correspond to those of quantum-chemistry-based values for $H_2SO_4$–$NH_3$ clusters (note that this is only a simplified example to illustrate the simulation set-up, and in reality the evaporation rates must be explicitly calculated from the free energies). The cluster sink follows a power-law size dependence, and the vapor source is of the order of magnitude of an atmospheric $H_2SO_4$ source. The initial cluster concentrations are set to zero. It must be noted that this is a simple example, and the simulation system does not correspond quantitatively to any real atmospheric cluster set.

```
clear variables

% Set the number of different cluster sizes (i.e. the number of
    molecules in the largest cluster)
nmax = 5;

% Determine the rate constants
[K,E,S] = rate_constants_1comp(nmax);

% Set the vapor monomer source and no sources for other clusters
Q = zeros(1,nmax);
Q(1) = 5e9;                    % (molecules m^-3 s^-1)

% Set the initial cluster concentrations to zero
c_init = zeros(1,nmax);
```

**Listing 6.2** Initialization of the simulation for a simple one-component system.

```
function [K,E,S] = rate_constants_1comp(nmax)

% This function calculates the rate constants for simulating the
    cluster concentrations
% Output:
%    K = collision rate coefficient matrix (m^3 s^-1)
%    E = evaporation rate coefficient matrix (s^-1)
%    S = cluster scavenging coefficient vector (s^-1)

% Physical constants
kB = 1.3806504e-23;            % Boltzmann constant (J K^-1)
amu2kg = 1e-3/6.02214179e23;   % Conversion from atomic mass units to
    kg

% Temperature
temp = 280;                    % (K)
```

```
% Cluster properties
m1 = 100*amu2kg;              % Mass of one molecule (kg)
rho = 1500;                   % Assumed density (kg m^-3)

% Cluster mass, volume and diameter
m = m1.*(1:nmax);             % (kg)
V = m./rho;                   % (m^3)
dp = (6/pi.*V).^(1/3);        % (m)

% Allocate the matrices and vectors for the rate constants
K = zeros(nmax,nmax);
E = zeros(nmax,nmax);
S = zeros(1,nmax);

for i = 1:nmax
    for j = i:nmax
        % Hard-sphere collision rate of clusters i and j
        K(i,j) = (3/4/pi)^(1/6)*(6*kB*temp*(1/m(i)+1/m(j)))^(1/2)*(V
(i)^(1/3)+V(j)^(1/3))^2;
        K(j,i) = K(i,j);
    end
    % Loss rate constant to external sinks
    S(i) = 1e-3*(dp(i)/dp(1))^-1.6;
end

% Evaporation rate of molecules from clusters (no fissions assumed)
% A hypothetical example where the smallest clusters evaporate at
    the order of magnitude of
% quantum-chemistry-based rates for H2SO4-NH3 clusters
% NOTE: When calculated properly, E is dependent on T and also on K
    due to the detailed balance
E(1,1) = 2e-3;
E(2,1) = 6e-3;
E(3,1) = 3e-3;
E(4,1) = 2e-4;
% Assume a low evaporation rate for the larger sizes
E(5:nmax,1) = 1e-5;
for i = 1:nmax
    E(1,i) = E(i,1);
end

% Divide the constants by two for collisions and evaporations of two
    identical clusters to
% avoid counting the same process twice
for i = 1:nmax
    K(i,i) = 0.5*K(i,i);
    E(i,i) = 0.5*E(i,i);
end

end
```

Listing 6.3 Rate constants for a simple one-component system. The collision and coagulation sink rate constants are calculated according to Eq. (6.6) and Lehtinen et al. [8], respectively.

Different numerical methods can be applied to solve the set of coupled differential equations: the simplest approach would be to call function "dgde_1comp" in a loop over the simulation time steps, and use, for example, the Euler or the Runge-Kutta method to obtain the concentrations at the next time step based on the derivatives. For instance, a simple Euler forward method can be implemented as

```
% Set the time for which the cluster dynamics are simulated
% Note: Larger systems might take longer to reach the steady state
t_final = 3600;              % (s)

% Solve the equations by simple Eulerian iteration

% Set the time step for the Euler method - no guarantee if this
    gives correct results...
t_step = 1;                  % (s)
% Number of time steps
imax = floor(t_final/t_step);

T = zeros(1,imax);
C = zeros(imax,nmax);
C(1,:) = c_init;
for i = 2:imax
    dcdt = dgde_1comp(T(i-1),C(i-1,:),K,E,S,Q);
    C(i,:) = C(i-1,:) + dcdt*t_step;
    T(i) = T(i-1) + t_step;
end
```

**Listing 6.4** Application of the simple Euler method to solve the discrete GDE.

Such approach is common in aerosol dynamics models that simulate the distribution of particles larger than a couple of nanometers (Chapter 7.5). However, realistic, multi-component molecular cluster systems (see Section 6.4.2) typically exhibit large variations in the rate constants, making simple integration methods unreliable. Instead, the discrete GDE is conveniently solved by automatic solvers designed for stiff problems. The above set of equations can be solved, for example, by the Matlab solver ode15s by

```
% Use a solver - accurate and fast
[T,C] = ode15s(@(t,c) dgde_1comp(t,c,K,E,S,Q),[0,t_final],c_init);

% See that the concentrations are not negative - this might happen
    for very stiff systems
if min(min(C)) < -1e-50
    error('Negative concentrations')
else
    C(C<0) = 0.0;
end
```

**Listing 6.5** Application of an ODE solver to solve the discrete GDE.

The formation rate of clusters growing out of the simulation system can be calculated from the resulting concentrations by

```
% Calculate the formation rate out of the simulation system
J = zeros(length(T),1);      % (m^-3 s^-1)
for i = 1:nmax
    for j = i:nmax
        if i+j > nmax
            J = J + K(j,i)*C(:,i).*C(:,j);
        end
    end
end
```

**Listing 6.6** Calculation of the formation rate out of the simulation system (Eq. (6.3)).

Finally, the cluster concentrations and the formation rate can be visualized as below, with the output shown in Figure 6.5.

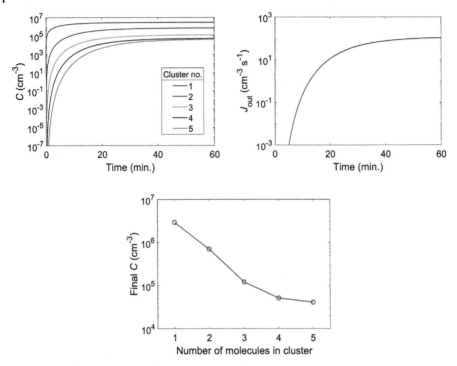

**Figure 6.5** Output given by the plotting script 6.7.

```
% Set the line style for the figures - handy when plotting different
    simulations
% in the same figure by running the script again when the figures
    are open
lstyle = '-';                    % '-', '--', '-.', or ':'

% Concentrations as a function of time
figure(1)
semilogy(T/60,C*1e-6,'LineStyle',lstyle)
hold on; set(gcf,'Color','white'); set(gca,'ColorOrderIndex',1)
xlabel('Time (min.)')
ylabel('{\itC} (cm^{-3})')
lgd = legend(string(1:nmax),'Location','best');
title(lgd,'Cluster no.','FontWeight','normal')

% Final concentrations as a function of cluster size
figure(2)
semilogy(1:nmax,C(end,:)*1e-6,'Marker','o','LineStyle',lstyle)
hold on; set(gcf,'Color','white'); set(gca,'ColorOrderIndex',1)
xlabel('Number of molecules in cluster')
xlim([0.5 nmax+0.5]); xticks(1:nmax)
ylabel('Final {\itC} (cm^{-3})')

% Formation rate as a function of time
figure(3)
semilogy(T/60,J*1e-6,'LineStyle',lstyle)
hold on; set(gcf,'Color','white'); set(gca,'ColorOrderIndex',1)
xlabel('Time (min.)')
ylabel('{\itJ}_{out} (cm^{-3} s^{-1})')
```

**Listing 6.7** Plotting the simulation results.

Since small atmospheric clusters typically evaporate significantly, it is critical to ensure that the size $i_{max}$ of the simulation system is large enough, corresponding to relatively low evaporation rate constants for the largest included clusters. For collisions $i, j \rightarrow (i + j)$ where $(i + j) > i_{max}$, the evaporation term $\gamma_{(i+j)\rightarrow i,j}C_{(i+j)}$ is omitted, meaning that it must be small compared to the collision term $\beta_{i,j}C_iC_j$ for the results to be robust. If the system is too small so that the excluded clusters $> i_{max}$ evaporate significantly, the omission of the backward evaporation fluxes leads to erroneous simulation results. When thermodynamic data is available for a large set of clusters, the sufficient system size can be assessed by varying $i_{max}$ to examine if concentrations $C_i$ are independent of it.

It is a good practice to perform consistency controls to ensure that the implementation of the equations is correct. These can include the following:
For a steady-state situation (where by definition $dC_i/dt = 0$),

- Check that the fluxes that lead to a given cluster composition $i$ (positive terms in Eq. (6.1)) equal the fluxes away from $i$ (negative terms in Eq. (6.1)).
- In the absence of cluster sinks and external sources, check that the total number of molecules (of each type in case of multi-component systems) remains constant when collisions leading out of the simulation system are disabled.
- In the absence of cluster sinks and cluster–cluster collisions, check that the net flux from cluster $i$ to the subsequent size $(i + 1)$ is independent of the composition $i$.

## 6.3 Multi-component Systems: Need for an Equation Generator

The model for a one-component cluster system considered in Section 6.2 is relatively easy to construct and understand. However, atmospheric particle formation typically involves two or more components, such as in an acid–base mixture. Constructing the GDE for a set of clusters consisting of multiple components is substantially more challenging than for single-component clusters, as the following aspects must be considered:

- Cluster indexing is not straight-forward anymore, as the indices $i$ (Eq. (6.1), code listing 6.1) cannot be simply set to correspond to the number of molecules in the cluster. Instead, each index $i$ corresponds to a different molecular composition, but these compositions do not need to be in any specific order.

  In addition, the cluster set does not need to include all possible molecular compositions, if certain types of compositions can be assumed highly unstable. For example, for acid–base systems, compositions for which the acid:base molar ratio differs significantly from one (that is, there is a large excess of either acid or base molecules) are typically unstable and thus do not need to be included in the simulation set (Figure 6.2). Therefore, the set of cluster compositions cannot simply be constructed by looping over all combinations of molecule numbers from $0 \ldots n_{acid,\,max}$ and $0 \ldots n_{base,\,max}$, where $n_{acid,\,max}$ and $n_{base,\,max}$ are the numbers of acid and base molecules in the largest included cluster.

- The stability criteria for outgrowing clusters also becomes more complex: unlike for the one-component system, for which the criterion is in practice simply $i > i_{max}$ (code listing 6.6), the criteria here must be set according to the composition of the outgrowing clusters. That is, if a collision results in a cluster composition that is outside of the simulation system but cannot be expected to be stable, the cluster cannot be let to grow out.

This also introduces the question of how to treat such unstable clusters that attempt to grow outside of the simulation set. In general, highly unstable compositions can be expected to undergo fast, subsequent evaporations until the resulting composition is stable. Clusters that end up outside of the simulated set but fail to meet stability criteria are thus brought back into the set by evaporation. The stability criteria are given as minimum numbers of molecules of different types that the outgrowing clusters must contain (Section 6.4.1).

- Some components may also become charged by atmospheric ionizing species, leading to ionic molecules and clusters. For example, acids are proton donors and thus tend to become negatively charged by removal of a $H^+$ ion, while bases are proton acceptors and can become positively charged by addition of $H^+$.

Charged species require some special considerations when constructing the GDE: due to electrostatic repulsion, small molecules or clusters of the same charge cannot collide, and a collision between species of opposite charges must result in a neutral cluster. In addition, the GDE must include equations also for the generic ionizing species in order to model their collisions with the simulated clusters.

- Finally, the multi-component equations can become numerically difficult to solve if the concentrations of different vapor species—and thus their collision rates—differ by several orders of magnitude. Also evaporation rates may exhibit such extremely large differences. However, if the rate coefficients of all processes related to a specific compound are orders of magnitude higher than those related to other compounds, the clusters can be expected to be in equilibrium with respect to this compound. This means that the collision and evaporation processes occur in such short time scales that they reach an equilibrium state instantaneously compared to all other processes, and thus do not need to be modeled explicitly. Note that the equilibrium assumption implies that the clusters cannot grow solely through addition of such compound. For an equilibrated compound, the collision and evaporation fluxes are balanced and thus there is no net growth along the corresponding coordinate on the "composition matrix" (Figure 6.2).

For atmospheric clusters, a prime example of such a compound is water, the concentration of which is approximately ten orders of magnitude higher than that of other condensable compounds. Water is also more loosely bound to the clusters, evaporating at much higher rates. In the presence of water, molecular clusters can become hydrated by taking up one or more water molecules, and clusters containing different amounts of water can exist simultaneously in an equilibrium hydrate distribution. The effect of water molecules on the collision and evaporation rate coefficients of the other compounds can be taken into account by calculating the rate coefficients as weighted averages over the hydrate distributions [12]. That is, water is not simulated explicitly as one of the compounds in the multi-component system, but its effect is included implicitly in the rate constants. It can be noted that such averaging approach is not an approximation, but corresponds to the exact solution for the explicit multi-component system when the equilibrium assumption is valid.

Based on the aspects discussed above, it is clear that the discrete GDE cannot be trivially written for arbitrary multi-component cluster sets. In order to conveniently and reliably construct the cluster equations, some kind of automatic equation generator is needed.

Such a tool is the Atmospheric Cluster Dynamics Code [ACDC; 13], which was developed to flexibly build and solve the cluster GDE for a given molecular cluster set and input for the corresponding rate constants. While the discrete GDE has been solved for specific cluster chemistries by different model approaches [see e.g. [14, 15]], ACDC can be applied to any chemical components and selected sets of cluster compositions. Therefore, we focus on ACDC for the rest of this chapter, introducing the model features, usage and examples of how to apply the model for different purposes.

It is important to note that ACDC is not a particle formation model as such, but instead a tool that enables the simulation of particle formation for given chemical systems and environments. The user must provide the cluster properties and ambient conditions as input, and ACDC generates and integrates the cluster equations. Therefore, it needs to be kept in mind that expressions such as "formation rate modeled by ACDC" does not actually give enough information on what has been modeled. The model results depend on which cluster set and conditions are considered, and how the rate constants are obtained, especially which quantum chemistry data set has been used for the evaporation rates.

## 6.4 Brief Introduction to Atmospheric Cluster Dynamics Code

The most fundamental part of ACDC is the main source code, written in the Perl language, that generates the code corresponding to the GDE for the given input system. The output files containing the GDE code can be printed out in either Matlab or Fortran, depending on the user's preference and on the application (Section 6.4.4). Perl is suitable for handling input and output text files and for working with strings, thus enabling convenient construction and printing of the equations. The code is named acdc_date.pl, where the date corresponds to the code version.

For the Matlab option, the Perl code also prints out a driver file including an ODE solver application that can be run directly to solve the generated equations. For the Fortran option, the open-source solver VODE [16] has been applied. In addition to the Perl code, additional Matlab and Fortran code packages are available for "standard" simulations (Section 6.4.4), such as for obtaining steady-state formation rates over ranges of input conditions.

Briefly, using the ACDC model consists of the following steps:

1. Define the set of simulated molecular clusters, and prepare input files that give the cluster compositions and formation free energies $\Delta G$ (or $\Delta H$ and $\Delta S$; Section 6.4.1.1), and possibly also other input depending on the simulation set-up.
2. Run the Perl code to generate the GDE files in either Matlab or Fortran. When running the Perl code, one can define various options related to the ambient conditions (e.g. temperature, relative humidity, ...) and to the included dynamic processes (which types of sinks are used, disabling some collisions or evaporations, ...).

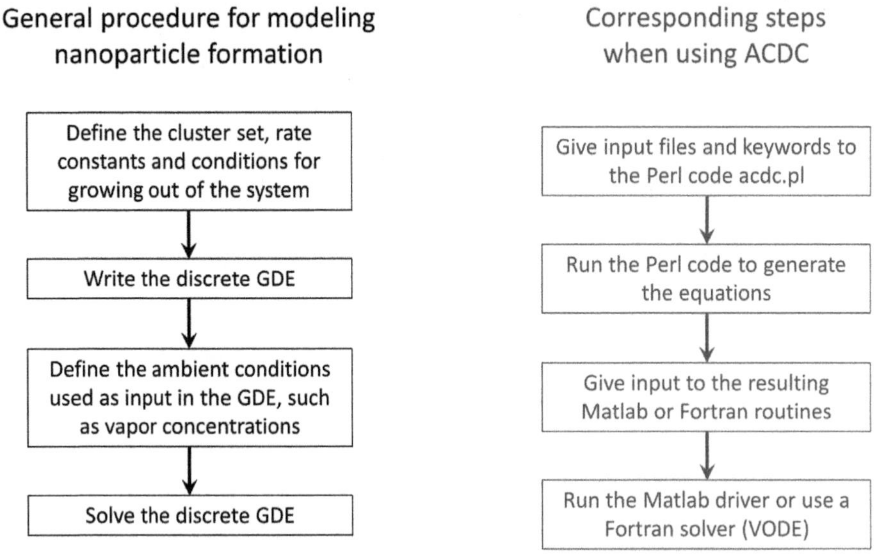

| General procedure for modeling nanoparticle formation | Corresponding steps when using ACDC |
|---|---|
| Define the cluster set, rate constants and conditions for growing out of the system | Give input files and keywords to the Perl code acdc.pl |
| Write the discrete GDE | Run the Perl code to generate the equations |
| Define the ambient conditions used as input in the GDE, such as vapor concentrations | Give input to the resulting Matlab or Fortran routines |
| Solve the discrete GDE | Run the Matlab driver or use a Fortran solver (VODE) |

**Figure 6.6** Implementation of the general steps for modeling cluster formation when applying the ACDC model.

3. Solve the equations by numerical integration using the Matlab or Fortran solver that comes with ACDC. Readymade code packages for this are available.

These steps and their relation to the general procedure of modeling cluster formation dynamics, listed in Box 6.1.1, are illustrated by the flowchart in Figure 6.6. The following sections introduce the essentials of modeling particle formation by ACDC; detailed information on all available features can be found in the ACDC Technical Manual [13].

### 6.4.1 ACDC Input

#### 6.4.1.1 Primary ACDC Input Files and Simulated Cluster Set

The minimum number of input files needed for ACDC is two: one for defining which clusters are included, and (unless evaporation is omitted) one for the free energies $\Delta G$ (or alternatively $\Delta H$ and $\Delta S$) which are used to calculate the evaporation rate constants ($\gamma$ in Eq. (6.1); Eq. (6.7)). The collision and sink rate constants ($\beta$ and $S$ in Eq. (6.1)) are by default calculated according to built-in formulae, but some or all rate constants can also be set in separate input files if the user wishes to vary them (Section 6.4.3). When charged species are included, default parameterizations that consider attractive forces are used to calculate the collision rate constants between charged and neutral species. In this case, the dipole moments and polarizabilities of the neutral molecules and clusters, available from the same quantum chemical calculations that yield the free energies, need to be also given as input.

```
#                          sulfuric_acid   ammonia
name:                      A               N
charge:                    0               0
mass [g/mol]:              98.08           17.04
density [kg/m^3]:          1830.0          696.0
base strength:             -1              1
```

```
acid strength:                    1                    -1
#########################################################
# Pure acid clusters:
                              1-2                 0
# Acid-ammonia clusters:
                              0-3                 1
                              2-4                 2
                              3-5                 3
                              4-5                 4
                              4-5                 5
#########################################################
# Criteria for growing out of the simulation:
out neutral                       6                    5
```

**Listing 6.8** ACDC input file for the simulated cluster set (see Figure 6.2).

Listing 6.8 gives an example of a cluster set input file for a system of electrically neutral clusters that contain up to five sulfuric acid and five ammonia molecules. More information on the cluster set file, for example settings for including charged species, can be found in the ACDC Technical Manual. The columns correspond to different molecule types, and the first line is a title line that is used to deduce the number of the molecule types in the file. It can contain for example the chemical names of the molecules. Comment lines begin with a hashtag (except for the first line, which is read in). The next lines give information on the molecular properties, defined by the following keywords:

| | |
|---|---|
| **name:** | One or more letters, used to label the molecule and construct cluster name labels |
| **charge:** | Charge of the molecule (0, −1, or 1) |
| **mass, density:** | Molecular mass and liquid-phase density, used to calculate the cluster diameters and the collision coefficients |
| **acid strength, base strength:** | Numbers describing the relative acid and base strengths, used to determine how the unstable clusters outside of the system are evaporated back into the simulation when several acidic or basic species are included |

After these definitions, the included clusters are listed according to their molecular content. Each line gives one or more cluster compositions in terms of numbers of molecules of each type in the cluster. For each line, one column can contain a range of molecule numbers instead of a single number for a neater input. For instance, the first composition line gives clusters that consist of 1–2 sulfuric acid molecules and contain no ammonia, that is, all in all two different cluster compositions. The next line gives clusters that consist of 0–3 sulfuric acid and one ammonia molecule, i.e., four different clusters, and so on. The clusters listed in the file form the set illustrated in Figure 6.2. The selection of clusters depends naturally on the available thermodynamic data (listing 6.9): all compositions for which data exists should be included, unless some compositions evaporate at such high rates that they in practice do not exist and can thus be excluded for computational efficiency. Such possibility can be readily tested by performing simulations using either the full or the reduced cluster sets and comparing the results (for assessing which clusters to exclude, see Section 6.4.3.3).

Finally, lines beginning with keyword "out" give the criteria for allowing clusters to grow out of the system. As these criteria are often different for different charging states, the criteria for neutral, negative and positive clusters are given separately as "out neutral", "out negative," and "out positive," respectively. In listing 6.8, clusters that contain *at least* 6 acid and 5 ammonia molecules are let to grow out. This means that clusters of, for example, 6 and 5, 8 and 6, or 10 and 10 acid and ammonia molecules are able to grow out. One can define more than one condition for growing out; for example, we could assume that also clusters containing at least 5 acid and 6 ammonia molecules are stable, and add this condition to the input file.

When a collision results in a cluster that is outside of the simulation system, but does not satisfy the outgrowth criteria, the cluster is treated as follows:

- Excess molecules are removed from the cluster, corresponding to fast evaporations. For example, if a cluster of 6 acid and 2 ammonia molecules is formed, 2 acids are removed to yield a $(H_2SO_4)_4 \cdot (NH_3)_2$ cluster that is included in the simulation (Figure 6.2).
- The cluster and the removed free molecules are returned into the simulation.
- For more complex molecular systems, namely those containing different types of acids and bases, there can be several possible evaporation chains and evaporation products. In ACDC, the evaporation route is in this case chosen according to the acid and base strengths, given in the cluster set input file. The removal of molecules is started from the weakest acids and bases, and continued until the cluster has reached a composition that is included in the system.

The above approach is chosen in order to deal with the finite size of the cluster set based on our best chemical understanding. It is reasonable to assume that highly unstable compositions evaporate instantaneously, and weaker acids and bases are expected to be more loosely bound to the clusters than stronger acids and bases. The terms corresponding to these processes are added automatically to the GDE. If the products that return to the system after removing the excess molecules are the same as the original colliding parties, as for e.g. $(H_2SO_4)_5 \cdot (NH_3)_4 + H_2SO_4 \longrightarrow (H_2SO_4)_6 \cdot (NH_3)_4 \longrightarrow (H_2SO_4)_5 \cdot (NH_3)_4 + H_2SO_4$, the terms are redundant and thus not included.

An example of the cluster formation free energy input file is given in listing 6.9. The first two lines give the reference pressure and temperature at which the free energies are calculated (note that the reference pressure is a parameter used in the quantum chemical calculations for the entropic contribution; it cancels out when calculating the evaporation rates and does not need to correspond to the actual ambient pressure). The following lines give the free energies of each cluster, with the cluster labels giving the composition of the cluster according to the molecule name labels given in the cluster set file (listing 6.8). For example, "3A2N" refers to the cluster consisting of 3 acid and 2 ammonia molecules. Note that the file can contain data also for clusters not included in the simulation; only the data corresponding to the cluster set file is used.

The free energy file can contain either the Gibbs free energies of formation $\Delta G$, or, as in listing 6.9, the corresponding enthalpies $\Delta H$ and entropies $\Delta S$, which give the Gibbs free energy as $\Delta G = \Delta H - T\Delta S$. The latter is beneficial for simulating clustering at different temperatures, as the temperature-dependence of the enthalpies and entropies of formation is normally negligible in atmospherically relevant temperature range.

```
101325
298.15

# Units:
# Pa and K for reference pressure and temperature, respectively (the
    first lines)
# kcal/mol for Gibbs free energy DeltaG and enthalpy DeltaH
# cal/(mol K) for entropy DeltaS

#         DeltaH                    DeltaS

2A        -17.8487481487            -33.418352

1A1N      -15.99849325487           -28.136935
2A1N      -44.9962620190            -71.024601
3A1N      -66.0568351924            -107.724444

2A2N      -64.4605106074            -104.450256
3A2N      -92.0891545072            -143.176618
4A2N      -115.1292995691           -183.339004

3A3N      -116.59903394959          -177.985175
4A3N      -145.1669199422           -222.332473
5A3N      -168.7911001831           -260.547804

4A4N      -164.3486235509           -251.025092
5A4N      -191.8609959824           -291.053716

4A5N      -186.4720560274           -296.507159
5A5N      -221.6506206689           -332.489463
```

Listing 6.9 ACDC input file for the formation free energies, here given as $\Delta H$ and $\Delta S$. The data are used to obtain the evaporation rate constants (Eq. (6.7)) through the Gibbs free energy $\Delta G = \Delta H - T\Delta S$.

### 6.4.1.2 Other ACDC Input Options

In addition to the input files containing information on cluster properties and/or rate constants, more input options can be given to the Perl code through keywords. These options include, for instance, the temperature and selection of sink terms (see also Section 6.4.3). To generate the GDE for the system given by the input files in Section 6.4.1.1 at 280 K and including a coagulation sink, the Perl code can be called as follows:

```
perl acdc_2020_11_06.pl \
--i input_AN_neutral_example.inp \
--e HS298.15K_example.txt \
--temperature 280 \
--use_cs \
--cs exp_loss \
--exp_loss_coefficient 0.001
```

Here, the input files for the cluster set and energies are given through keywords "--i" and "--e", respectively, followed by the other input options. "--use_cs" tells ACDC to use the coagulation sink (while other sinks such as the sink on chamber walls are not switched on), and the abbreviation "exp_loss" which comes after the keyword "--cs" gives the wanted functional form for the coagulation sink. "exp_loss" refers to a sink that decreases with cluster diameter according to a power law [8]. "exp_loss_coefficient" gives the reference sink rate constant corresponding to vapor molecules, and the sink rates of clusters are scaled according to the power law based on the reference rate (unit $s^{-1}$; if several vapors are available and $H_2SO_4$ is one of them, "exp_loss_coefficient" is assumed to give the $H_2SO_4$ sink rate).

Comprehensive information on the available keywords can be found in the ACDC Technical Manual. By default, the equations are printed out in Matlab unless keyword "--fortran" is given. Note that not all ambient parameters of the simulation are set by the Perl code, but can instead be varied after the generation of the equations. Such parameters include the vapor concentrations, which are included in the concentration array $C$, and are thus inserted in the solver call as initial or constant values without the need to re-generate the GDE. Furthermore, some parameters, for example the coagulation sink rate, can be re-set or varied within the Matlab or Fortran simulation by changing their absolute values while keeping the functional form constant (Section 6.4.3).

## 6.4.2 Running an ACDC Simulation

The Perl code generates the source code files that contain the birth–death equations for the given set of clusters, the rate constants, and other information on the system, such as cluster compositions. For atmospheric cluster chemistries, the rate constants within the equation set—mainly the evaporation rate constants—typically exhibit large variations. This makes the equations numerically stiff, meaning that very short integration steps may be required also for slowly varying solutions. Too long integration steps can give distorted results, and thus simplified numerical methods such as the Euler method are normally not feasible. Instead, the cluster equations are conveniently solved by ODE solvers suitable for stiff problems that employ efficient adaptive time-stepping.

First, the initial conditions, including vapor and cluster concentrations and possible vapor sources, and simulation time are set according to the simulated environment. Often the quantity of interest is the steady-state particle formation rate (Section 6.5), and thus the simulation is simply run until the concentrations do not change anymore. In this case, the vapor concentrations can be fixed to constant values instead of using vapor source terms, as one typically wishes to obtain the steady-state formation rate for given vapor concentrations. The simulation can also be run for a given time using specific vapor source terms, for instance, for simulating the evolution of a cluster population in an aerosol chamber experiment. Also time-dependent vapor concentration profiles or other parameters can be applied (Section 6.4.3).

We will now demonstrate conducting simulations with the example cluster set using the Matlab version due to its easy-to-use syntax. After generating the equations (Section 6.4.1), written out in "equations_acdc.m," and other related files, the driver that solves the equations can be called as shown in the code below. The driver applies solver ode15s, and gives various output parameters, including the simulation time vector and the time-dependent cluster concentrations and corresponding cluster name labels. If the user wishes to exclude some unnecessary output parameters, they can be replaced by the tilde character in Matlab, as demonstrated below. Information on all available in- and output parameters can be obtained by the command "help driver_acdc."

In the simplest form, the driver can be called by setting the initial concentration array "C0", corresponding to the simulated set of clusters, and the simulation time "t_sim" to yield the solved concentrations "C" as a function of time "T":

```
C0 = zeros(1,16);    % Set the concentrations of all 16 clusters to
    zero
C0(1) = 1e7;         % Set the H2SO4 concentration
C0(3) = 1e9;         % Set the NH3 concentration
t_sim = 10*60;       % Set the simulation time to e.g. 10 min.
[C,T] = driver_acdc(t_sim,C0);
```

The number of clusters and the indices corresponding to the $H_2SO_4$ and $NH_3$ monomers (i.e. single vapor molecules) can be found in "equations_acdc.m". Here, "C0" gives the initial concentrations, and no source terms or constant concentrations are used.

However, normally we want to set (at least) the vapor concentrations in a convenient manner, which can be done by an input text file:

```
constant 1A 1e7
constant 1N 1e9
```

Here, the $H_2SO_4$ and $NH_3$ monomer concentrations are set to constant values of $10^7$ and $10^9$ cm$^{-3}$, respectively. In addition, we may want to ensure that the simulation is run until a steady state is reached by setting a long simulation time and using the keyword "repeat" as shown below. Finally, we wish to obtain some additional output parameters. In the example below, the argument "ok" gives information on the convergence to the final steady state: a value of 1 corresponds to steady state, 0 to evolving concentrations, and $-1$ is a flag for negative concentrations, meaning that an error has occurred. "clust" is an array of cluster names corresponding to the columns in "C", and "J_out" is the time-dependent formation rate out of the system. We name the vapor concentration file "driver_input.txt", and call the driver as follows:

```
[C,T,ok,clust,~,~,~,J_out]=driver_acdc(1e5,'Sources_in','driver_input
    .txt','repeat');
```

This gives the simulation results for a single set of ambient conditions. If we want to obtain, for example, the formation rate as a function of $H_2SO_4$ concentration, the driver can be called in a loop as shown below. Here, "driver_input.txt" is re-written within the loop for each $[H_2SO_4]$ value.

```
clear variables

% Define vapor concentrations (cm^-3)
% NOTE: H2SO4 vapor concentration is here assumed to equal the
    monomer concentration - this may not
% hold for strong H2SO4-base dimer formation
Ca_vector=10.^(6:0.1:8);                    % Logarithmically evenly
    spaced H2SO4 values
Cb = 1e9;

J = zeros(size(Ca_vector));

% Loop over the H2SO4 concentrations
for nCa = 1:length(Ca_vector)

    Ca = Ca_vector(nCa);

    % Set the vapor concentrations
    fid=fopen('driver_input.txt','w');
    fprintf(fid,'constant 1A %e\n',Ca);     % Assuming H2SO4 monomers
```

```
    fprintf(fid,'constant 1N %e\n',Cb);
    fclose(fid);

    % Call the driver
    [C,T,ok,clust,~,~,~,J_out]=driver_acdc(1e5,'Sources_in','
    driver_input.txt','repeat');

    % Check if the steady state has been reached
    if ok ~= 1
        fprintf('Did not reach steady state for Ca = %.2e cm^-3, Cb
    = %.2e cm^-3\n',Ca,Cb)
        % Check for negative concentrations
        if ok == -1
            fprintf('Negative concentrations for Ca = %.2e cm^-3, Cb
    = %.2e cm^-3\n',Ca,Cb)
        end
        continue
    end

    % Save the formation rate
    J(nCa) = J_out(end);

end

figure(1)
loglog(Ca_vector,J)
hold on; set(gcf,'Color','white')
xlabel('[H_2SO_4] (cm^{-3})')
ylabel('{\itJ} (cm^{-3} s^{-1})')
```

**Listing 6.10** Calling ACDC in a loop in Matlab to obtain the formation rate as a function of $H_2SO_4$ concentration.

The ACDC repository contains templates for automatically looping over vapor concentrations at different conditions, including e.g. different temperatures, sinks, and ion production rates (see Section 6.4.4). The Fortran version of ACDC includes an equations file similar to that of the Matlab version, but no driver is generated by the Perl code. Instead, a readymade driver that uses the open-source solver VODE is available, and outputs the cluster concentrations and formation rate similarly to the Matlab version.

### 6.4.3 Code Features Useful for Studying Clustering Mechanisms

In addition to constructing and solving the cluster equations using standard formulae for rate constants, optional code features that enable, for instance, switching on or off specific dynamic processes or investigating cluster growth mechanisms, are useful for flexible modeling and analysis of the clustering processes. Such features, available in the ACDC model, are listed in Sections 6.4.3.1–6.4.3.3.

#### 6.4.3.1 Options for Processes and Rate Constants

In order to determine the significance of different dynamic processes, it is beneficial to include options for easily enabling or disabling them. The role of, for example, processes between two clusters (as opposed to cluster–monomer processes) for a given chemical system and conditions can be studied by ACDC through flags to include or omit these processes. Keyword "--disable_nonmonomers", given to the Perl code, disables all collisions and evaporations involving two clusters, and "--disable_nonmonomer_evaps" disables cluster fission (non-monomer evaporation) while omitting cluster–cluster collisions.

The sensitivities of the cluster concentrations and formation rates to the quantitative rate constants can be assessed by changing the values of all or individual rate constants. For instance, so-called sticking factors can be incorporated to the collision coefficients, corresponding to either imperfect attachment upon collisions due to e.g. cluster structure (factors < 1), or increased attachment probability compared to the default collision constants (hard-sphere collision rates for neutral–neutral collisions) due to attractive forces (factors > 1). Similarly, evaporation rates can be modified by making changes to the input formation free energies $\Delta G$. This is convenient for assessing the effects of uncertainties estimated for the $\Delta G$ data on the simulated concentrations and formation rates, as the response is typically non-linear. Modifications to the rate constants can be given to ACDC either as keywords to the Perl code (e.g. "--sticking_factor 2.0" to modify all collision coefficients between electrically neutral species), or in separate input files (e.g. "--sticking_factor_file_name file.txt" to modify specific collision coefficients, with the file listing the individual collisions parties and the corresponding sticking factors).

The effect of water on the rate coefficients can be taken into account by giving the relative humidity as an input parameter in the Perl call as e.g. "--rh 20", corresponding to a RH of 20%. All rate coefficients for all included processes are in this case calculated as effective rates over the hydrate distributions. The included hydrates are automatically deduced from the free energy input file (listing 6.9): for all cluster compositions listed in the cluster set file (listing 6.8), clusters that have additional water molecules, but otherwise the same composition, are detected in the free energy file and included in the hydrate distribution. For this, water molecules need to be called "W", and the name labels of hydrated clusters are thus e.g. "3A2N2W". The main challenge for including the hydration effects is, however, the availability of quantum chemical data: including water is cumbersome and computationally expensive, and thus not many comprehensive datasets of hydrated clusters have been computed.

Finally, pre-determined concentrations can be given for specific molecules or clusters instead of solving their differential equations. In ACDC, constant concentrations can be set for given species (Section 6.4.2), or time-dependent concentration profiles can be given as input to the Matlab driver. This is mainly relevant for vapor monomers, and is useful if, for example, a vapor profile is known from an experiment. Furthermore, constant concentrations can be set also with respect to other simulated species' concentrations. For example, a measured $H_2SO_4$ concentration may consist of free $H_2SO_4$ monomers and $H_2SO_4$ molecules clustered with one or more base molecules [17]. In this case, the $H_2SO_4$ monomer concentration in the simulation can be set to the measured concentration minus the sum of concentrations of clusters consisting of one $H_2SO_4$ and any number of base molecules.

### 6.4.3.2 Tracking the Cluster Growth Mechanisms

Cluster concentrations and formation rates do not give direct information on the clustering mechanisms. To understand the clustering dynamics, it should be determined (1) through which molecular growth steps the net cluster growth proceeds, (2) which clusters are the key steps on the main cluster formation pathways, and (3) if and how these depend on ambient conditions. In order to track the clustering pathways, ACDC can give the net fluxes $I_{i \to (i+j)}$ between all different cluster compositions:

$$I_{i \to (i+j)} = \beta_{i,j} C_i C_j - \gamma_{(i+j) \to i,j} C_{(i+j)}. \tag{6.8}$$

These fluxes are given by the output parameter "flux" of the Matlab driver. To extract and visualize the main growth pathways from single molecules to growth out of the simulation system, an automatic tracking program that utilizes the flux matrix is available in the ACDC Matlab package. Also fluxes out of the system to unstable compositions that are forced to evaporate back can be included in the tracking to examine the behavior of the cluster fluxes at the system boundaries.

### 6.4.3.3 Ensuring That the Cluster Set Is Sufficient

As discussed in Section 6.1.1, it must be ensured that clusters that are allowed to grow out of the simulation system can truly be expected to be stable. In practice, this means that also the largest clusters included in the system must be relatively stable; otherwise it is not realistic to assume that the clusters are suddenly stabilized as they cross the system boundary. If the largest clusters are unstable, the cluster set is too small and should be extended to sizes at which the clusters become stable.

The stabilities of the cluster compositions within the system are conveniently examined by presenting, for example, the ratios of collision and evaporation rate constants on a "matrix" similar to that of Figure 6.2. In ACDC, the rate constants are readily available in the files generated by the Perl code (for Matlab, in files "get_coll.m" and "get_evap.m"), and a Matlab program for drawing such matrices is available in the ACDC package (Section 6.4.4). If the collision rates of vapor molecules with the largest simulated clusters exceed the evaporation rates of the clusters, it is reasonable to assume that the subsequent clusters outside of the system are also stable. Note that the relative stabilities of the clusters, that is, the rate constant ratios, depend on the vapor concentrations and on the temperature. Therefore, the stabilities must be examined at the lowest simulated vapor concentrations and at the highest simulated temperature to ensure that the criteria are satisfied at all simulation conditions.

### 6.4.4 ACDC Applications

The ACDC model can be used in different ways for different purposes. For studying the details of cluster formation for given chemical systems, the Matlab version is an easy-to-use tool that does not require the user to have much programming experience. For implementing ACDC results in aerosol box or transport models, the Fortran version can be used to obtain particle formation rates over large sets of atmospheric conditions. Examples of ACDC applications are listed below, and the implementation of cluster formation modeling in aerosol dynamics frameworks is discussed more in Section 6.5.

In fundamental cluster formation studies, the topics of interest include the effects of various factors on the formation rate and the clustering mechanisms, such as the effects of different chemical compounds and dynamic processes, as well as the quantum chemical input data and other parameters related to the rate constants. For this, simulations with different input can easily be run and visualized with the Matlab version. The ACDC package includes readymade applications for

- plotting the free energies and rate constants for a given cluster set to examine the cluster stability (Section 6.4.3),

- running and plotting steady-state formation rates and cluster concentrations at different conditions, and
- plotting the cluster formation pathways molecule-by-molecule.

Figure 6.7 shows an example of the output of the Matlab templates for the formation rate and clustering pathways.

While the benefits of the Matlab tools include easy inspection of data for given input, Matlab is not efficient for running extremely large sets of simulations. For such tasks, the Fortran version can be applied. The Fortran version is suitable for

- generation of very large data tables, and
- explicitly combining time-dependent molecular clustering simulations with an aerosol dynamics model (Section 6.5.2.2).

Finally, there are a couple of remarks worth making regarding the computational efficiency. In general, the more different compositions the cluster set covers, that is, the more coupled differential equations there are, the heavier the simulation becomes regardless of the programming language used. Moreover, the selection of dynamic processes to be included in the model adds to the code complexity if it is done during the run time. In order to optimize the simulation, executable codes should not include redundant contents.

In ACDC, this is considered by generating the equations by the Perl code, which controls that only the processes and features that are needed for the given input case are included in the output Matlab or Fortran source code files. This means that if a specific dynamic process is not included, the related terms and rate constants are not printed out at all. Also, the equations do not include other unnecessary terms: for example, a process where an unstable cluster is brought back from outside of the system is not printed out if the collision parties and resulting evaporation products are the same (Section 6.4.1.1).

In principle, there is no limit for how many compositions can be included in the simulation. As atmospheric clusters typically stabilize at sizes consisting of at least a few to ~10 molecules, ACDC is primarily designed for relatively small cluster sets

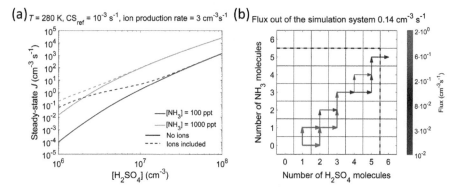

**Figure 6.7** Example of the output given by the ACDC Matlab package. Panel (a): Formation rate for the $H_2SO_4$–$NH_3$ system at different vapor concentrations. Panel (b): Clustering pathways without ions at $[H_2SO_4]=5 \times 10^6$ cm$^{-3}$ and $[NH_3]=100$ ppt, obtained by following the main net fluxes (Eq. (6.8)).

consisting of ~15–100 compositions. For substantially larger sets of clusters, the simulation becomes slower and eventually infeasible. It can be noted that the Fortran version includes an option for simulating very large clusters sets with simplified approaches for the rate constants (keyword "--loop"). The option is for studying the dynamics of large cluster populations rather than detailed clustering chemistry, and can be applied to simulate at least ~1000–2000 compositions, depending on the values of the rate constants (i.e. the system stiffness). In any case, for all simulation systems, it is always beneficial to exclude redundant, extremely unstable compositions. Such compositions can be detected by investigating the evaporation rates, and by ensuring that the simulation results do not change—even at high vapor concentrations and low temperatures—when the unstable clusters are excluded from the input set.

## 6.5 From Clustering to Particle Growth: Implementation of Initial Particle Formation in Aerosol Dynamics Models

While the molecular cluster dynamics resulting in a net formation of new particles are complex, aerosol dynamics models typically implement the formation process through a single parameter, namely the formation rate of particles of ~1–3 nm in diameter, commonly referred to as the nucleation rate. This is generally a robust approach, but involves a few assumptions that must hold in order to avoid inaccuracies in the implementation. The conventional approach and the assumptions are discussed below in Section 6.5.1.

Implementing the initial particle formation as an incoming flux of new particles in the smallest size bin of an aerosol dynamics model is a simplification that works well for many atmospheric conditions. However, the complete particle formation and growth process can also be modeled explicitly by a combination of molecular cluster and aerosol dynamics models, avoiding the uncertainties that may arise from the assumptions made in the conventional approach for some simulation systems and conditions. Such combination approaches, discussed in Section 6.5.2, can be applied in computationally less demanding applications, including box and trajectory models. For large-scale models, embedding an explicit molecular cluster model in the aerosol dynamics framework becomes too heavy, and the conventional formation rate approach must be used. These standard and explicit approaches are illustrated by Figure 6.8.

### 6.5.1 The Default Approach: Particle Formation Rate as an Input Parameter

The standard approach to include the formation of new nanoparticles from vapors in atmospheric aerosol modeling is to insert the given number of particles, obtained from the formation rate $J$ and the model time step $\Delta t$ as $\Delta C = J \times \Delta t$, at the lowest end of the aerosol size distribution (Chapters 1 and 7). The formation rate is obtained from theoretical or empirical parameterizations or data tables, and is normally given as a function of vapor concentrations, temperature and possibly also the ambient coagulation sink caused by the aerosol population.

In other words, the initial formation process is modeled or parameterized separately from the rest of the aerosol distribution, although the fundamental dynamic processes involved are similar to aerosol dynamics (Section 6.1.1). The fact that the formation rate is generally given as a function of ambient conditions implies that it is assumed to be

## (a) Standard approach

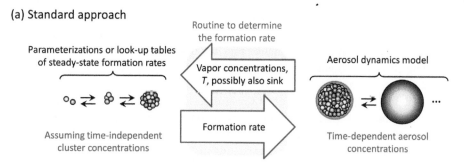

## (b) Explicit approach

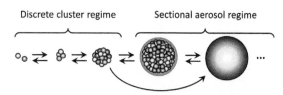

**Figure 6.8** Schematic presentation of approaches to implement initial particle formation in an aerosol dynamics model. The standard (a) and the explicit (b) approaches are discussed in Sections 6.5.1 and 6.5.2, respectively.

time-independent as no time dimension is included. This corresponds to assuming that the concentrations of small clusters are always in a steady state. The reasons for such implementation are that modeling the initial clustering is computationally complex – and impossible for large-scale models– and the assumed simplifications are normally justifiable. However, the simplifications may be a source of error at some conditions, and the reasoning behind them must thus be understood in order to assess their validity. The assumptions involved in the standard approach are listed and briefly discussed below [18, 19].

- The formation rate is in steady state, which is based on the assumption that the time scales of the clustering dynamics are substantially shorter than the time scales in which the ambient atmospheric conditions vary. That is, the cluster concentrations reach the steady state instantaneously at every modeled time step. Conditions at which this assumption may become uncertain include remote, clean environments where the clustering time scales are longer due to low vapor concentrations and low scavenging sink.
- The initial molecular clusters and the larger aerosol particles interact principally through the formation and growth of the clusters to the size range covered by the aerosol model. Coagulation between the cluster and aerosol size regimes is either not considered, or it is included in a simplified manner in the input formation rate through the effect of scavenging of the small clusters on the formation rate (e.g. Sections 6.4.1.2 and 6.5.2.2).

  The scavenging effect is typically non-negligible and should thus be included in the input formation rate; input data that normally do not give the formation rate as a function of scavenging sink include classical nucleation models and experimental parameterizations. The effect of coagulation of small clusters on aerosol growth is inherently omitted as only the formation rate, not the cluster distribution, is considered. The effect is normally expected to be minor, but it may be non-negligible in polluted environments.

- The formation rate, normally given at the size of ~1–2 nm, can also be scaled to a larger size, typically ~3–20 nm, before insertion in the aerosol model by considering the assumed nanoparticle growth and scavenging rates [8] [Chapter 7.5]. The scaling reduces the computational burden of modeling the very smallest sizes, but similarly to the initial formation scheme, applies the steady-state assumption. The assumption becomes worse as the nanoparticle size increases since the changes that occur at the very initial sizes take time to propagate along the size axis. Therefore, the larger the scaled size is, the more uncertain the scaling becomes.

The above simplifications cannot be avoided in regional and global models. The factors affecting their validity should, however, be understood in order to assess if and under which conditions the initial formation scheme involves potential uncertainties.

The steps to incorporate the steady-state formation rates in an aerosol dynamics framework are summarized in Algorithm 6.1. The formation rate data should cover atmospherically relevant conditions at which particle formation from the given species is expected to occur. The parameters that determine the formation rate must include at least the concentrations of the simulated vapor species, the temperature and the scavenging sink rate, and depending on the simulated cluster set, possibly also the ion production rate and/or the relative humidity.

---

**Algorithm 6.1:** The standard approach to incorporate particle formation in an atmospheric model in a nutshell.

---

1 **Result:** Nucleation rates in an aerosol dynamics model
2 – *In the particle formation model* –
3 Select the ranges of atmospherically relevant conditions for which formation
    rates will be generated (Sections 6.4.2 and 6.5.1);
4 **for** *each combination of parameter values* **do**
5   |   Solve GDE until steady state reached, write out formation rate;
6 **end**
7 – *In the aerosol dynamics model* –
8 Incorporate the formation rate data tables in the "nucleation" routine of the
    aerosol dynamics model:
9 **Function** formation rate (*parameters defining conditions*):
10   |   Interpolate the formation rate for given conditions from the data table[a];
11   |   Multiply the formation rate by the model time step to yield the number of
       new particles;
12   |   **return** Number of new particles to add to the corresponding size bin of the
       aerosol dynamics model[b] at each time step;

---

[a] Also parameterizations may be derived based on the data; computational-burden-wise the choice between look-up tables or parameterizations is not likely to be critical, but robust parameterizations may be challenging to derive.
[b] In practice the smallest size bin, which needs to cover the size of the new particles modeled by the particle formation scheme.

## 6.5.2 The Dynamic Approach: Combination of Molecular Cluster and Aerosol GDEs

For computationally cheaper model frameworks, such as aerosol box models applied for local-scale or laboratory investigations, a more explicit treatment of particle formation is possible. The simplifications discussed in Section 6.5.1 can be avoided by combining the discrete and the continuous GDE models for molecular clusters and aerosol particles, respectively, to simulate the time-dependent particle formation dynamics starting from single vapor molecules. Such combination models have not been widely used in atmospheric applications thus far, but they are briefly introduced here for completeness.

### 6.5.2.1 Discrete-sectional Models

The most explicit approach to model the whole cluster and particle size range are so-called discrete-sectional models, which apply the discrete, molecule-by-molecule GDE for the smallest particle sizes and the continuous GDE for larger particles [18, 20]. The benefits of this approach are that no steady-state assumptions are applied, and the interactions between small clusters and larger particles are treated explicitly, including the coagulation of the clusters onto the particles. This enables accurate simulation of particle population dynamics at all ambient conditions.

The drawback of discrete-sectional models is that they apply very simplified chemistry, typically a one-component substance. Including arbitrary chemical species is complex for both the "cluster" and the "aerosol" size regimes (Section 6.3; Chapters 2-4); specifically, including a multi-component cluster regime would in practice require an ACDC-type of equation generator and explicit coupling of the cluster–aerosol coagulation terms to the aerosol regime. The restrictions in the description of chemistry limit the applicability of such models on ambient atmospheric particle formation and growth processes, in which the role of different species typically varies with nanoparticle size; that is, the first clustering steps, the initial sub-10 nm growth and the growth of larger particles are likely to be driven by different compounds. However, rather than detailed chemical studies, the purpose of discrete-sectional models is to accurately simulate the size distribution dynamics, thus helping to understand and interpret nanoparticle size distribution measurements.

### 6.5.2.2 Combination of Separate Molecular Cluster and Aerosol Models

It is also possible to combine cluster and aerosol size distribution dynamics by coupling separate models to each other. Such approach has not been widely used, but with currently available tools to simulate the cluster population dynamics, it is a feasible solution for avoiding the steady-state approximation commonly applied to the initial cluster formation. The ACDC model has been coupled to an aerosol dynamics model by replacing the nucleation subroutine that determines the steady-state formation rate with a call to an ACDC routine [21], which keeps track of the time-dependent cluster concentrations during the simulation as summarized in Algorithm 6.2.

The workflow of Algorithm 6.2 corresponds to panel (a) in Figure 6.8, but with the time-independent formation rates replaced by an in-situ cluster dynamics simulation. In this way, the initial formation is modeled dynamically as a molecular cluster extension to the lower end of the aerosol size distribution.

---

**Algorithm 6.2:** A non-steady-state approach to incorporate particle formation in an atmospheric model.

---

1 **Result:** Non-steady-state nucleation rates in an aerosol dynamics model
2   – *In the aerosol dynamics model* –
3   At every model time step, the model calls the ACDC routine:
4 **Function** `formation rate` (*parameters defining conditions[a]*) **:**
5   |   – *In the particle formation model (ACDC)* –
6   |   ACDC simulates the time evolution of the cluster concentrations for the model time step at the given conditions; the initial concentrations are the values saved at the end of the previous time step;
7   |   The formation rate is obtained as the number of stable molecular clusters that grow out of the ACDC system during the time step;
8   |   **return** Number of new particles to add to the corresponding size bin of the aerosol dynamics model at each time step;

---

[a] These include current values for vapor concentrations, temperature, scavenging sink caused by particles simulated by the aerosol model, and possibly also ion production rate.

As a final remark, it can be noted that the division into nucleation and particle growth is largely artificial—in essence the same physics apply. If there were no limitations at all in computational resources and thermodynamic data, particle formation from individual molecules and molecular clusters to larger aerosols could be modeled with a single model for arbitrary chemistries. However, due to the different roles of different chemical species at different stages of the particle formation process, and the behavior of the rate constants of the general dynamic equation as a function of particle size, modeling the initial formation and further growth can be split into molecular cluster and aerosol dynamics simulations, respectively.

As new sets of quantum chemical data for different molecular cluster chemistries are becoming available, particle formation schemes in atmospheric models can be updated with more versatile chemical mechanisms. At the same time, adequate treatment of cluster and particle size distribution dynamics and cluster–aerosol feedbacks is needed for interpreting and predicting particle formation phenomena and their effects in diverse environments, from very clean to heavily polluted conditions. Detailed dynamics modeling and high-level input data are needed not only for fundamental understanding, but also for deriving and evaluating simplified approaches in order to reduce code complexity and computational time.

## Bibliography

[1] Tinja Olenius, Lukas Pichelstorfer, Dominik Stolzenburg, Paul M. Winkler, Kari E. J. Lehtinen, and Ilona Riipinen. Robust metric for quantifying the importance of stochastic effects on nanoparticle growth. *Scientific Reports*, 8(1):14160.

**[2]** C. F. Clement and M. H. Wood. Equations for the growth of a distribution of small physical objects. *Proceedings of the Royal Society of London. A. Mathematical and Physical Sciences*, 368(1735):521–546.

**[3]** Tinja Olenius, Oona Kupiainen-Määttä, Kari E. J. Lehtinen, and Hanna Vehkamäki. Extrapolating particle concentration along the size axis in the nanometer size range requires discrete rate equations. *Journal of Aerosol Science*, 90:1–13, 2015.

**[4]** Anna Shcherbacheva, Tracey Balehowsky, Jakub Kubeka, Tinja Olenius, Tapio Helin, Heikki Haario et al. Identification of molecular cluster evaporation rates, cluster formation enthalpies and entropies by Monte Carlo method. *Atmospheric Chemistry and Physics*, 20(24):15867–15906.

**[5]** Roope Halonen, Evgeni Zapadinsky, Theo Kurtén, Hanna Vehkamäki, and Bernhard Reischl. Rate enhancement in collisions of sulfuric acid molecules due to long-range intermolecular forces. *Atmospheric Chemistry and Physics*, 19(21):13355–13366.

**[6]** J. Leppä, S. Gagné, L. Laakso, H. E. Manninen, K. E. J. Lehtinen, M. Kulmala et al. Using measurements of the aerosol charging state in determination of the particle growth rate and the proportion of ion-induced nucleation. *Atmospheric Chemistry and Physics*, 13(1):463–486.

**[7]** Hanna Vehkamäki and Ilona Riipinen. Thermodynamics and kinetics of atmospheric aerosol particle formation and growth. *Chemical Society Reviews*, 41(15):5160–5173.

**[8]** Kari E. J. Lehtinen, Miikka Dal Maso, Markku Kulmala, and Veli-Matti Kerminen. Estimating nucleation rates from apparent particle formation rates and vice versa: Revised formulation of the Kerminen-Kulmala equation. *Journal of Aerosol Science*, 38(9):988–994, 2007.

**[9]** A. Kürten, C. Williamson, J. Almeida, J. Kirkby, and J. Curtius. On the derivation of particle nucleation rates from experimental formation rates. *Atmospheric Chemistry and Physics*, 15(8):4063–4075.

**[10]** Jonas Elm, Jakub Kubeka, Vitus Besel, Matias J. Jääskeläinen, Roope Halonen, Theo Kurtén et al. Modeling the formation and growth of atmospheric molecular clusters: A review. *Journal of Aerosol Science*, 149:105621, 2020.

**[11]** Lauri Partanen, Hanna Vehkamäki, Klavs Hansen, Jonas Elm, Henning Henschel, Theo Kurtén et al. Effect of Conformers on Free Energies of Atmospheric Complexes. *Journal of Physical Chemistry*, 120(43):8613–8624.

**[12]** Henning Henschel, Theo Kurtén, and Hanna Vehkamäki. Computational Study on the Effect of Hydration on New Particle Formation in the Sulfuric Acid/Ammonia and Sulfuric Acid/Dimethylamine Systems. *Journal of Physical Chemistry*, 120(11):1886–1896, 2016.

**[13]** Tinja Olenius. Atmospheric Cluster Dynamics Code: Software repository. https://github.com/tolenius/ACDC, January 2021.

**[14]** Coty N. Jen, Peter H. McMurry, and David R. Hanson. Stabilization of sulfuric acid dimers by ammonia, methylamine, dimethylamine, and trimethylamine. *Journal of Geophysical Research: Atmospheres*, 119(12):7502–7514, 2014. _eprint: https://agupubs.onlinelibrary.wiley.com/doi/pdf/10.1002/2014JD021592.

**[15]** Fangqun Yu, Alexey B. Nadykto, Jason Herb, Gan Luo, Kirill M. Nazarenko, and Lyudmila A. Uvarova. $H_2SO_4$-$H_2O$-$NH_3$ ternary ion-mediated nucleation (TIMN): kinetic-based model and comparison with CLOUD measurements. *Atmospheric Chemistry and Physics*, 18(23):17451–17474.

**[16]** Peter N. Brown, George D. Byrne, and Alan C. Hindmarsh. VODE: A Variable-Coefficient ODE Solver. *SIAM Journal on Scientific and Statistical Computing*, 10(5):1038–1051.

**[17]** L. Rondo, S. Ehrhart, A. Kürten, A. Adamov, F. Bianchi, M. Breitenlechner et al. Effect of dimethylamine on the gas phase sulfuric acid concentration measured by Chemical Ionization Mass Spectrometry. *Journal of Geophysical Research: Atmospheres*, 121(6):3036–3049, 2016. _eprint: https://agupubs.onlinelibrary.wiley.com/doi/pdf/10.1002/2015JD023868.

**[18]** Fred Gelbard and John H Seinfeld. The general dynamic equation for aerosols. Theory and application to aerosol formation and growth. *Journal of Colloid and Interface Science*, 68(2):363–382, 1979.

**[19]** Tinja Olenius and Ilona Riipinen. Molecular-resolution simulations of new particle formation: Evaluation of common assumptions made in describing nucleation in aerosol dynamics models. *Aerosol Science and Technology*, 51(4):397–408.

**[20]** Chenxi Li and Runlong Cai. Tutorial: The discrete-sectional method to simulate an evolving aerosol. *Journal of Aerosol Science*, 150:105615, 2020.

**[21]** Pontus Roldin, Mikael Ehn, Theo Kurtén, Tinja Olenius, Matti P. Rissanen, Nina Sarnela et al. The role of highly oxygenated organic molecules in the Boreal aerosol-cloud-climate system. *Nature Communications*, 10(1):4370.

# 7

## Box Models

Box models represent aerosol physical and chemical processes occurring in an air parcel using a zero dimensional framework. An air parcel is a hypothetical unit of mass of air which is fully elastic and does not exchange mass with its surroundings. The air parcel can be considered to represent a moving or a stationary point in space. A box model can bring together all the aerosol processes described in previous chapters and simulate the evolution of an aerosol or hydrometeor population. Aerosol box models are also used to simulate laboratory chamber experiments or in situ observations [e.g. [1–3]]. They can also be implemented in three-dimensional frameworks where they solve aerosol microphysics in each grid box [4].

In this chapter we will construct a box model as an example on how different aerosol processes can be brought together and how they are solved in the box model framework. In Chapter 2, absorptive uptake of semivolatile organic species to a population of different sized particles was presented using a sectional framework. The code in Chapter 2 can in itself be already considered a box model and in this chapter we expand that model to include more species, microphysical processes, and techniques for treating the evolution of an aerosol population in a sectional framework. As shown in Chapter 2, microphysical processes are typically described using ordinary differential equations (ODE) which can be solved using an ODE solver. Here we use the same ODE solver as in Chapter 2 and add ODEs for new particle formation by nucleation. We also modify the model that instead of using it for simulating a stationary air parcel it can also be used in simulating an ascending cloud parcel.

Additions to the model framework require extending the amount of variables to be solved by the ODE solver (see Figure 2.7). Additional variables include number concentrations, concentrations of water, concentrations of sulfate in each bin as well as in the gas phase. In addition, the box model described in this chapter will solve differential equations for temperature and pressure. Thus the array shown in Figure 2.7 is modified to that shown in Figure 7.1. We include aerosol number concentration of each bin in the array so that the first cells of the array are occupied by the number concentrations. For cloud parcel modelling we will also include water volume concentrations in each bin which follow the number concentrations in the array. Following water concentrations, we add molar concentrations of the core particles which in this case are assumed to be ammonium sulfate. Next in the array are the condensing organic species which were already introduced in Figure 2.7. Gas phase concentrations for individual condensing gases occupy array elements after the particle phase species. The last two elements in

*Introduction to Aerosol Modelling: From Theory to Code.*
First Edition. Edited by David Topping and Michael Bane.
© 2022 John Wiley & Sons Ltd. Published 2022 by John Wiley & Sons Ltd.

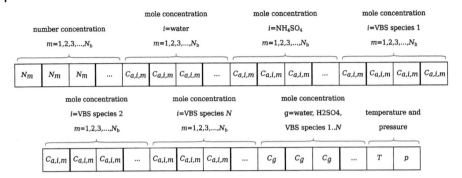

**Figure 7.1** Schematic illustration of the numerical array to include solved variables in ODE solver.

the array are temperature $T$ and pressure $p$ to allow for simulating their changes in expanding air parcels.

## 7.1 box_model.py

The box model used in this chapter is built around the *box_model.py* main routine and its parameters are set in the file *parameters.py*. The box model has currently been setup so that by default it can simulate four cases:

1. Dry aerosol with semivolatile organic compounds partitioning between the gas and the aerosol phase. This setup is selected by setting `model_setup=1` in the file *parameters.py*.
2. Dry aerosol with semivolatile organic species and sulfuric acid partitioning between the gas and the aerosol phase. Sulfuric acid is also involved in nucleation of new particles. This setup is selected by setting `model_setup=2`.
3. Cloud parcel simulation with wet aerosol. The air parcel ascends with a constant updraft velocity $w$ and with water partitioning between the gas and the liquid phase. This setup is selected by setting `model_setup=3`.
4. Cloud parcel simulation with wet aerosol. The air parcel ascends with a constant updraft velocity $w$ and with water and semivolatile organic species partitioning between the gas and the liquid phase. This setup is selected by setting `model_setup=4`.

Note, however, that the model setups are not restrict to these and can be freely modified.

The first setup is identical to the simulation of the absorptive condensation case shown in Section 2.4.1. Setup 2 adds more complexity to the model and is useful in demonstrating how remapping of the size bins affect the simulated evolution of aerosols. Setups 3 and 4 are used in Section 7.4.1 where we present the modifications to the model which allow for modelling cloud formation.

Routine *box_model.py* first initializes the size distribution by calling a function `log_normal_distribution` in *distribution.py* (see Listing 7.1). This function calculates a log normally distributed aerosol number concentrations using volume-ratio sectional size bins as shown in Section 1.3.2. The size distribution is calculated for given geometric standard deviation `sigma_g`, mean diameter `mu`, and total number of particles `N_L` in a mode for which the routine provides number concentrations $n_{sect,i}$, diameter $d_{sect,i}$, volume $v_{sect,i}$, and the lower and higher volume boundaries $v_{sect,i,lo}$ and $v_{sect,i,hi}$ of each size bin $i$. Note that in the example cases presented in this chapter, the log normal size distribution is the size distribution of the dry aerosol.

```python
import numpy as np

def log_normal_distribution(mu, sigma_g, N_T):

    # Define lower and upper limit of the size distribution [m]
    d_1 = 10.0e-9
    d_Nb = 1.0e-6

    # Volume ratio between bins, Equation (1.17)
    v_rat = (d_Nb/d_1)**(3.0/(Nb-1.0))
    # Diameter ratio between bins,, Equation (1.18)
    d_rat = v_rat**(1.0/3.0)

    # diameters of each size bin [m]
    d=np.zeros((Nb), dtype=float)
    d[0] = d_1

    for step in range(Nb):
        if step > 0:
            d[step]=d[step-1]*d_rat

    # volumes of each size bin [m3], Equation (1.3)
    v = 4.0 / 3.0 * np.pi *(d / 2.0)**3

    # lower limit of each size bin [m], Equation (1.22)
    v_lo = 2 * v / (1.0 + v_rat)

    # upper limit of each size bin [m], Equation (1.21)
    v_hi = v_rat * v_lo

    # width of the size bin [m], Equation (1.23)
    delta_d=d*2**(1.0/3.0)*(v_rat**(1.0/3.0) - 1)/\
        (1 + v_rat)**(1.0/3.0)

    # log normal distribution p(x), Equation (1.16)
    n = N_T*delta_d/(d*np.sqrt(2.0*np.pi)*np.log(sigma_g))*\
        np.exp(-(np.log(d)-np.log(mu))**2/(2.0*np.log(sigma_g)**2))

    return n, d, v, v_lo, v_hi
```

**Listing 7.1** Determining a log-normal sectional size distribution in routine *distribution.py*.

Once the size distribution has been defined, the composition distribution is calculated. Concentrations of individual species are first stored in a two-dimensional array `c` in which each row represents one compound and the columns represent size bins. The first row `c[0, :]` is allocated for number concentrations. The dry aerosol is assumed to be initially composed of ammonium sulfate which is denoted as `SO4`. In all the model setups presented in this chapter, all semivolatile organic species are assumed to be in the gas phase in the beginning of the simulation and thus their concentrations in the aerosol phase are set to zero.

```
# initialize number size distribution
# using the module log_normal_distribution
n, d, v, v_lo, v_hi = log_normal_distribution(mu, sigma_g, N_T)

# initialize the vector aerosol composition size distribution
c = np.zeros((2+num_vbs_species,n.shape[0]))

# concentration of sulfate [mol/m3]
c[1,:] = n * np.pi/6.0*d**3 * rho['SO4'] / M['SO4']
```

Listing 7.2 Initializing the number, volume, and the composition size distribution for dry aerosol (*box_model.py*).

For cases with dry aerosol (`model_setup = 1 and 2`) we use the gas phase concentration values from Section 2.4.1 for semivolatile organics. In model setups 1 and 2, initial water and sulfuric acid / sulfate concentrations are zero.

```
# set the gas phase concentrations for dry aerosol cases
if model_setup <= 2:

    C_gases=[
        0.0,                                    # water
        0.0,                                    # sulfuric acid
        abundance['VBS0']/M['VBS0']*1.e-09,     # VBS species
        abundance['VBS1']/M['VBS1']*1.e-09,
        abundance['VBS2']/M['VBS2']*1.e-09,
        abundance['VBS3']/M['VBS3']*1.e-09,
        abundance['VBS4']/M['VBS4']*1.e-09,
        abundance['VBS5']/M['VBS5']*1.e-09,
        abundance['VBS6']/M['VBS6']*1.e-09,
        abundance['VBS7']/M['VBS7']*1.e-09,
        abundance['VBS8']/M['VBS8']*1.e-09,
        abundance['VBS9']/M['VBS9']*1.e-09,
    ]
```

Listing 7.3 Initializing the water composition size distribution to zero in each bin and initializing the gas phase concentrations for cases `model_setup=1 and 2` (*box_model.py*).

Once the initial conditions have been set, the non-equilibrium partitioning of semivolatile species between the gas and the particle phase is solved using the absorptive update approach. The initial values for number concentrations, molar concentrations both in particle and gas phase, temperature, and pressure are assigned to `array` (see Figure 7.1).

```
# Determine the initial values for ODEs
array=n.tolist()                         # number concentration
array.extend(c[0,:].tolist())            # water
array.extend(c[1,:].tolist())            # sulfate
for i in range(2,2+num_vbs_species):     # condensing organics
    array.extend(c[i,:].tolist())
array.extend(C_gases)                    # gas phase species
array.extend([T0, p0])                   # temperature and pressure
```

Listing 7.4 Assigning initial values of ODEs to `array` (*box_model.py*).

The array provides the initial conditions for which ODEs are solved over the integration time.

```
# set the boundary conditions for the time step
t = np.linspace(1, integration_time, num=integration_time)

solution = odeint(dy_dt, array, t,
                  rtol=rtol, atol=atol,
                  tcrit=None, mxstep=5000)
```

**Listing 7.5** Solving ODEs dy_dt with initial values in array (*box_model.py*).

Differential equations for aerosol concentrations, temperature, and pressure are now defined in Function dy_dt in the *differential_equations.py* file. In order to make the function more concise, functions which calculate the Kelvin effect, diffusion coefficient, and hygroscopicity (kappa) are provided in a separate Python file *auxiliary_functions.py*.

The function dy_dt first initializes variables and retrieves variables from vector y to a more readable form using the initialize_variables function which is also provided in the *auxiliary_functions.py* file.

```
R_dry, R_wet, T, N_m, mol_tot, k_i_m, C_g, X, dydt, Vs, Vw, kappa\
    = initialize_variables(y)
```

**Listing 7.6** Converting array to separate variables (*differential_equations.py*).

After this, differential equations for condensation/evaporation of semivolatile species are implemented.

```
for species in Condensing_species:

    # Calculate the Kelvin effect
    K = Kelvin_effect(R_wet, T, species)

    # find the indices of each species in Array y
    # particle phase
    bins = range(Nb*index[species],Nb*index[species]+Nb)
    # gas phase
    gas = Nb+len(bins)*len(index)+index[species]-1

    # mole fraction of an individual species in a bin
    X[species] = y[bins]/np.maximum(mol_tot, 1.e-30)

    # gas phase concentration of a species
    C_g[species]=y[gas]

    # gas phase concentration above the droplet surface,
    # see Eq (Equation 2.26)
    C_surf = K * X[species]*C_star[species](T)

    # Calculate the diffusion coefficient
    # for each individual species
    D_eff = diffusion_coefficient(N_m, R_wet, T, S_w, species)

    # condensation equation for individual particle species,
    # see Eq (2.43)
    dydt[bins] = N_m*4.0*np.pi*R_wet*D_eff * (C_g[species]-C_surf)

    # condensation equation for gases
    dydt[gas] = - sum(dydt[bins])
```

**Listing 7.7** Condensation equations for semivolatile species (*differential_equations.py*).

This model setup solves absorptive uptake of semivolatile species and can be chosen by setting `model_setup=1` in *parameters.py*. The initial conditions such as aerosol and gas phase concentrations, size distribution parameters, integration time, and ambient conditions are also set in *parameters.py*. The schematic of *box_model.py* is shown in Figure 7.2. Other model setups will be presented later in this chapter.

## 7.2 Remapping Size Distribution When Using the Sectional Method

In Section 2.4, the full-moving method [5] was used for describing the droplet size for mono-disperse and poly-disperse aerosol populations. In the full-moving method, particles are allowed to grow and shrink freely without remapping aerosol mass or number between the size bins. When simulating condensation and evaporation in zero dimensional frameworks, the full-moving sectional method has high accuracy and is the most suitable method for the task. However, when for example coagulation or formation of new particles by nucleation is added to the modeling system the full-moving method may not be the optimal for describing the evolution of particle sizes and compositions. When using the full-moving method, particles can grow significantly due to microphysical processes. Condensation is especially effective in growing the smallest particles since they have the highest surface to volume ratio. If such growth has happened, once nucleation initiates, newly formed particles will be merged in the smallest size bin

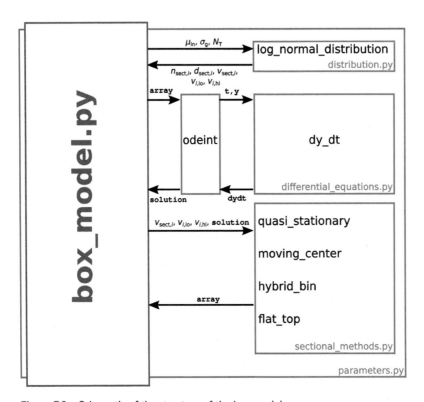

**Figure 7.2** Schematic of the structure of the box model.

which may have grown to a much larger in size than the nucleating particles. Since nucleated particles will have to be added to the smallest available size bin, they are added to a too large size in the model size distribution. In addition, when simulating aerosol in three dimensional atmospheric models, the advection of aerosol populations from one grid box to another requires restricting the size of the particles in each size section to reside within a certain size range. This enables that the advected particles will have a size bin of a correct size in the grid box they are transferred to. There are several ways to using such fixed size sections to better accommodate solving coagulation and nucleation as well as simulating the advection of aerosol in 3D. Here we present three methods: quasi-stationary sectional, moving center, and hybrid-bin sectional methods (with two different approaches).

### 7.2.1 Quasi-Stationary Sectional Method

One option for remapping the size distribution is the quasi-stationary sectional method where aerosol particles have fixed sizes. If particles in a size bin grow or shrink, their number and volume concentrations are remapped between two adjacent size bins to this fixed size. The basic principle of the remapping algorithm is illustrated in Figure 7.3.

In practice, the evolution of the size distribution due to microphysical processing is calculated for a given time step after which the aerosol particles are remapped to the original fixed size grid. The quasi-stationary method is especially well suited for new particle formation and emissions because the method guarantees that pre-existing sizes for newly formed particles or primary emissions exist. This method is also convenient when simulating aerosol transport in three-dimensional models since the transferred particles have the same sizes as those already in the grid cell they are transferred to.

In the quasi-stationary method, number and volume concentrations are conserved. If a size bin $j$ has grown to a size between two adjacent size bins $k$ and $l$, the number concentration is divided to $\Delta n_{\text{sect},k}$ and $\Delta n_{\text{sect},l}$ according to the equation

$$n_{\text{sect},j} = \Delta n_{\text{sect},k} + \Delta n_{\text{sect},l}. \tag{7.1}$$

From this it follows that the volume $v_{\text{sect},j}$ is divided between the same size bins according to equation

$$n_{\text{sect},j} v_{\text{sect},j} = \Delta n_{\text{sect},k} v_{\text{sect},k} + \Delta n_{\text{sect},l} v_{\text{sect},l}. \tag{7.2}$$

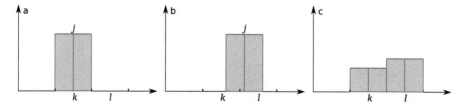

**Figure 7.3** Illustration of how a growing or a shrinking particle size bin $j$ is remapped using the quasi-stationary method. Panel (a) shows the initial state for particles $j$ in size bin $k$, (b) particles $j$ have grown to a size between two size bins $k$ and $l$, (c) particles $j$ remapped to the fixed sizes of $k$ and $l$.

From these two equations, we obtain the change in the number concentration in size bin $l$

$$\Delta n_{\text{sect},l} = \frac{v_{\text{sect},l} - v_{\text{sect},j}}{v_{\text{sect},l} - v_{\text{sect},k}} n_{\text{sect},j} = x_l n_{\text{sect},j} \tag{7.3}$$

and in size bin $k$

$$\Delta n_{\text{sect},k} = (1 - x_l) n_{\text{sect},j}. \tag{7.4}$$

In turn, the volume concentrations $V_{\text{sect},i,j}$ of individual compounds $i$ in bin $j$ are divided in size bins $k$ and $l$ following

$$\Delta V_{\text{sect},i,l} = x_l V_{\text{sect},i,j} \tag{7.5}$$

and

$$\Delta V_{\text{sect},i,k} = (1 - x_l) V_{\text{sect},i,j}. \tag{7.6}$$

Although the quasi-stationary method is widely used and is suitable for particular aerosol processes, the drawback is that it exhibits numerical diffusion and tends to make the size distribution wider.

When modelling only kinetic absorptive partitioning with the full-moving method in Section 2.4.1, the ODE solver solved the time evolution gas-to-particle partitioning with one call. When using the quasi-stationary method (and the following remapping methods) the size distribution needs to be remapped throughout the simulation. Here we do the remapping every one second and thus the time integration in *box_model.py* is split into 1 s intervals.

```
# split the time interval for remapping schemes
for k in range(integration_time):

    # set the boundary conditions for a 1 sec time step
    t = [float(k), float(k+1)]

    # solve ODE's over one time step
    solution[range(k-1,k+1),:] = odeint(dy_dt, array, t,
                        rtol=rtol, atol=atol,
                        tcrit=None, mxstep=5000)
```

Listing 7.8 Looping the droplet growth one second at a time. (*box_model.py*).

The algorithm `quasi_stationary.py` uses the volumes of the center of each bin as well as the array set up for the ODEs as inputs for remapping the aerosol size distribution. First, the variables are initialized, and retrieved from array y to a more readable form. Molar concentrations are converted to volume concentrations in order to use the remapping methods.

```
def quasi_stationary(v, y):

    # initialize variables
    delta_n, delta_V, n, V, v_j, bins = initialize_remapping(y)
```

Listing 7.9 Initializing and retrieving the variables from array y. (Function `quasi_stationary` in *sectional_methods.py*).

Initialization and retrieving variables is done in function *initialize_remapping.py*. Volume concentrations are also used in the rest of the remapping methods of this chapter, so the initialization routine is a common routine for all these methods.

```python
def initialize_remapping(y):

    # initialize variables
    v_j = np.zeros(Nb)          # Volumes of grown/evaporated bins
    delta_n = np.zeros(Nb)      # fraction of remapped number
                                # concentration in size bins
    delta_V = dict()            # fraction of remapped volume
                                # concentrations in size bins
    for i in All_species:
        delta_V[i] = np.zeros(Nb)
    bins = dict()               # indices of each aerosol species
                                # in array y

    V = dict()                  # volume concentrations

    # number concentration in each bin
    n = y[range(Nb)]

    # Calculate the total volume of each bin
    for i in All_species:

        # find the indices of each species i
        bins[i] = range(Nb*index[i],Nb*index[i]+Nb)

        # convert molar concentration of individual i
        # to volume concentrations [m3/m3]
        V[i] = y[bins[i]]*M[i]/rho[i]

        # volume of individual particles in grown/evaporated bins
        v_j+=V[i]/np.maximum(n, N_low_limit)

    return delta_n, delta_V, n, V, v_j, bins
```

**Listing 7.10** Initializing the variables and converting the molar concentrations to volume concentration. (*sectional_methods.py*).

After the initialization, the routine loops over the size bins. First, the bins $k$ and $l$ between which the grown/evaporated bin $j$ lies (see Figure 7.3) are identified. After this, the number concentrations of each size bin and volume concentrations of each compound in each bin are remapped using Equations (7.4)–(7.6). If the smallest size bin has shrunk or the largest size bin has grown, they are not remapped since they cannot be remapped between two bins which is a requirement for this scheme

```python
for j in range(Nb):

    # find the index of bins between which the size
    # of each bin has evolved: k an l in Equation (7.1)
    l = np.argmax(v_j[j] < v)
    k = l - 1

    # if the index was found and the bin is not empty
    if l > 0 and n[j] > N_low_limit:
```

```
        # calculate the fraction of n_j remapped to n_l (Equation
7.3)
        x_l = (v_j[j] - v[k])/(v[l] - v[k])

        x_k = 1.0 - x_l

        # fraction of n_j to be remapped to n_l (Eq (7.3))
        delta_n[l] += x_l*n[j]
        # fraction of n_j to be remapped to n_k (Eq (7.4))
        delta_n[k] += x_k*n[j]

        # loop over individual species
        for i in All_species:
            # fraction of V_i,j to be remapped to V_i,k (Eq (7.6))
            delta_V[i][l] += x_l*V[i][j]*v[l]/v_j[j]
            # fraction of V_i,j to be remapped to V_i,l (Eq (7.5))
            delta_V[i][k] += x_k*V[i][j]*v[k]/v_j[j]

    # if the size bin does not fall between any bins
    else:

        delta_n[j] += n[j]

        # loop over individual species
        for i in All_species:
            delta_V[i][j] += V[i][j]
```

Listing 7.11 Searching for indices of bins between which particles have grown or evaporated then remapping them to the fixed size of each bin. (Function `quasi_stationary` in *sectional_methods.py*).

Once the number concentrations after remapping have been calculated, they are substituted to array y. Volume concentrations are converted back to molar concentrations and substituted to array y.

```
    # substitute the remapped values to number concentrations
    n = delta_n

    # substitute these values to array y
    y[range(Nb)] = n

    # volume concentrations of individual species
    for i in All_species:

        V[i] = delta_V[i]

        # convert volume concentration
        # back to molar concentration
        # and substitute the values in array y
        y[bins[i]] = V[i]*rho[i]/M[i]

    return y
```

Listing 7.12 Substituting the values to array y. (Function `quasi_stationary` in *sectional_methods.py*).

To use this scheme in the box model, the string `quasi-stationary` must be added to the `remapping_scheme` list in *parameters.py*.

### 7.2.2 Moving Center Method

In the moving center method, aerosol size bins do not have fixed sizes. Instead, their size is bound between the boundaries of the size bin which are fixed. The principle of this method is shown in Figure 7.4. The aerosol size is allowed to evolve freely until their size exceeds either the lower ($v_{sect,l,lo}$ in Figure 7.4) or the upper boundary ($v_{sect,i,hi}$) of the size bin $i$. When the boundaries are exceeded, all of the aerosol number and volume of all species in the size bin are merged with the size bin with bin boundaries between which the grown or shrunk size bin lies.

The benefits of the method are that it is efficient, simple to implement, and it exhibits less numerical diffusion than e.g. the quasi-stationary method. One disadvantage in using the moving center method is that when it is used in zero-dimensional (box model) simulations, remapping a size bin which has exceeded its bin boundaries to another size bin, can leave an empty size bin. In a box model simulation, unless new primary particles are emitted to the system, the number of empty size bins can never decrease. However, this can be avoided by smoothing or remapping techniques [6]. In three-dimensional models this is not a problem since the advection of aerosol between grid boxes tends to fill up empty size bins.

The routine applying moving-center will use the upper bin boundaries $v_{sect,i,hi}$ (see Figure 7.4) and array $y$ as inputs. The initialization of the routine is identical to the quasi-stationary method and uses the `initialize_remapping` routine.

```python
def moving_center(v_hi, y):

    delta_n, delta_V, n, V, v_j, bins =initialize_remapping(y)
```

**Listing 7.13** Initialization of the moving center method. (*sectional_methods.py*).

After the initialization, the routine will loop over all size bins and locates the fixed size bin $l$ for which $v_{sect,l,lo} < v_{sect,j} < v_{sect,l,hi}$ (see Figure 7.4). Once the new size bins have been identified, all number and mass are transferred from bin $j$ to bin $l$. Similarly to the quasi-stationary scheme, the smallest and the largest particle size bins may not fall between the boundaries of any of the fixed size bins and thus such bins cannot be remapped.

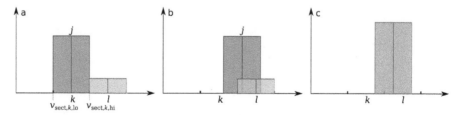

**Figure 7.4** Illustration of how growing particles in size bin $j$ are remapped using the moving center method. Panel (a) shows the initial state for particles in size bins $k$ and $l$, (b) particles $j$ have grown a part of the bin has exceeded the lower boundary of size bin $l$, (c) a fraction of particles $j$ have been merged with pre-existing particles in size bin $l$.

```
for j in range(Nb):

    # find the index of the size bin
    # where v_lo < v_j < v_hi
    l = np.argmax(v_j[j] < v_hi)

    # check if the index was found,
    # the size class has grown/evaporated
    # out of its initial size bin,
    # and the bin is not empty
    if l != 0 and j != l and n[j] > N_low_limit:

        # particle numbers are transferred
        # from bin j to bin l
        delta_n[l] += n[j]
        # the same amount is subtracted from bin j
        delta_n[j] -= n[j]

        # loop over individual species
        for i in All_species:
            # all volume of compound i in bin j
            # is transferred to size bin l
            delta_V[i][l] += V[i][j]
            # this amount is subtracted from bin j
            delta_V[i][j] -= V[i][j]

# transfer the remapped values to number concentrations
n = n + delta_n

# substitute these values to array y
y[range(Nb)] = n

# volume concentrations of individual species
for i in All_species:

    V[i] = V[i] + delta_V[i]

    # convert volume concentration back
    # to molar concentration
    # and substitute the values in array y
    y[bins[i]] = V[i]*rho[i]/M[i]

return y
```

**Listing 7.14** Remapping the size distribution using the moving center method. (Function `moving_center` in *sectional_methods.py*).

To use the moving center scheme with the box model, the string `moving-center` needs to be added to the `remapping_scheme` list in *parameters.py*.

### 7.2.3 Hybrid Bin Method

A method that can be considered a mix between the fixed sectional and moving center methods is the hybrid bin (or moving bin) method [7]. Instead of merging the whole size bin to the size bin it has grown or shrunk to, it is remapped to the adjacent size bins between which it lies. However, in this method the particles are not remapped to fixed size bins as is done in the quasi-stationary method. Instead, the size of the remapped size bins are allowed to vary in size between given size bin boundaries. This method assumes a width for the size bin. When a fraction of the size bin $j$ exceeds its boundaries, the exceeding fraction is merged to the size bin $l$ for which $v_{\text{sect},l,\text{lo}} < v_{\text{sect},j} < v_{\text{sect},l,\text{hi}}$. In the hybrid bin method, the size bin can be either assumed to have a "flat-top" or it can be assumed to have a linear size distribution within the size bin. The principle of remapping the size bin using the hybrid bin, when assuming a "flat-top" for the size bin, is illustrated in Figure 7.5.

An alternative method is to assume that the number concentration $y$ has a linear dependency on the particle size $x$ within a size bin according to the folllowing:

$$y(x) = k(x - x_0) + y_0, \tag{7.7}$$

where $k$ is the slope of the linear distribution and $y_0$ is the value of $y$ at the center of the size bin (see Figure 7.6). The size $x$ can be e.g. mass, volume, or a log of radius.

Here we assume a sub-grid linear dependence of number concentration as a function of particle volume $v$. Based on the conservation of total number $n_{\text{sect},i}$ and volume $V_{\text{sect},i}$ concentration in bin $j$ (of all species), when integrating from the lower edge of the size bin $v_1$ to the upper edge $v_2$, we get

$$n_{\text{sect},j} = \int_{v_1}^{v_2} k(v - v_0) + n_0 \, dv \tag{7.8}$$

$$V_{\text{sect},j} = \int_{v_1}^{v_2} kv(v - v_0) + kn_0 \, dv, \tag{7.9}$$

where $v_0 = (v_2 + v_1)/2$ and $n_0$ is the number concentration at $v_0$. From these equations we get

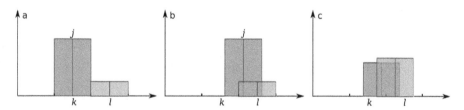

**Figure 7.5** Illustration of how growing particles in size bin $j$ are remapped using the hybrid bin method. Panel (a) shows the initial state for particles in size bins $k$ and $l$, (b) particles $j$ have grown and their size has exceeded the lower boundary of size bin $l$, (c) particles $j$ have been merged with pre-existing particles in size bin $l$.

**Figure 7.6** Linear size distribution within one size bin.

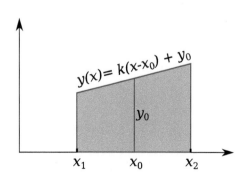

$$n_0 = \frac{n_{\text{sect},j}}{v_2 - v_1}, \quad k = \frac{12(v - v_0 n_{\text{sect},j})}{(v_2 - v_1)^3}. \tag{7.10}$$

Figure 7.7 visualizes, how this sub-size bin assumption works if a sine function $n(x) = \sin(x) + 1$ is discretized using this method. From the figure, we can see that the shape of the sine function is well captured. However, there are discontinuities at the points where the linear approximations meet. In addition, linear approximations exhibit negative values for example near $x = 5$. Although discontinuities could be avoided using higher order polynomials, here we use only linear functions. We will show how to deal with these negative values and discontinuities later in this section.

After the size bin has grown or shrunk during a time step, a fraction of the size bin exceeding the boundaries of the size bin is transferred to the other size bin taking into account the sub-grid shape of the size bin. There are two alternatives for the assumption for the width of the bin and the shape of the sub-grid size distribution: we can either assume that the size bin maintains the width and the shape during growth (the pre-growth linear method) or the width and the shape of the size bin are allowed to change during growth (the post-growth linear method).

The number and volume after a time step from $t$ to $t + \Delta t$ are

$$n_{t+\Delta t} = n_t + \Delta n, \quad v_{t+\Delta t} = v_t + \Delta v \tag{7.11}$$

Similarly, the volumes of the lower end ($v_1$) and the upper end ($v_2$) of the size bin have shifted by $\Delta v$. When using the pre-growth linear method, both $v_1$ and $v_2$ have both shifted the same amount

$$v_{1,t+\Delta t} = v_{1,t} + \Delta v, \quad v_{2,t+\Delta t} = v_{2,t} + \Delta v, \tag{7.12}$$

where

$$\Delta v = \frac{V_{t+\Delta t}}{n_{t+\Delta t}} - \frac{v_t}{n_t}. \tag{7.13}$$

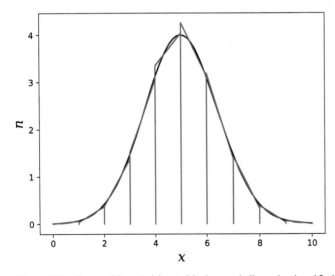

**Figure 7.7** Curve $n(x) = \sin(x) + 1$ (black curve) discretized to 10 size bins assuming linear size distribution within size bins.

When using the post-growth linear method, $v_1$ and $v_2$ shift by a different amount and the width of the size bin can change. In this method, the change in the volumes at the edges of the size bin can be written as

$$v_{1,t+\Delta t} = v_{1,t} + \Delta v_1, \quad v_{2,t+\Delta t} = v_{2,t} + \Delta v_2. \tag{7.14}$$

The shift of the size bin edges depend on the microphysical processes affecting the particle size. For example, if condensation grows the particle size, the shift in the lower and upper edge of the size bin can be approximated to be

$$\Delta v_1 \approx \Delta v \left(\frac{v_{1,t} n}{v}\right)^{\frac{1}{3}}, \Delta v_2 \approx \Delta v \left(\frac{v_{2,t} n}{v}\right)^{\frac{1}{3}}. \tag{7.15}$$

As shown in Figure 7.5, a fraction of the size bin which has exceeded its boundaries is merged into the adjacent size bin. The number of aerosol particles $\Delta n_{\text{sect},j}$ to be merged can be calculated from

$$\Delta n_{\text{sect},j} = \int_a^b y(x)dx = (b-a)\left[n_0 - k\left(x_0 - \frac{a+b}{2}\right)\right], \tag{7.16}$$

where $y(x)$ is the function given in Equation (7.7). The volume concentration of particles $\Delta V_{\text{sect},j}$ to be merged is

$$\Delta V_{\text{sect},j} = \int_a^b y(x)dx = n_0 \frac{b^2 - a^2}{2} + k\left(\frac{b^3 - a^3}{3} - x_0 \frac{b^2 - a^2}{2}\right). \tag{7.17}$$

The integration range $a$ to $b$ depends on if the particles in the size bin have grown or shrunk (see Figure 7.8(a) and (b)). When the particles have grown, the integration starts from the fixed upper boundary of the size bin ending to the upper boundary of the grown size bin $j$ (Figure 7.8(a)) and thus

$$a = v_{\text{sect},j,\text{hi}}, \quad b = v_{2,t+\Delta t}. \tag{7.18}$$

For a shrinking size bin (Figure 7.8(b))

$$a = v_{1,t+\Delta t}, \quad b = v_{\text{sect},j,\text{lo}}. \tag{7.19}$$

As illustrated in Figure 7.7 and Figure 7.8(c), the linear fit can produce negative values. Thus, a positivity check has to be made for both size bin edges. In case the positivity check fails, the sub-grid distribution in Equation (7.7) is modified to be of the form

$$y(x) = k_*(x - x_*), \tag{7.20}$$

where

$$v_* = \frac{3v}{n} - 2v_2, \ k_* = \frac{2n}{(v_2 - v_*)^2}, \text{ for } y(v_1) < 0, \tag{7.21}$$

$$v_* = \frac{3v}{n} - 2v_1, \ k_* = \frac{-2n}{(v_1 - v_*)^2}, \text{ for } y(v_2) < 0. \tag{7.22}$$

The routine that implements the hybrid bin method requires the mean size of each size bin $v$, the volume boundaries of each bin $i$ ($v_{\text{sect},i,\text{lo}}$ and $v_{\text{sect},i,\text{lo}}$) as well as the array $y$ as inputs. Again, the initialization of variables is identical to the quasi-stationary and moving center routines. In addition, the volumes of grown or shrunk particles are calculated.

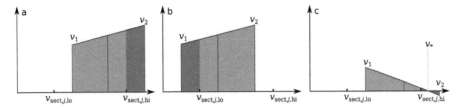

**Figure 7.8** Visualization of (a) a growing, (b) shrinking size bin, and (c) a growing size bin for which the linear sub-grid assumption produces negative values.

```
def hybrid_bin(v, v_lo, v_hi, y):

    # initialize variables for remapping
    delta_n, delta_V, n, V, v_t_dt, bins =initialize_remapping(y)

    # volume concentration of all species
    # in the grown/evaporated size bins
    V_t_dt = v_t_dt * np.maximum(n, N_low_limit)
```

**Listing 7.15** Initialization of the hybrid bin method. (*sectional_methods.py*).

After the initialization, the routine runs a loop over all size bins and checks that the bin to be remapped is not empty.

```
for j in range(Nb-1):

    # If bin is not empty
    if n[j] > N_low_limit:
```

**Listing 7.16** Looping over size bins. (Function `hybrid_bin` in *sectional_methods.py*).

Following this, the routine calculates the change in volume $\Delta V$ (see Equation (7.13)) and the volume boundaries of the shifted size bin (Equation (7.12)). For simplicity, we assume the pre-growth method where both volume boundaries shift by the same amount.

```
    # Calculate the change in volume
    # during a time step (Eq (7.13))
    delta_v = v_t_dt[j] - v[j]

    # Volumes of the boundaries
    # of size bins (Eq (7.13))
    # assuming pre-growth linear method
    v_1 = v_lo[j] + delta_v
    v_2 = v_hi[j] + delta_v
```

**Listing 7.17** Change in the volume of the size bin and the volume boundaries of the size bin $j$. (Function `hybrid_bin` in *sectional_methods.py*).

Next, the parameters for representing the linear size distribution within the size bin are calculated according to Equation (7.10)

```
# calculate n_0, k, (Eq (7.10))
# and x_0 for integrating Eqs (7.16) and (7.17)
n_0 = n[j] / (v_2-v_1)
v_0 = (v_2+v_1) / 2.0
k = 12.0 * (V_t_dt[j]-v_0*n[j])/(v_2-v_1)**3
x_0 = v_0
```

Listing 7.18 Calculating the paramaters for integrating Equations (7.16) and (7.17). (Function hybrid_bin in *sectional_methods.py*).

Once the sub-grid distribution for the size bins has been determined, remapping is carried out. When the size bin has grown, a fraction of the bin is remapped to the larger bin. The number of particles to be remapped is calculated using Equation (7.16) and the volume of particles to be remapped is based on Equation (7.17). The lower limit of the integral is the upper bound $v_{sect,i,hi}$ of the fixed bin (see Figure 7.8) and the upper limit is given by $a$ in Equation (7.18). In addition, the positivity check is done for the shifted bin edges and if the check fails, integration limits are adjusted according to Equations (7.21) and (7.21)

```
# if the particles have grown and
# we are not remapping the largest bin
if delta_v > 0 and j != Nb:

    # set the limits a and b
    # for integrating Eqs (7.16) and (7.17)
    a = v_hi[j]
    b = v_2
    # the index of the bin to which
    # particles are transferred to
    m = j+1

    # positivity check
    # y(v_1) in Eq (7.21)
    if k * (v_1 - v_0) + n_0 < 0:

        # Equation (7.21)
        # Notice that the volume concentrations
        # of individual species are summed
        v_ast = 3.0*V_t_dt[j]/n[j] - 2.0*v_2
        n_0 = 0.0
        x_0 = v_ast
        k = k_ast = 2.0*n[j]/(v_2-v_0)**2
        a = v_hi
        b = v_2


    # v(v_2) in Eq (7.22)
    if k * (v_2 - v_0) + n_0 < 0:
        # Equation (7.22)
        v_ast = 3.0*V_t_dt[j]/n[j] - 2.0*v_1
        n_0 = 0.0
        x_0 = v_ast
        k = k_ast = -2.0*n[j]/(v_1-v_0)**2
        a = v_2
        b = v_ast
```

Listing 7.19 Remapping using hybrid bin method when the bin has grown. (Function hybrid_bin in *sectional_methods.py*).

Similarly, if the bin has shrunk in size, a part of the bin is remapped to the smaller size bin. Now the lower limit of the integration is the lower limit of the shrunken bin while the upper limit of integration is the fixed lower limit of the size bin $v_{\text{sect},i,\text{lo}}$. The positivity check is done also for a shrunken bin and if it fails, bin limits are recalculated according to Equations (7.21) and (7.22)

```python
# if the particles have shrunk in size
elif delta_v < 0 and j != 0:

    # Equation (7.19)
    a = v_1
    b = v_lo[j]
    # the index of the bin to which
    # particles are tranferred to
    m = j-1

    # positivity check
    # y(v_1) in Eq (7.21)
    if k * (v_1 - v_0) + n_0 < 0:

        # Equation (7.21)
        # Notice that the volume concentrations
        # of individual species are summed
        v_ast = 3.0*V_t_dt[j]/n[j] - 2.0*v_2
        n_0 = 0.0
        x_0 = v_ast
        k = k_ast = 2.0*n[j]/(v_2-v_0)**2
        a = v_ast
        b = v_lo[j]

    # v(v_2) in Eq (7.22)
    if k * (v_2 - v_0) + n_0 < 0:

        # Equation (7.22)
        v_ast = 3.0*V_t_dt[j]/n[j] - 2.0*v_1
        n_0 = 0.0
        x_0 = v_ast
        k = k_ast = -2.0*n[j]/(v_1-v_0)**2
        a = v_1
        b = v_lo[j]
```

**Listing 7.20** Remapping using hybrid bin method when the bin has shrunk in size. (Function `hybrid_bin` in *sectional_methods.py*).

```python
else:
    # nothing is remapped
    k = 0.0
    a = v[j]
    b = v[j]
    m = j
```

**Listing 7.21** If the size bin limits of the evolved particles are within the fixed bin limits, particles are not remapped. (Function `hybrid_bin` in *sectional_methods.py*).

If either the upper limit $a$ or the lower limit $b$ of the integral in Equation 7.16 exceeds the fixed boundaries $v_{\text{sect},j,\text{lo}}$ or $v_{\text{sect},j,\text{hi}}$ of the size bin $j$, remapping is carried out.

```
        if b > v_hi[j] or a < v_lo[j]:
            # Equation (7.16)
            delta_n[m] += (b-a) * (n_0-k*(x_0-(a+b)/2.0))
            delta_n[j] -= (b-a) * (n_0-k*(x_0-(a+b)/2.0))

            # Equation (7.17)
            delta_V_all = n_0 * (b**2-a**2)/2.0 \
                + k*((b**3-a**3)/3.0 - x_0*((b**2-a**2)/2.0))

            # loop over individual species
            for i in All_species:
                # V[i][j]/np.sum(V[:][j]) is the volume fraction
                # of species i in size bin j
                delta_V[i][m] += delta_V_all \
                    * V[i][j]/(v_t_dt[j]*n[j])
                delta_V[i][j] -= delta_V_all \
                    * V[i][j]/(v_t_dt[j]*n[j])

    # substitute the remapped values to number concentrations
    n = np.maximum(n + delta_n, 0.0)

    # substitute these values to array y
    y[range(Nb)] = n

    # volume concentrations of individual species
    for i in All_species:

        V[i] = np.maximum(V[i] + delta_V[i], 0.0)

        # convert volume concentration back to molar concentration
        # and substitute the values in array y
        y[bins[i]] = V[i]*rho[i]/M[i]

    return y
```

**Listing 7.22** Remapping of a size bin using hybrid bin method. (Function `hybrid_bin` in *sectional_methods.py*).

To use the hybrid bin scheme in the box model, the string `hybrid-bin` needs to be added to the `remapping_scheme` list.

### 7.2.3.1 Flat-Top Size Bins with Hybrid Bin Method

The hybrid bin method requires several if statements, in particular related to the positivity checks, which can decrease the computational efficiency. One way to avoid this is to assume the size bin to have a flat-top linear distribution, i.e. $k$ in Equation (7.7) is set to zero. This allows for omitting the positivity checks thus simplifying the remapping routing. With this approach when using the pre-growth method, codes in Listings 7.19–7.22 can be replaced by the following Listing 7.23.

```
def flat_top(v, v_lo, v_hi, y):

    # initialize variables for remapping
    delta_n, delta_V, n, V, v_t_dt, bins =initialize_remapping(y)

    # volume concentration of all species
    # in the grown/evaporated size bins
```

```
    V_t_dt = v_t_dt * np.maximum(n, N_low_limit)

for j in range(Nb-1):

    # If bin is not empty
    if n[j] > N_low_limit:

        # Calculate the change in volume
        # during a time step (Eq (7.13))
        delta_v = v_t_dt[j] - v[j]

        # Volumes of the boundaries
        # of size bins
        v_1 = v_lo[j] + delta_v
        v_2 = v_hi[j] + delta_v

        # Equation (7.10)
        n_0 = n[j] / (v_2 - v_1)

        # if the particles have grown and
        # we are not remapping the largest bin
        if delta_v > 0 and j != Nb:

            b = v_2
            a = v_hi[j]
            m = j + 1

        # if the particles have shrunk in size
        elif delta_v <= 0 and j != 0:

            b = v_lo[j]
            a = v_1
            m = j - 1

        else:
            # nothing is remapped
            k = 0.0
            a = v[j]
            b = v[j]
            m = j

        delta_n[j] -= np.minimum(n[j], (b-a)*n_0)
        delta_n[m] += np.minimum(n[j], (b-a)*n_0)

        # loop over individual species
        for i in All_species:

            # V[i][j]/np.sum(V[:][j]) is the volume fraction
            # of species i in size bin j
            delta_V[i][j] -= np.minimum(V[i][j],n_0 \
                * (b**2-a**2)/2.0 * V[i][j]/V_t_dt[j])
            delta_V[i][m] += np.minimum(V[i][j],n_0 \
                * (b**2-a**2)/2.0 * V[i][j]/V_t_dt[j])

# substitute the remapped values to number concentrations
n = n + delta_n

# substitute these values to array y
y[range(Nb)] = n

# volume concentrations of individual species
for i in All_species:

    V[i] = np.maximum(V[i] + delta_V[i], 0.0)
```

```
             # convert volume concentration back to molar concentration
             # and substitute the values in array y
             y[bins[i]] = V[i]*rho[i]/M[i]

        return y
```

**Listing 7.23** Remapping of a size bin using hybrid bin method assuming a flat subgrid size distribution. (*sectional_methods.py*).

This method can be used in the box model by adding the string `flat-top` in the `remapping_scheme` list in *parameters.py*.

## 7.3 Simulating Absorptive Uptake and New Particle Formation Simultaneously

In Section 7.1 we introduced ODEs for absorptive uptake. Now we add ODEs for simulating new particle formation (NPF) due to nucleation of sulfuric acid particles to demonstrate how different microphysical processes can be combined in a box model.

As explained in Section 6.5, simulating new particle formation by nucleation accurately is complex and typically atmospheric models implement new particle formation using formation rate of a give sized particles [8–10]. Here we use a very simplified equation for the formation rate of 10 nm particles

$$\left(\frac{dn_{\mathrm{sect},1}}{dt}\right)_{\mathrm{NPF}} = AC_{g,\mathrm{H_2SO_4}},\tag{7.23}$$

where $A$ is the NPF rate constant and $C_{g,\mathrm{H_2SO_4}}$ is the gas phase concentration of sulfuric acid. Here we use $A = 10^{20}\,\mathrm{mol^{-1}s^{-1}}$. The value is not physically-based, but only selected for demonstration purposes. However, any new particle formation rate parameterization can be implemented similarly in the box model.

Since the formed particles are 10 nm in diameter, we obtain the formation rate of sulfuric acid in the condensed phase from

$$\left(\frac{dC_{a,\mathrm{SO_4},1}}{dt}\right)_{\mathrm{NPF}} = \frac{v_{\mathrm{NPF}}\rho_{\mathrm{SO_4}}}{M_{\mathrm{SO4}}}\left(\frac{dn_{\mathrm{sect},1}}{dt}\right)_{\mathrm{NPF}},\tag{7.24}$$

where $C_{a,\mathrm{H_2SO_4},1}$ is the molar concentration of sulfate in the smallest size bin, $v_{\mathrm{NPF}}$ is the volume of a particle formed in NPF, $\rho_{\mathrm{SO_4}}$ is the density of sulfate, and $M_{\mathrm{SO4}}$ is the molar mass of sulfate. The change in the gas phase concentratio is

$$\left(\frac{C_{g,\mathrm{H_2SO_4}}}{dt}\right)_{\mathrm{NPF}} = -\left(\frac{dC_{a,\mathrm{SO_4},1}}{dt}\right)_{\mathrm{NPF}}.\tag{7.25}$$

In this model setup, we use the same initial conditions for gas and aerosol phase concentrations as in `model_setup=1`. Once the simulation has been started, gas phase $\mathrm{H_2SO_4}$ is injected at a constant rate of $0.001\,\mu\,\mathrm{g\,m^{-3}s^{-1}}$ for the first 50 s which initiates the formation of new particles. Equations (7.23)–(7.25) are added to `dy_dt` in the model in order to include NPF,

```
# Change in the number concentration and volume concentration
# of the smallest size bin due to nucleation
        if model_setup == 2 and species == 'SO4':

            if t < 50.:
                # inject sulfuric acid in the gas phase
                # for 50 seconds
                dydt[gas] = dydt[gas] + \
                    abundance['SO4']/M['SO4']*1.e-9

                # formation rate of new particles, Eq(7.23)
                dydt[0] = dydt[0] + 1.0e20*C_g[species]

                # volume of the new particles
                v_NPF = np.pi / 6.0 * 10.0e-9**3

                # moles of sulfate in the new particles in Eq (7.23)
                C_NPF = v_NPF * rho[species] / M[species]

                # change in molar concentration, Eq (7.24)
                dydt[bins[0]] = dydt[bins[0]] + dydt[0] * C_NPF

                # change in gas phase concentration, Eq (7.25)
                dydt[gas] = dydt[gas] - dydt[0] * C_NPF
```

Listing 7.24 Calculating the formation rate of new particles, formation rate of sulfate in condensed phase in the smallest size bin, and the change in gas phase concentration of sulfuric acid. (*differential_equations.py*).

This model setup can be selected by setting `model_setup=2`.

## 7.4 Cloud Parcel Models

In the previous sections, we assumed that the ambient conditions in the air parcel stay constant. Cloud parcel models typically simulate cloud droplet formation following an air parcel in which temperature change drives condensation and evaporation of water to and from hydrometeors. The main process driving the partitioning of water (and other semivolatile species) between the gas and particle/liquid phase is the temperature change which changes the saturation ratio of the condensing compounds. Typically cloud parcel models simulate an ascending air parcel in which the parcel does not exchange heat with its surroundings. As the parcel ascends, it expands and cools. Thus, the saturation ratio of water in air increases. Simultaneously, condensation warms the air parcel through latent heat. When considering these two opposing effects, the temperature of air in a rising air parcel is given by

$$\frac{dT}{dt} = -\frac{gw}{c_{tot}} + \frac{1}{\rho c_{tot}} \sum_{j=1}^{n} L_{e,i} \frac{dc_{i,j}}{dt}, \tag{7.26}$$

where the first term in the equation accounts for the temperature change caused by vertical motion. In the equation, $g$ is the gravitational acceleration, $w$ is the updraft velocity of the air parcel, and $\rho$ is the density of air. The second term in the equation gives the temperature change due to latent heat released in condensation. In the second term, $L_{e,i}$ is the latent heat of evaporation for species $i$. In Equation (7.26), $c_{tot}$ is the total heat capacity of the air parcel, defined as

$$c_{tot} = \frac{m_{air}c_{p,air} + \sum_{j=1}^{n_{gas}} m_j c_{p,j} + \sum_{k=1}^{n_{liq}} m_k c_{p,k}}{m_{tot}}, \tag{7.27}$$

where $c_{p,i}$ are the heat capacities of individual species, $m_i$ is the mass for individual species $i$.

The equation for the total pressure is

$$\frac{dP}{dt} = -gPM_{air}wRT \tag{7.28}$$

where $g$ is the gravitational acceleration, $M_{air}$ is the average molar mass of air, $w$ is the horizontal updraft velocity of the air parcel, $R$ is the gas constant and $T$ is the air temperature. The vertical position of the air parcel is

$$\frac{dz}{dt} = w. \tag{7.29}$$

The adiabatic expansion of the air parcel changes the volume of the air parcel. The rate of change in volume can be calculated from ideal gas law to obtain

$$-\frac{dV_{parcel}}{dt} = V_{parcel}\left(\frac{1}{P}\frac{dP}{dt} - \frac{1}{T}\frac{dT}{dt}\right) = V_{parcel}\gamma. \tag{7.30}$$

Since the air parcel expands, the change in the volume of the air parcel affects the concentration of gases, the number concentration of the aerosol droplets and the concentration of the species in the droplets. Using the ideal-gas law, we get the equations for the gas phase concentration of species $i$ as

$$\left(\frac{dC_{g,i}}{dt}\right)_{volume} = C_{g,i}\gamma, \tag{7.31}$$

for the mole concentration of the species $i$ in size bin $j$ in the liquid phase as

$$\left(\frac{dC_{a,i,j}}{dt}\right)_{volume} = C_{a,i,j}\gamma, \tag{7.32}$$

and the number concentration $n_{sect,j}$ of particles in size bin $j$

$$\frac{dn_{sect,i}}{dt} = n_{sect,i}\gamma. \tag{7.33}$$

### 7.4.1 Sectional Cloud Parcel Model

Next, we extend the box model to include the differential equations accounting for adiabatic expansion of the air parcel. In addition, since rapid condensation of water affects the temperature of the air parcel which have to be taken into account in condensation equations, we will have separate differential equations for condensation of water.

For the cloud parcel setup, the model is initialized with wet aerosol so that the water activity equals to the saturation vapor pressure of water in air. The volume of water is calculated according to Equation (2.66).

```
# calculate the volume of water [m3]
# based on Kappa K\"ohler, Eq (2.66)
a_w = RH0 / 100.0
v_w = a_w / (1 - a_w) * kappa_i['SO4'] * np.pi/6.0*d**3
```

```
# convert to molar concentration [mol/m3]
c[0,:] = n * v_w * rho['H2O'] / M['H2O']

C_gases=[
    a_w * C_star['H2O'](T0),          # water
    0.0,                              # sulfuric acid
    0., 0., 0., 0., 0., 0., 0., 0., 0.,0.,  # VBS species
]
```

Listing 7.25 Initializing the aerosol water using Kappa Köhler theory and the gas phase water based on the initial saturation ratio. (*box_model.py*).

Note that we omit the Kelvin effect in initializing aerosol water making the water content too high especially for the smallest particles. Accounting for the Kelvin effect would require an iterative approach to obtain a complete equilibrium between the aerosol water and saturation ratio of water vapour. However, water reaches equilibrium between the gas and aerosol extremely fast and this can be considered a reasonably good initial guess for the amount of water.

For semivolatile organic species we assumed that their surface concentration is a function of their mole fraction and equilibrium vapor pressure. However, for water we use the Kappa Köhler equations to calculate the surface concentration of water. To use Kappa Köhler in modeling condensation and evaporation of water, we will add separate calculation for surface concentration of water in the dy_dt function in *differential_equations.py*.

```
# Saturation ratio of water, Eq (2.68)
S_w = (R_wet**3 - R_dry**3)/np.maximum((R_wet**3-R_dry**3*(1.0-kappa)),
    1.e-30) * K
# Equilibrium concentration on the surface
C_surf = S_w * C_star[species](T)
```

Listing 7.26 Calculating the saturation ratio of water at the droplet surface. (*differential_equations.py*).

In addition, when water condenses on droplets, the latent heat affects the condensation at the droplet surface and needs to be taken into account (see Equation (2.36)).

```
    # temperature change due to
    # condensation /evaporation of water
    if species == 'H2O':

        # pressure
        p = y[-1]
        # moles of water in all size bins per m3 air
        C_water = sum(y[bins])
        # density of the air parcel
        rho_air = p * M_air / (R_gas * T)
        # mass density of the air parcel
        m_tot = rho_air + (C_water + C_g[species]) * \
            M[species]
        # heat capacity of the air parcel, Eq (7.27)
        c_tot = (rho_air * c_p_air +                    \
                (C_water + C_g[species]) *              \
                M[species] * c_p_water)/m_tot
```

```
        # temperature change, the 2nd term in Eq (7.26)
        dydt[-2] = 1.0 / (rho_air * c_tot) *               \
            L_e[species](T) * (-dydt[gas])
```

**Listing 7.27** Differential equation for temperature change due to latent heat of evaporation / condensation. (*differential_equations.py*).

Next, we add ODEs accounting for adiabatic expansion of the air parcel.

```
    # Ordinary differential equations for an ascending air parcel
    if model_setup >= 3:
        # temperature change due to adiabatic
        # expansion of the air parcel, the 2nd term in Eq (7.26)
        dydt[-2] = dydt[-2] - g * w / c_tot

        # pressure change due to adiabatic
        # expansion of the air parcel, Eq (7.28)
        dydt[-1] = -g * p * M_air * w / (R_gas * T)
        # term gamma accounting for adiabatic expansion
        # in Eq (7.30)
        gamma = dydt[-1] /p - dydt[-2] / T
        for species in Condensing_species:

            # gas phase
            gas = Nb+len(bins)*len(index)+index[species]-1
            # add the term for air parcel volume change
            # for gas phase concentrations, Eq (7.31)
            dydt[gas] = dydt[gas] + y[gas] * gamma

        for species in All_species:

            # find the indices of each species in Array y
            # particle phase
            bins = range(Nb*index[species],Nb*index[species]+Nb)
            # add the term for air parcel volume change
            # for liquid phase concentrations, Eq (7.32)
            dydt[bins] = dydt[bins] + y[bins] * gamma

        # for number concentrations, Eq (7.33)
        dydt[0:Nb] = dydt[0:Nb] + y[0:Nb] * gamma
```

**Listing 7.28** Differential equations accounting for the adiabatic expansion of the cloud parcel. (*differential_equations.py*).

Now solving these ODEs we can simulate the changes in the gas phase concentration, condensed phase concentrations, and number concentrations. Other differential equations describing changes in gas phase, aerosol, and ambient conditions due to microphysical and chemical processes can be added to the differential equation in a similar fashion.

This model setup can now simulate activation of aerosol particles to cloud droplet and this setup can be selected by setting `model_setup=3` in *parameters.py*. In its current configuration, this setup can only use the full-moving method, because the bin limits $v_{sect,i,lo}$ and $v_{sect,i,hi}$ have been defined for dry aerosol and should be defined for wet aerosol in cloud parcel setups.

#### 7.4.1.1 Sectional Cloud Parcel Model with Co-condensing Organic Species

In addition to water, semivolatile species can also condense simultaneously during cloud droplet activation adding to the amount of hygroscopic material in droplets. The increase in hygroscopic mass supresses the maximum supersaturation required for the droplets to activate to cloud droplets [11].

Co-condensation can be simulated without additional code modifications in the model setup described in previous sections. The box model will use the routines for solving condensation and evaporation of semivolatile species as describe in Listing 7.7. For this model setup, the model is initialized with VBS compounds in the gas phase. We use one-tenth lower concentrations for gas phase concentrations of VBS species than in the aerosol setup of the box model.

```
if model_setup == 4:
    C_gases[2:-1]=[
        abundance['VBS0']/M['VBS0']*0.1e-09,    # VBS species
        abundance['VBS1']/M['VBS1']*0.1e-09,
        abundance['VBS2']/M['VBS2']*0.1e-09,
        abundance['VBS3']/M['VBS3']*0.1e-09,
        abundance['VBS4']/M['VBS4']*0.1e-09,
        abundance['VBS5']/M['VBS5']*0.1e-09,
        abundance['VBS6']/M['VBS6']*0.1e-09,
        abundance['VBS7']/M['VBS7']*0.1e-09,
        abundance['VBS8']/M['VBS8']*0.1e-09,
        abundance['VBS9']/M['VBS9']*0.1e-09,
    ]
```

**Listing 7.29** Initialization of gas phase concentrations for cloud parcel simulation with co-condensing organic gases. (*differential_equations.py*).

Using the box model with co-condensation of water and VBS species can be done selecting model_setup=4.

### 7.5 SALSA

One example of an aerosol model that can be used in a box setup as well as in two- and three-dimensional frameworks is SALSA (Sectional Aerosol module for Large-Scale Applications). It was originally developed to be implemented in the aerosol-climate model ECHAM5-HAM [12, 13]. Later, it has been implemented in a chemical transport model MATCH [14], Regional climate model RCA4 [15], aerosol-chemistry-climate model ECHAM-HAMMOZ [16], and two large eddy simulations (LES) models UCLALES [17] and PALM [18]. Although, SALSA has been mainly designed for large-scale models, it can be run in a box setup with high resolution and has been used to simulate aerosol chamber experiments [1].

SALSA was aimed to be implemented in computationally heavy atmospheric models and was designed to be computationally efficient while optimizing the numerical accuracy. This was achieved by minimizing the number of size bins and the number of chemical included in each size bin. In the default setup, it describes aerosol size distribution using 10 bins in size space ranging from 3 nm to 10 $\mu$m. It includes model species sulfate ($SO_4$), organic carbon (OC), semivolatile species (VBS), sea salt (SS), black carbon (BC), and mineral dust (DU).

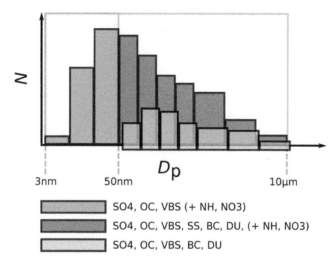

Figure 7.9 Schematic of the aerosol number and composition size distribution used in SALSA.

The size distribution includes two parallel externally mixed size bins: one set of bins ranging over the whole size range and one ranging from 50 nm to 10 $\mu$ m. The width of the bins varies so that the width is smallest in the size range where the size is sensitive with respect to cloud activation, i.e. 50 nm – 700 nm. The details of which species are included in each size bin can be seen in Figure 7.9.

SALSA solves the following aerosol processes:

- Water uptake is calculated using the Zdanovskii, Stokes, and Robinson mixing assumption presented in Section 2.5.1 (see Equation (2.66)).
- Condensation: Since SALSA is designed to be computationally fast, it does not allow for using detailed ODE solvers which were used in the box model. Instead, condensation is calculated over one atmospheric model time step which can be minutes in global scale models. To achieve this, SALSA uses Analytical Predictor of Condensation [19] for water and VBS, and Analytical Predictor of Dissolution [19] in combination with Analytical Predictor of Condensation for nitric acid and ammonia.
- Coagulation is calculated using the semi-implicit method presented in Section 5.3.2 (see Algorithm 5.7)
- New particle formation: Nucleation of new particles can be calculated using different nucleation parameterizations [20]. As explained in Section 6.5, atmospheric models often use parameterizations of new particle formation rates of particularly sized particles. In the case of SALSA using the nucleation rates, aerosol size distribution, and ambient conditions, the new particle formation rates of particles of 3 nm in diameter are calculated using the parameterization of Kerminen and Kulmala (2002) [8], Lehtinen et al. (2004) [9], or Anttila et al. (2010) [10].

In addition to aerosol processes, SALSA also accounts for cloud and precipitation droplets as well as ice particles. Cloud droplets include size bins in the range of 50 nm to 10 $\mu$ m, precipitation droplet bins range from 50 $\mu$ m to 2 mm, and ice particle bins range from 2 $\mu$ m to 2 mm. In SALSA, the activation of cloud droplets can be calculated using a cloud activation parameterization [21] which provides the activated droplets per bin using the aerosol size and composition distribution and the updraft velocity.

Optionally, cloud droplet activation can also be calculated in SALSA by solving the condensation equations similarly to what was described in Section 7.4.1.

## Bibliography

[1] H. Kokkola, P. Yli-Pirilä, M. Vesterinen, H. Korhonen, H. Keskinen, S. Romakkaniemi et al. The role of low volatile organics on secondary organic aerosol formation. *Atmospheric Chemistry and Physics*, 14(3):1689–1700, 2014.

[2] S. P. O'Meara, S. Xu, D. Topping, M. R. Alfarra, G. Capes, D. Lowe et al. Pycham (v2.1.1): a python box model for simulating aerosol chambers. *Geoscientific Model Development*, 14(2):675–702, 2021.

[3] L. Huang and D. Topping. Jlbox v1.1: a julia-based multi-phase atmospheric chemistry box model. *Geoscientific Model Development*, 14(4):2187–2203, 2021.

[4] G. W. Mann, K. S. Carslaw, C. L. Reddington, K. J. Pringle, M. Schulz, A. Asmi et al. Intercomparison and evaluation of global aerosol microphysical properties among aerocom models of a range of complexity. *Atmospheric Chemistry and Physics*, 14(9):4679–4713, 2014.

[5] F. Gelbard. Modeling Multicomponent Aerosol Particle Growth by Vapor Condensation. *Aerosol Science and Technology*, 12:399–412, 1990.

[6] Adam J Mohs and M Bowman. Eliminating numerical artifacts when presenting moving center sectional aerosol size distributions. *Aerosol and Air Quality Research*, 11(1):21–30, February 2011.

[7] Jen-Ping Chen and Dennis Lamb. Simulation of cloud microphysical and chemical processes using a multicomponent framework. Part I: Description of the microphysical model, *Journal of the Atmospheric Sciences* 51(18):2613–2630, 1994.

[8] V. M. Kerminen and M. Kulmala. Analytical formulae connecting the "real" and the "apparent" nucleation rate and the nuclei number concentration for atmospheric nucleation events. *Journal of Aerosol Science*, 33:609–622, 2002.

[9] Kari E. J. Lehtinen, U. Rannik, Markku Kulmala, and Pertti Hari. Nucleation rate and vapour concentration estimations using a least squares aerosol dynamics method, *Journal of Geophysical Research* 109(D21209), 2004.

[10] Tatu Anttila, Veli-Matti Kerminen, and Kari E.J. Lehtinen. Parameterizing the formation rate of new particles: The effect of nuclei self-coagulation. *Journal of Aerosol Science*, 41(7):621–636, 2010.

[11] A. Laaksonen, P. Korhonen, M. Kulmala, and R. J. Charlson. Modification of the Köhler equation to include soluble trace gases and slightly soluble substances. *Journals of the Atmospheric Sciences*, 55:853–862, March, 1998.

[12] H. Kokkola, H. Korhonen, K. E. J. Lehtinen, R. Makkonen, A. Asmi, S. Järvenoja et al. SALSA: a Sectional Aerosol module for Large Scale Applications. *Atmospheric Chemistry and Physics*, 8(9):2469–2483, 2008.

[13] T. Bergman, V.-M. Kerminen, H. Korhonen, K. J. Lehtinen, R. Makkonen, A. Arola et al. Evaluation of the sectional aerosol microphysics module SALSA implementation in ECHAM5-HAM aerosol-climate model. *Geoscientific Model Development*, 5(3):845–868, 2012.

**[14]** C. Andersson, R. Bergström, C. Bennet, L. Robertson, M. Thomas, H. Korhonen et al. MATCH-SALSA - Multi-scale Atmospheric Transport and CHemistry model coupled to the SALSA aerosol microphysics model - Part 1: Model description and evaluation. *Geoscientific Model Development*, 8(2):171–189, 2015.

**[15]** M. A. Thomas, M. Kahnert, C. Andersson, H. Kokkola, U. Hansson, C. Jones et al. Integration of prognostic aerosol-cloud interactions in a chemistry transport model coupled offline to a regional climate model. *Geoscientific Model Development*, 8(6):1885–1898, 2015.

**[16]** H. Kokkola, T. Kühn, A. Laakso, T. Bergman, K. E. J. Lehtinen, T. Mielonen et al. Salsa2.0: The sectional aerosol module of the aerosol–chemistry–climate model echam6.3.0-ham2.3-moz1.0. *Geoscientific Model Development*, 11(9):3833–3863, 2018.

**[17]** J. Tonttila, Z. Maalick, T. Raatikainen, H. Kokkola, T. Kühn, and S. Romakkaniemi. Uclales–salsa v1.0: a large-eddy model with interactive sectional microphysics for aerosol, clouds and precipitation. *Geoscientific Model Development*, 10(1):169–188, 2017.

**[18]** M. Kurppa, A. Hellsten, P. Roldin, H. Kokkola, J. Tonttila, M. Auvinen et al. Implementation of the sectional aerosol module salsa2.0 into the palm model system 6.0: model development and first evaluation. *Geoscientific Model Development*, 12(4):1403–1422, 2019.

**[19]** M. Z. Jacobson. *Fundamentals of Atmospheric Modeling, Second Edition*. Cambridge University Press, New York, 2005.

**[20]** P. Paasonen, T. Nieminen, E. Asmi, H. E. Manninen, T. Petäjä, C. Plass-Dlmer et al. On the roles of sulphuric acid and low-volatility organic vapours in the initial steps of atmospheric new particle formation. *Atmospheric Chemistry and Physics*, 10(22):11223–11242, 2010.

**[21]** Hayder Abdul-Razzak and Steven J. Ghan. A parameterization of aerosol activation 3. Sectional representation, *Journal of Geophysical Research* 107(D3)(4026), 2002.

# 8

# Software Optimization

In the previous chapters, we have discussed a number of scientific techniques and how to translate these in to computer programs. As part of this translation we have also covered numerical methods and explained the importance of choosing an appropriate solver, whether for accuracy or speed. We have codes written in a number of programming languages including Fortran, C, Java, Julia, Python, and MATLAB. Each of these can run "as is" on a CPU architecture, such as the microprocessor computing chips sold by Intel and AMD (of the x86_64 architecture), and those designed by IBM (of the Power architecture) and ARM (of their own architecture).

The topics of software engineering and the software development life cycle are key to writing quality software. Regrettably, there is too much of interest to those implementing the codes discussed in this book than we can fit in this chapter. We will first highlight two key topics and then apply to some of the codes previously covered, and in Section 8.4 we will pick up again some of the more general principles of good coding.

The main topics and opportunities that atmospheric scientists should be aware of when programming can be grouped as "portability" and "performance":

1. Portability: ensuring your coding efforts can run anywhere, while giving good results
2. Performance: ensuring your code can run efficiently, while giving good results

As you can see, the primary concern is to ensure you maintain the correct answers whether you and your collaborators are looking to run your simulation on different hardware, or whether you are optimizing and parallelizing your programs to get the results more quickly or to access larger memory systems. We cover Portability and Performance in the following sections.

## 8.1  Portability

As we noted a little earlier in this chapter, one of the main challenges of good programming is ensuring portability across different architectures. But what does this mean? And how much of a challenge is it?

The good news is that for compiled languages we write the code and then compile on our chosen machine. And we can copy the code to different machines of choice, merely re-compiling on each one. As long as we have kept to the programming language's standards this should ensure portability so that we can run our code in different places

*Introduction to Aerosol Modelling: From Theory to Code.*
First Edition. Edited by David Topping and Michael Bane.
© 2022 John Wiley & Sons Ltd. Published 2022 by John Wiley & Sons Ltd.

and obtain the same results. The results may not be identical in some cases and it is important to have an understanding of why this may be.

Computers use binary to store numbers and only have a fixed number of bits to store any given number. This is generally fine for small integers such as 2 ("0010" in binary with 4 bits) and 5 ("0101" in binary) but consider when we want 1/2 and 1/5. 1/2 would be 0.5 as a decimal and "0.1" in binary, neither of which raise any issue. However, 1/5 would be 0.2 in decimal (again, no issue) but in binary would be "0.00110011" as a recurring number and this cannot be represented in a finite number of bits.

So not only would we need the translation from our chosen programming language to the target machine code to be identical, we will require how the hardware handles representation of floating point numbers to be identical. This is where "IEEE754" comes in! This is a standard that describes how floating point numbers should be represented on computers and as long as both target architectures (and compilers used) adhere to the standard there should be no issue. But, for example, GPUs do not adhere to IEEE754 and thus you may see differing results from a long simulation, compared to CPU results, due to the differences in how they represent and process real numbers. These differences, particularly when as a result of differing compiler optimizations (see below) are known as **rounding error**. It will ultimately be in the hands of the aerosol scientist as to whether results are "good enough" or too different, and it is important to realize this decision depends heavily on what is being simulated. Rounding error when predicting future global average temperatures over timescales of centuries may be less important than when determining whether a critical temperature is being exceeded in a nuclear reactor.

We will cover software maintenance of codes used on many machines in Section 8.4.

## 8.2 Performance

The other question is how well our code runs on different machines. This will be a function of the target hardware, the compiler (if appropriate), and our code.

In this section we restrict our attention to traditional CPU processors, holding off discussion of accelerators (such as GPU and FPGA technologies) until Section 8.3.3. Even with just CPU processors we still have a wide variety of potential hardware configurations: Which ISA applies? How many cores per processor? How many sockets per node? We may even wonder if we have more than one node in our machine whether we can make use of them all! Indeed, we can write parallel programs to make good use of many processor cores across these different levels, but the key is to ensure good performance on each core. We will cover parallelization further in Section 8.3 but for this section we focus on just a single processor core.

### 8.2.1 Compiler Optimization

For compiled languages, recall that we use a compiler and that it can view all the source code during compilation. All modern compilers support "optimization" which allows transformations of the source code during compilation to machine code in order to improve performance. Compilers typically offer different levels of compiler optimization with higher optimization levels providing the compiler more and more freedom to be more and more aggressive in their transformation of the original code to machine code that the processor core will execute.

These compiler optimizations include re-ordering of loops to ensure data is fed in to the processor core "engine" as fast as is needed (and minimizing the number of times any given variable needs to be fetched from memory), and the re-organization of the statements to minimize the effort required to reach a similar arithmetical result. The compiler can also, for example, make use of approximate trigonometric function values, rather than the more time consuming task of computing to higher accuracy. A more in-depth discussion of compiler optimizations can be found at [1] and [2], and on floating point arithmetic at [3].

Typically, increasing the compiler optimization level will give substantially quicker run times, but you have to ensure you maintain the required accuracy of the simula-tion. The scientist will need to decide how to balance between decreased run times and potentially decreased accuracy.

As we discuss in Section 8.4, good programming practice has many aspects includ-ing modularity - having discrete functions (or subroutines) written for specific tasks required by the simulation. Each of these discrete functions should be in its own dis-crete file. What this allows, and is often overlooked, is the opportunity to use different compiler optimizations for different files. So, if some functions are numerically unsta-ble they would still be compiled with a low optimization level. However, we do not need to reduce the optimization level for all of the code to that level, and we could use high optimization levels on those files containing a function that has no numerical sensitivity.

## 8.2.2  Profiling

Each chapter in this book has at least one code to cover one specific aspect of aerosol dynamics. Each of these is not insubstantial in length. You may already be considering that a comprehensive model of aerosol physics would need to incorporate several such codes, leading to a rather large piece of software. For example, the UK MetOffice "Uni-fied Model" has over a million lines of code [4] (and processes over 200 billion global daily data points [5]).

For such large codes, a process is required to determine which regions of the code should be considered for some form of optimization. By use of **profiling** we can identify which routines or functions, or even lines of code, are taking the most time. These form the best candidates to make faster since they have more significant impact than, say, halving the time for a region of code whose time contribution to the whole run time is already negligible.

Profiling involves using a **profiler** as a wrapper to obtain either deterministic or sam-pling data regarding the running code. The sampling approach records which part of your code is currently executing at a series of given intervals (usually much less than a second). This is usually sufficient to illustrate which parts of the code are consuming most time (for the given data set). You can then tackle those areas of code, to try and help the compiler to optimize further, or for the interpreter to handle more efficiently.

Having made any changes, your first next step should always be to check whether the code outputs are still satisfactory. There is little point having a faster version of your code that produces rubbish! You can then repeat this process, optimizing the next region of code consuming most time, and so on. You will get diminishing returns – taking more coding effort to save run time, but the time saved getting smaller and smaller – so will draw a line at some point, at which point you will have a faster and correct code but only for the data set you have been using. If this is a typical data set, then all is well; otherwise you should undertake the above process with a selection of data sets so you can ensure you do not inadvertently slow down the code when a certain type of data set exercises a different path of control within your code.

We mentioned deterministic profiling. This is applicable to interpretted languages and relates to recording what happens when each code statement is interpreted. Rather than sampling we therefore obtain a full picture of time spent during execution, albeit potentially with a higher overhead (of the time taken by the profiler itself). Similar to deterministic profiling is the concept known as **tracing**. Rather than showing where time is spent, tracing gives a time plot of what is happening when. Typically, one uses tracing for compiled languages when requiring to know the order of events and is of particular use when optimizing parallel programs (e.g. to see why messages may be delayed).

Many compilers are also capable of more than just translating a programming language in to machine code. They can report on what optimization transformations they have performed but also on what they have not been able to perform. You can use this part of the report to identify potential areas to tweak manually to aid the compiler in performing the optimization transformation. First, we look at using a profiler to speed up one of the codes from Chapter 7.

### 8.2.3   Case Study: Speeding-up Box Model

We examine the code from Chapter 7 that runs a box model simulation for the same physical situation but applies a number of different computer models (flat top etc.). We remove the plotting statements (lines 180-184 of "*box_model.py*") so there's no interactive element which distracts from the profiling process. Rather than run the box model to simulate an hour, we have chosen to simulate 30 seconds. This is a long enough run time to give meaningful data each time we undertake a profile, without having to expend unnecessary computing power and our time. The code is written in Python. While Python is an interpreted language, behind the scenes some caching takes place by the Python run time system as an aid to improving performance. We therefore run the code once before we do any profiling.

It is also important to realize that different hardware will give different timing data but that generally the pattern of where time is spent is likely to be similar for similar architectures. The data we present here is from running the code on an Intel i7-3770 (4 core "IvyBridge"), with 16 GB RAM, running Debian 10 with hyper-threading turned off. We do a base line run for comparison with our optimizations:

```
$ time python3 box_model.py
simulating using the hybrid-bin method
Only full-moving scheme can used with the parcel setup
simulating using the quasi-stationary method
Only full-moving scheme can used with the parcel setup
simulating using the moving-center method
Only full-moving scheme can used with the parcel setup
simulating using the flat-top method
```

```
Only full-moving scheme can used with the parcel setup
simulating using the full-moving method

real    0m53.025s
user    0m56.443s
sys     0m6.613s
```

This shows the wall clock time to run the box model with these settings is 53.0 seconds and the CPU ("user") time of 56.4 seconds (so there was a little use of more than 1 processor core).

There are various profilers available. For Python the popular choices are Profile and cProfile which will give profiling information but only at the function level. There are many ways to use the cProfiler, but to quickly profile the full "*box_model.py*" code from Chapter 7:

```
python3 -m cProfile -o box_model.prof box_model.py
```

which saves all the profiling statistics in the file "box_model.prof." We then employ "pstats" on this file to sort and print useful data. Let us consider the top ten "cumulative time" entries

```
>>> import pstats
>>> p = pstats.Stats('box_model.prof').strip_dirs().sort_stats('
    cumtime').print_stats(10)
Thu Jul 15 11:46:30 2021     box_model.prof

         3668909 function calls (3661083 primitive calls) in 56.389
    seconds

   Ordered by: cumulative time
   List reduced from 2991 to 10 due to restriction <10>

   ncalls  tottime  percall  cumtime  percall filename:lineno(
    function)
   477/1     0.003    0.000   56.389   56.389 {built-in method
    builtins.exec}
       1     0.000    0.000   56.389   56.389 box_model.py:4
    (<module>)
       1     0.000    0.000   56.053   56.053 odepack.py:26(odeint)
       1     0.858    0.858   56.053   56.053 {built-in method scipy
    .integrate._odepack.odeint}
   27569    27.799    0.001   55.195    0.002 differential_equations
    .py:6(dy_dt)
   27598    10.894    0.000   12.375    0.000 auxiliary_functions.py
    :56(initialize_variables)
   330828    8.100    0.000    8.203    0.000 auxiliary_functions.py
    :4(diffusion_coefficient)
   330828    3.666    0.000    3.666    0.000 auxiliary_functions.py
    :37(Kelvin_effect)
   358397    2.336    0.000    2.336    0.000 {built-in method
    builtins.sum}
   27598     1.260    0.000    1.307    0.000 auxiliary_functions.py
    :43(calculate_kappa)
```

Cumulative time relates to time spend within the named function *and* all the functions that it calls (and that they call). We can see that the top level "box_model" takes 56.4 seconds. (We bear in mind the statistical nature of the profiling so only quote to one decimal place.) Of this time, the "odeint" function (in a file "*odepack.py*") has a cumulative time of 56.0 seconds mainly to the built-in scipy function. (Note that we can repeat the statistics analysis but without the "strip_dirs()" method to get the full path file of source files for these scipy "ode" functions.) For all these functions discussed so far we can see that their internal time (from the "tottime" column) is negligible, so we need to keep diving down the table. We note that the calculation of "dy_dt" and the "initialize_variables" functions have internal time as a significant proportion of their cumulative times.

We interpret this that the time is spent not in ODE solver itself but within the forming of dy_dt, and that in turn, most of that time is actually within the function "initialize_variables" within the file "*auxiliary_functions.py*."

A limitation of Profile and cProfile is that they do not provide profiling at the line level. We can therefore also use "pprofile" [6] to gain a deeper understanding of the "initialize_variables" function. Note that on a Debian system running Python 3, the command line interface to run "pprofile" is "pprofile3" but it can also be installed via "pip" and run within the Python interpreter.

```
pprofile3 -o pprof.txt box_model.py
```

This runs deterministic profiling (by default) and will give a *long* (350K lines) output to the file "pprof.txt," with every statement of every executed function having profile data listed. Manipulating the output under Linux we can view the names of the top ten most time consuming files:

```
$ grep -E '^File' pprof.txt |grep -v '0.00%'|head -20
File: /home/mkb/HEC/book/chpt7-boxModels-Harri/
    differential_equations.py
File duration: 73.6149s (46.56%)
File: /home/mkb/HEC/book/chpt7-boxModels-Harri/auxiliary_functions.
    py
File duration: 73.1373s (46.25%)
File: /home/mkb/HEC/book/chpt7-boxModels-Harri/parameters.py
File duration: 3.92591s (2.48%)
File: /usr/lib/python3/dist-packages/scipy/integrate/odepack.py
File duration: 1.30699s (0.83%)
File: <frozen importlib._bootstrap_external>
File duration: 0.706139s (0.45%)
File: <frozen importlib._bootstrap>
File duration: 0.68939s (0.44%)
File: /usr/lib/python3.7/inspect.py
File duration: 0.670048s (0.42%)
File: /usr/lib/python3.7/sre_parse.py
File duration: 0.638671s (0.40%)
File: /usr/lib/python3/dist-packages/matplotlib/_cm_listed.py
File duration: 0.296832s (0.19%)
File: /usr/lib/python3/dist-packages/matplotlib/__init__.py
File duration: 0.252546s (0.16%)
```

It is key to note that there is more overhead with deterministic profiling as can be seen from the indicated time now being approximately 73 seconds in total (compared to 56 seconds). This illustrates that it is often the patterns (not the absolute timing data) that are important within the profiling process. From this output we see that the the "differential_equations" and "auxiliary_functions" functions are (as before) the key time-consuming functions within the box model simulation.

We can then examine line by line each of these and we see that line 84 of the file "*auxiliary_functions.py*" as taking most time within that given function. The columns are line number, number of calls to that statement, total time for that statement, average time for that statement, percentage contribution to total run time, and the statement itself:

```
76|          0|          0|          0| 0.00%|       # Sum up
the volumes of all species to obtain
77|          0|          0|          0| 0.00%|       # the
total volume of the droplet
78|     358774|    1.97225|  5.4972e-06| 1.25%|       for
species in All_species:
79|          0|          0|          0| 0.00%|
80|          0|          0|          0| 0.00%|
# find the indices of each species
81|     331176|    1.68868|  5.09903e-06| 1.07%|
bins = range(Nb*index[species],Nb*index[species]+Nb)
82|          0|          0|          0| 0.00%|
83|          0|          0|          0| 0.00%|
# volume of one soluble species in each bin, in Eq (2.70)
84|     331176|    7.14521|  2.15753e-05| 4.52%|
V_si[species]=y[bins]*M[species]/rho[species]/np.maximum(N_m,
N_low_limit)
85|          0|          0|          0| 0.00%|
86|          0|          0|          0| 0.00%|
# total volume of particles in each bin, Eq (2.67)
87|     331176|    2.65936|  8.03004e-06| 1.68%|
V_T+=V_si[species]
88|          0|          0|          0| 0.00%|
89|     331176|    1.61037|  4.86258e-06| 1.02%|
if species == 'H2O':
90|          0|          0|          0| 0.00%|
# volume of water in droplets
91|      27598|   0.171452|  6.21247e-06| 0.11%|
V_w+=V_si[species]
92|          0|          0|          0| 0.00%|
93|          0|          0|          0| 0.00%|
else:
94|          0|          0|          0| 0.00%|
# volume os the solid part of droplets (see Eq (2.70))
95|     303578|    1.86477|  6.14265e-06| 1.18%|
V_s+=V_si[species]
96|          0|          0|          0| 0.00%|
97|          0|          0|          0| 0.00%|
# total number of moles in individual bins
98|     331176|    5.89494|    1.78e-05| 3.73%|
mol_tot+=y[bins]
99|          0|          0|          0| 0.00%|
100|          0|          0|          0| 0.00%|
# Calculate the droplet radius
101|      27598|   0.348389|  1.26237e-05| 0.22%|
R_wet = (3.0/(4.0*np.pi)*V_T)**(1./3.)
```

From experience (and as could be determined from first principles) we know that repeating an operation often takes longer than doing it once, saving to an intermediate variable and then using that intermediate result thereafter. We also know that the cost of a division is expensive (relative to summation and multiplication).

By examination of "species" loop enveloping line 84 we see that the values of $N\_m$ and $N\_low\_limit$ do not vary during loop execution. Therefore we can pop the calculation of the reciprocal of their maximum to before the loop, save the result in an intermediate variable *recip* and then multiply by that value within the loop. We thus have

```
# HEC: move loop invariant factor outside of loop to save
recalculations
recip = 1.0/np.maximum(N_m, N_low_limit)

for species in All_species:

    # find the indices of each species
    bins = range(Nb*index[species],Nb*index[species]+Nb)

    # volume of one soluble species in each bin, in Eq (2.70)
##  V_si[species]=y[bins]*M[species]/rho[species]/np.maximum(N_m
    , N_low_limit)
    V_si[species]=y[bins]*M[species]/rho[species] * recip

    # total volume of particles in each bin, Eq (2.67)
    V_T+=V_si[species]


    if species == 'H2O':
        # volume of water in droplets
        V_w+=V_si[species]

    else:
        # volume os the solid part of droplets (see Eq (2.70))
        V_s+=V_si[species]

    # total number of moles in individual bins
    mol_tot+=y[bins]

# Calculate the droplet radius
R_wet = (3.0/(4.0*np.pi)*V_T)**(1./3.)
```

We need to run this modified version initially to set up the Python caches and then we can run again and observe the total run time is better than halved

```
$ time python3 box_model.py
simulating using the hybrid-bin method
Only full-moving scheme can used with the parcel setup
simulating using the quasi-stationary method
Only full-moving scheme can used with the parcel setup
simulating using the moving-center method
Only full-moving scheme can used with the parcel setup
simulating using the flat-top method
Only full-moving scheme can used with the parcel setup
simulating using the full-moving method

real    0m25.348s
user    0m26.952s
sys     0m2.883s
```

Furthermore, profiling the new version shows that the cumulative time taken within the file *"auxiliary_functions.py"* has better than halved (from 73.1 seconds to 34.1 seconds):

```
$ pprofile3 -o new-pprof.txt box_model.py
simulating using the hybrid-bin method
Only full-moving scheme can used with the parcel setup
simulating using the quasi-stationary method
Only full-moving scheme can used with the parcel setup
simulating using the moving-center method
Only full-moving scheme can used with the parcel setup
simulating using the flat-top method
Only full-moving scheme can used with the parcel setup
simulating using the full-moving method

$ grep -E '^File' new-pprof.txt |grep -v '0.00%'|head -20
File: /home/mkb/HEC/book/chpt7-boxModels-Harri/auxiliary_functions.
    py
File duration: 34.1449s (44.71%)
File: /home/mkb/HEC/book/chpt7-boxModels-Harri/
    differential_equations.py
File duration: 34.0418s (44.57%)
File: /home/mkb/HEC/book/chpt7-boxModels-Harri/parameters.py
File duration: 1.82516s (2.39%)
File: <frozen importlib._bootstrap_external>
File duration: 0.684319s (0.90%)
File: <frozen importlib._bootstrap>
File duration: 0.666788s (0.87%)
File: /usr/lib/python3.7/inspect.py
File duration: 0.649586s (0.85%)
File: /usr/lib/python3.7/sre_parse.py
File duration: 0.625644s (0.82%)
File: /usr/lib/python3/dist-packages/scipy/integrate/odepack.py
File duration: 0.364391s (0.48%)
File: /usr/lib/python3/dist-packages/matplotlib/_cm_listed.py
File duration: 0.296762s (0.39%)
File: /usr/lib/python3/dist-packages/pyparsing.py
File duration: 0.248438s (0.33%)
```

And finally, we run with sampling profiling that illustrates how the cumulative time for function "initialize_variables()" has fallen from 12.4 to 5.4 seconds.

```
python3 -m cProfile -o new_box_model.prof box_model.py
```

from which "pstats" shows us

```
>>> p = pstats.Stats('new_box_model.prof').strip_dirs().sort_stats('
    cumtime').print_stats(10)
Thu Jul 15 12:07:03 2021    new_box_model.prof

        1905833 function calls (1898007 primitive calls) in 25.420
    seconds

   Ordered by: cumulative time
   List reduced from 2991 to 10 due to restriction <10>

   ncalls  tottime  percall  cumtime  percall filename:lineno(
    function)
    477/1    0.003    0.000   25.420   25.420 {built-in method
    builtins.exec}
```

```
      1    0.000    0.000   25.420   25.420 box_model.py:4(<module
>)
      1    0.000    0.000   25.088   25.088 odepack.py:26(odeint)
      1    0.358    0.358   25.088   25.088 {built-in method scipy
.integrate._odepack.odeint}
  12876   12.577    0.001   24.730    0.002 differential_equations
.py:6(dy_dt)
  12905    4.773    0.000    5.402    0.000 auxiliary_functions.py
:56(initialize_variables)
 154512    3.506    0.000    3.555    0.000 auxiliary_functions.py
:4(diffusion_coefficient)
 154512    1.629    0.000    1.629    0.000 auxiliary_functions.py
:37(Kelvin_effect)
 167388    1.190    0.000    1.190    0.000 {built-in method
builtins.sum}
  12905    0.527    0.000    0.548    0.000 auxiliary_functions.py
:43(calculate_kappa)
```

A minor detail we observe from this output that the times for other functions has also changed. This is due to a combination of the statistical nature of this profiling and that it is possible some changes in one function may lead to effects in others (e.g. due to different memory access patterns).

To recap, we have shown how to use Python profiling to identify which statements to examine. We have applied our knowing of optimization and moved a single loop-invariant statement out of a loop and replaced a division within the loop by a multiplication. These minor changes have resulted in the same results for code running twice as fast. We have also seen that patterns are important when interpreting profiling results, and this is particularly true for Python.

### 8.2.4 Vectors

In recent years, processors have become multi-core and most cores now have wide vector units. For example, a Intel Skylake core supports AVX-512, AMD EYPC Rome supports AVX-256, and ARM has its configurable Scalable Vector Extension (SVE). These, in the Flynn Taxonomy [7], [8] can all be considered "Single Instruction, Multiple Data" (SIMD) architectures.

The "Advanced Vector eXtensions" (AVX) on Skylake is 512 bits wide, and for Rome is 256 bits wide. For single precision ("float" in C, C++) the typical implementation is 4 bytes which is 32 bits, so a 512 bits wide vector unit can concurrently process $512/32 = 16$ contiguous vector elements. For double precision ("double" in C, C++), a 512 bits wide vector unit can concurrently process 8 contiguous vector elements.

The ARM model allows the vendor of a specific processor fabrication to define the width of the implemented vector unit. However the ISA is written such that a code does not have to recompiled for different implementations and is portable across all processors supporting any SVE, although the performance will vary.

The potential increase in speed for the overall code will depend heavily on how much of the time-consuming elements can be vectorized as well as the width of the vector unit. While it is possible to write vector intrinsic function calls to employ use of the vector units, it is more usual to leave it to the compiler, with an appropriate optimization level. As discussed above, a compiler can re-arrange code statements and loop orderings, not only to maximize data re-use but also to use available vector units more effectively.

### 8.2.5 Hand Holding the Compiler

Compilers however are just a tool. A very smart tool, yes, but also very generalized. A compiler has to do the best job across all potential types of user codes with its optimizations speeding up results while avoiding the generation of wrong results. A specific example is that unless a compiler can deduce it is safe to optimize (or vectorize or parallelize) a section of code then it will not do so. Often the person implementing an algorithm will understand more that a compiler can deduce, and it may be logically safe to vectorize a loop for the given implementation but not in the generality that the compiler has to presume. For example, if we were updating contiguous elements of a vector, x, in this manner

```
x[i] = x[i+k] + y
```

for values of $i = s, s+1, s+2, ...., s+N$ then there may be relevant loop carried dependencies which means the updates to $x[i]$ cannot be carried out independently. A compiler may therefore refuse to vectorize this at all. However, the user may know that the practical values of $s, k$, and $N$ (for their given simulation) means that all updates could be carried out independently and thus the loop over $i$ could be vectorized. Various hints can be given to the compiler to encourage or force a loop to be vectorized [9].

We have seen in this section how the simple consideration of "portability" leads to a consideration of several important matters including hardware, language, and compilers. In the next section, we introduce the matters concerning parallelization with the goal of scaling up what we can from 1 to N "processing elements," where the term "processing elements" can mean a number of things.

### 8.2.6 Case Study: PartMC

As noted earlier, it is important to consider the hardware details of the platform upon which a code is running. Typically, if there is no memory contention, then comparing times between two processors (of the same ISA) the performance will be proportional to the clock speed (frequency) of the processor cores. It can also be shown[1] that the instantaneous power, $P$, consumed by a processor core is proportional to the frequency, $f^n$ where $1 \leq n \leq 3$. The import of the power consumed will dependent upon the setting e.g. for embedded use there may be a power cap and for HPC systems their size may impact upon system settings. To determine the magnitude of the number of processor cores in the most powerful supercomputers in the world, we can visit the biannual "Top 500" listing [10]. Governments and companies can run a LINPACK benchmark [11] on their machines and submit, with the "Top 500" then ranking these by the fastest (out of those submitted) by speed of performing this set of matrix-matrix operations. We will see that the number of processor cores in the top ten machines (as of the June 2021 listing) range from 448 thousand to 10.6 *million* processor cores. Many modern processors can run a range of frequencies and different sites will implement different policies,

---

1 Simplicistly, consider that power $P$ is proportional to $V * I$ where $V$ is the voltage and $I$ the current. The current $I$ would, by Ohm's Law, be proportional to $V/R$ where R is the resistance but a processor core is more complex and the current has some dependency on the voltage. Further, the voltage of a processor core usually varies (non-linearly) with the frequency.

such as "maximum compute power" (with each processor running at (fixed) maximum frequency) or a "balanced energy" (where processor frequencies fall when cores are not in use but rise to a "turbo" frequency when they are in use). These frequency-scaling technologies are present in most modern machines, whether laptops, desktops, or workstations. Depending upon how long a code runs, and what else may be occurring on a processor, the "balanced energy" approach may take a non-deterministic time to ramp up which may be non-negligible compared to the total run time of the simulation. For our exploration of optimizing the coagulation codes from Chapter 5, namely the "partmc" simulation code, we have therefore fixed the frequency of our platform (see Section 8.2.3) to 3.4 GHz. Power management of processor cores is rather complex and the details of how to alter a default set-up will vary between vendors. Our Intel i7-3770 "IvyBridge" system is running Linux and [12] gives a thorough overview. In summary, following [13] we first need to reboot to obtain a kernel using the ACPI power management kernel module since this allows a "userspace" configuration (that the default intel_pstate power management kernel does not support). We can then set the frequency to be fixed, and results in this section are for a 3.4 GHz frequency which matches the base frequency of the processor.

We will examine how to measure the performance of the stochastic particle-resolved aerosol model "PartMC" described in Section 5.5. We obtain PartMC by cloning the git repository:

```
git clone https://github.com/compdyn/partmc
```

which creates all the required source files and some example scenarios in to a new subdirectory called "partmc." First one needs to build the model and then empower the given scenario examples to the required model scenario.

As discussed in Section 8.4, reuse is a good software engineering principle. As Newton remarked [14], by standing on the shoulders of giants (those who have come before and laid some groundwork) we can see further (and code more efficiently). In this case, PartMC builds on the shoulder of a giant known as the "netCDF" library [15] – in turn building on shoulders of some NASA scientists who wrote the CDF libary. NetCDF is a format to save scientific data in a portable and space-efficient manner. The format is binary which means you will need to have access to a netCDF library on your computing platform in order to read or write such files.

Since PartMC uses netCDF to save its outputs, a netCDF installation is required in order to build PartMC. Full details of how to do this on all available platforms can be found on the official netCDF website at https://www.unidata.ucar.edu/software/netcdf/. To build on our Debian-10 platform we used "apt-get" to install the libnetcdff6 libnetcdff-dev libnetcdf13 and libnetcdf-dev packages. This is one example of the process know generally as "resolving dependencies" for installing software. It is an important and useful skill to learn if you are installing, using, or developing, software across systems (see Section 8.1).

The official PartMC web page [16] describes how to do a generic install of PartMC (and lists further dependencies for some of the advanced uses of the simulation suite). This includes setting some environment variables; for example, on Debian 10 using the netCDF packages listed above:

```
export NETCDF\_C\_LIB=/usr/lib/x86\_64-linux-gnu/libnetcdf.so.13
export NETCDF\_FORTRAN\_LIB=/usr/lib/x86\_64-linux-gnu/libnetcdff.so
    .6.2.1
export NETCDF\_INCLUDE\_DIR=/usr/include
```

Note that on systems that make use of "modules" (e.g. most HPC systems) these absolute paths will vary, and relate to the module loaded; but loading the module may also set these for you. If you are unsure of the absolute paths on a non-"modules" system, then the command "locate" (when supported) is useful to determine these (and likely to be very similar to above). On a "modules" system, you can use the "module show xxx" to see what paths are used when actually loading module xxx.

There are various systems that support building of software, and PartMC uses the "cmake" system. A key component of the cmake system is the separation of source files in one directory to where the build itself takes place. On a Linux system you should make a new subdirectory of the "partmc" directory created when you performed the "git clone" (above), such as "BUILD-partmc" (via command "mkdir BUILD-partmc") and change in to this directory ("cd BUILD-partmc"). From this subdirectory you can run the GUI-based "ccmake .." which then configures the build for you given system (making use of the environment variables you set above). Within ccmake you will use "c" to configure (and check and fix any errors) and "g" to generate the actual makefiles to use in the next step.

As per the PartMC instructions you then run "make" (to build all the executable files) and "make test" (to ensure all "unit tests" are passed, as discussed in Section 8.4.5).

Using a build using the Fortran compiler "f95" (version 8.3.0) on the platform described above, we can then construct an appropriate set of input files. As per Section 5.5.1, we use a particle-resolved run with Brownian coagulation with 1M particles ($n\_part$) for a simulation time ($t\_max$) of 86,400 seconds with $del\_t$ set to 60 seconds. Since we are interested in its performance we can ignore the extraction and processing steps of the generic scripts provided in the "partmc/scenarios" directories, and we turn off file output ($t\_output = 0$) and limit the information displayed on the screen by setting $t\_progress$ to be half of $t\_max$ (i.e. $t\_progress = 43200$).

As discussed previously, PartMC is a *stochastic* model. This means it makes use of random numbers, namely to set initial conditions and determine whether coagulation occurs for a specific collision. Let us run the example three times to see how long it takes on our system. As we can see from Figure 8.1, the number of particles ($n\_part$) and number of coagulations ($n\_coag$) vary on each run. We also see that the total time varies. The reason that the numbers vary is related to the behavior of the "random number generator" (RNG) used. It is important to understand the principles of how RNGs work so you can ensure you apply them appropriately (and how you apply them will vary depending upon your aims, as we will see). The RNGs that we encounter in our programming generate a sequence of pseudo-random numbers, with the $i^{th}$ value being calculated from the $(i - 1)^{th}$ value, as discussed in the likes of chapter 3 of the seminal book by Knuth [17]. The actual sequence will therefore depend on some initial value, which is known as the "seed" (although often the initial value is a function of the seed, not the seed itself). We can therefore deduce that the RNG used within f95 must use a different seed each time the RNG is initially started within a given execution of the code - as shown by $n\_part$ at $t\_sim = 0$ varying (this is not exactly 1,000,000 due to the Poisson sampling, as discussed in Chapter 5). For timing purposes we require bit-reproducible results. We can do this by manually setting the RNG seed (via the 'spec' file we set "rand_init 1" or any other integer), and re-running we will get the same numerical results, as seen in Figure 8.2.

It is important to consider whether you require to always use the same seed and sequence (for example when optimizing so you only time changes due to the optimization, not due to the use of different random numbers) or require a different seed and sequence (for example if undertaking ensemble modelling). PartMC typically requires the latter and you can see from Figure 8.1 that the *pattern* of the number of coagulations is the same for all three runs (namely, a peak at $t\_sim = 43,200$ seconds).

It is beyond this book to discuss in detail how to use RNG within parallel programs, but the same considerations should be made. For example, if you have several concurrent threads wanting a sequence of random numbers, how would you do this? Do you have a different seed per parallel processing element, or do you create one sequence on a 'root' thread and then distribute and use a subset for each parallel processing element? Similarly, different compilers may use a different internal RNG algorithm so comparing, say, outputs produced by running executables compiled by GNU and Intel compilers may not be straightforward since they produce different sequences even when given the same integer seed.

To illustrate the importance of compiler optimizations (as discussed in Section 8.2.1), we do a number of builds with no optimization flags set, with "-O0," "-O3," and "-Ofast." We run each of these and ensure we get the same output, and then we compare the times; see Table 8.1.

Compared to the time without any compiler optimization flag we see that using "-O3" saves 17% and using "-Ofast" saves 35% of that time. That is, by using "-Ofast" we

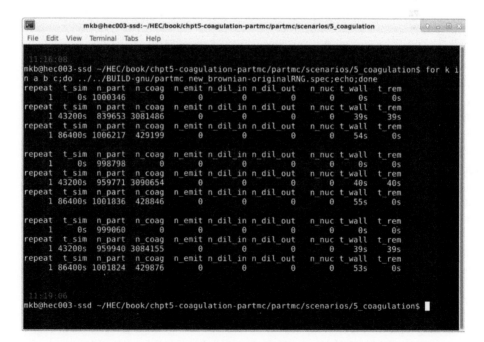

**Figure 8.1** Initial run of PartMC for Brownian coagulation. *Source*: The Linux Foundation.

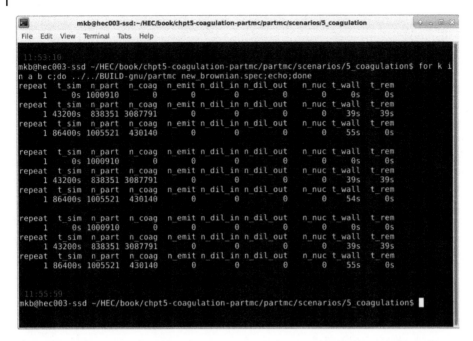

**Figure 8.2** Initial run of PartMC for Brownian coagulation with fixed RNG seed. *Source*: The Linux Foundation.

**Table 8.1** PartMC timings for various optimizations.

| Compiler flags | Time (seconds) |
|---|---|
| no flag | 54 |
| -O0 | 54 |
| -O3 | 43 |
| -Ofast | 35 |

go 1.5 times faster than not setting the flags at all. Note that when running "ccmake" you can switch between a set of pre-determined targets such as "None," "Debug," and "Release." If you do not set the value of "CMAKE_BUILD_TYPE" it defers to building the "None" target which will compile with no optimization flags. However, setting within "ccmake" the value of "CMAKE_BUILD_TYPE" to "Release" will use the "-O3" flags. Thus to get the optimal speed you will need to choose the advanced mode then set "CMAKE_Fortran_FLAGS" explicitly to "-Ofast" (leaving CMAKE_BUILD_TYPE undefined). You can confirm what flags are actually used by setting the "VERBOSE" variable for the "make" command:

We can use the "-fopt-info-optimized" flag for the f95 compiler to show what the compiler has optimized, and the vast majority is vectorization (see Section 8.2.4) and loop transformations (see Section 8.2.1).

### 8.2.7 Interpretted Languages

Not all languages have to be compiled – the process of taking a high-level language that humans can read and translating it to (efficient) low-level machine code that runs on a specific architecture. Many popular languages are "interpreted," which can be thought of as a basic "compile each line of code as we want to run it." From a functional viewpoint there is nothing wrong with this. Many interpreted languages are highly popular, due to their ease of learning, including Python.

## 8.2.8  Case Study: Droplet Growth Equation

As we have discussed the choice of programming language and the specific implementation choices made during coding of your algorithm, can have a significant impact on computational efficiency. We have indicated that interpretted languages are generally not as fast as compiled languages, due to the lack of optimizing transformations. To tackle this issue for Python, we can look at the Numba package [18]. On the first pass of a Python function, Numba translates it to optimized machine code (and vectorizes where possible), which can then be used in future invocations of the function. This offers a relatively simple route, for supported functions, to improvements in performance. Numba works well for loops over numpy arrays, functions, and loops, but not so well for distributions it has no knowledge of (such as pandas).

To examine the usefulness of Numba, let us consider the droplet growth equation examples given in Section 2.4.1. Recall that we used the "odeint" solver and plotted the output. Here we are now interested in the time taken so we remove the plotting functionality. We also introduce use of the inbuilt "timeit" Python function to time 100 runs of the ODE solver (which calls the "dy_dt" function we aim to accelerate). We can ensure any caching happens by doing an initial call to the ODE solver before we do this timing. The revised version of *chap2_alg4_Droplet_growth_single_bin.py* using the native Python only is given as *chap8_alg1_Droplet_growth_single_bin_speedtest.py*. The first change is the import of "timeit"

```
import timeit # for timing
```

Listing 8.1 Importing the timeit function.

We also remove the plotting and related statements, replace the single call to the "odeint" function as a function call of "find_solution," perform a single call to "find_solution" (to handle any caching) and then time 100 calls to "find_solution" and output how long that takes.

```
# Call the ODE solver with reference to our function, dy_dt, setting
    the
# absolute and relative tolerance.

# Do timing for native python: run once and then time next 100
    instances
def find_solution():
    solution = odeint(dy_dt, array, t, rtol=1.0e-6, atol=1.0e-4,
    tcrit=None)
find_solution
print("Time using native Python: ", timeit.Timer(find_solution).
    timeit(number=100))
# ignore output since we continually integrating forward over t
# no plot required
```

Listing 8.2 Timing the native implementation.

We run this three times, since there will be some small variation in time, as shown in the top half of Figure 8.3. We observe that the time taken to do 100 runs of the ODE solver, using the native Python implementation of "dy_dt," takes 7.7 seconds (with all remaining code taking approximately 0.3 seconds).

To invoke the Numba "just in time" (JIT) compilation we use the "@jit" decorator. We shall apply this to the "dy_dt" function for the single bin example. In order to use Numba we have to make three sets of changes which you can find in file *chap8_alg2_Droplet_growth_single_bin_numba_speedtest.py.* Firstly we have to import numba itself:

```
import timeit # for timing
import numba as nb
```

Listing 8.3 Importing the numba function.

Secondly, we include the "@jit" decorator

```
@nb.jit(nb.float64[:](nb.float64[:],nb.float64), nopython=True,
    cache=True)
def dy_dt(array,t):
```

Listing 8.4 Importing the numba function.

In addition to adding the @jit decorator we cast the input and output variables to be type "nb.float64" that Numba can optimize.

Numba has two compilation modes: "nopython" or "object" mode. While the former produces much faster code, any errors in the function definition can force a fallback to the (slower) object mode. To prevent Numba from falling back, and instead raise

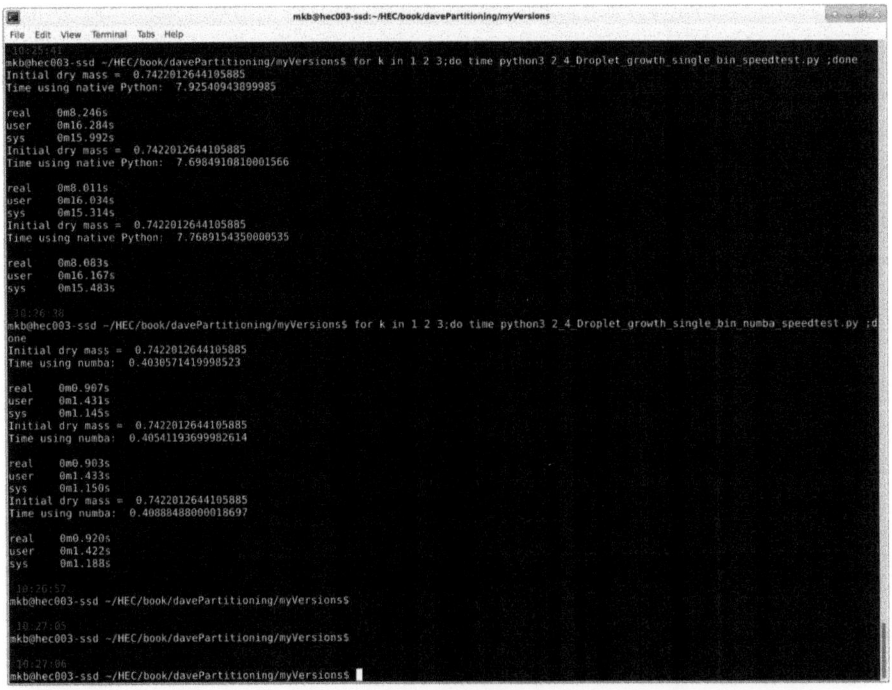

Figure 8.3 Single bin droplet simulation accelerated by use of Numba. *Source*: The Linux Foundation.

an error (so we can amend the source code accordingly), we pass "nopython=True."
We also use the option of caching of compiled functions by passing "cache=True"
which instructs Numba to write the result of function compilation into a file-based
cache.

Finally, we ensure we use the type "nb.float64" (rather than "float") for the variables we have declared of that type (via the "@jit" decorator). This also allows us to use numpy (sic) ufuncs ("universal functions") such as numpy.sum and numpy.exp. For example, in initializing an empty array for storing the mass concentration of all condensates in a given bin, we have used:

```
dy_dt_array = np.zeros((num_species+num_species*num_bins), dtype
    =nb.float64)
```

Listing 8.5 Using numpy ufunc to zero array.

Finally, we update the timing (merely comment and print output):

```
t = np.linspace(0, 10000, num=1000)

# Call the ODE solver with reference to our function, dy_dt, setting
    the
# absolute and relative tolerance.

# Do timing for numba: run once to compile and then time next 100
    instances
def find_solution():
    solution = odeint(dy_dt, array, t, rtol=1.0e-6, atol=1.0e-4,
    tcrit=None)
find_solution
print("Time using numba: ", timeit.Timer(find_solution).timeit(
    number=100))
```

Listing 8.6 Timing the Numba implementation.

We observe the timings shown in the lower half of Figure 8.3. We note that using
Numba the time taken to do 100 runs of the ODE solver now takes 0.4 seconds compared to using the native Python implementation of "dy_dt" taking 7.7 seconds. That
is a speed-up factor of 19 times. We also note for the Numba implementation that all
remaining code takes approximately 0.5 seconds (compared to native of 0.3 seconds)
showing that even with the overhead of compiling the "dy_dt" function that Numba
accelerates the time to solution.

For the multibin case we can also wrap our function using the @jit decorator and
compare the simulation time using Numba to that using native Python. In this case we
observe an increase in performance of $\approx 26x$.

There are many more options on using Numba, including parallelization that we
encourage the reader to explore to maximize the performance of their Python code.

## 8.3 Parallelization

As we discussed above, a CPU processor may have several cores, and a motherboard,
or node, may have one or more CPU processors. So, whether it is a modern laptop,

desktop, or workstation, you are likely to have several cores available for running your calculations. In a "cluster" or an "High Performance Computing" (HPC) facility, you would have a number of nodes connected together, and we discuss this situation below. Whether you have a single node or many nodes, you should consider seriously how to make effective use of these resources, and which is most appropriate to your modelling.

In terms of "thinking parallel," the key point to understand is that you have to identify at a logical or physical level where you can distribute independent segments of the work that can be carried out independently of the other work segments (which also means performing those work segments in any order). In such cases, these independent segments of work can be computed in a concurrent manner. If we have a number of "parallel processing elements" (we discuss what these may be shortly) then we can distribute the work segments over these parallel processing elements. In theory, if we can divide all of the work between $N$ parallel processing elements, and there is no overhead of doing so, then the work can be completed $N$ times more quickly. In practice, there will be inherently serial proportions of a code and there will be a number of overheads so we would likely process the work $1 < M < N$ times more quickly. This $M$ value is known as the "speed-up factor" (of the implementation on $N$ parallel processing elements with respect to the time on 1 parallel processing element).

As discussed in Section 8.1, it is always important to ensure correctness of the implemented solution over the speed of obtaining results.

Let us take a simple 2D example to illustrate. We have $N$ aerosol particles and wish to model how each particle is transported by the wind, and we presume particles would collide elastically were they to hit each other. So each particle is moved by a wind function, $w(x, y, t)$ and if the distance between particle $i$ and $j$ is zero we reverse the velocity in each of the $x$ and $y$ directions for each particle. Theoretically we need to simultaneously update all particles' new positions, based upon the wind function, and capture when any two particles would hit, amending their velocities (and taking account of this for all other particles). In practice, we *approximate* the solution by taking very small timesteps (so that we do not have to trace a long path for each particle). Further, we impose some ordering by looping over the particles according to an index – and this we have arbitrarily given them for computational purposes. So our serial implementation of this algorithm would be written as pseudo-code given in Algorithm 8.1

Note that we have no physical reason to process each particle one at a time, nor in the order shown in the algorithm. Indeed, in real life each particle will have a continuous influence on each other particle. However, by making $t\_delta$ sufficiently small, this algorithm will give a satisfactory approximate solution. Similarly, as $t\_delta \to 0.0$ we can rewrite Algorithm 8.1 to take a copy of the *velocity* values and then update using these saved values, as shown in Algorithm 8.2. If we now consider the outermost $i$ loop carefully, we can see that we could update each *velocity*[$i$] value independently and in any order we wish. These are the conditions we require in order to be able to do these updates concurrently. So, we can distribute these updates to different parallel processing elements in order to process the simulation more quickly.

There are some dangers of parallel programming that we should warn you about! By parallelizing a code that produces a global sum (for example) from partial sums we will affect the ordering of the summations undertaken, which in turn may affect the accumulation of rounding error and thus produce a (slightly) different global sum. The impact of this difference should be carefully considered. Also, if parallelization is

---

**Algorithm 8.1:** Serial particle algorithm.

---

1 **Data:** *wind*() function used to update each particle velocity
2 **Data:** *t_delta* is set appropriately small
3 **Result:** *update_positions*() updates (x,y) position for each updated velocity
4 **for** $t \leftarrow t\_start, t\_start + t\_delta, \dots, t\_final$ **do**
5      **for** $i \leftarrow 1, \dots, N$ **do**
6          $velocity[i]+ = wind(x, y, t)$;
7          **for** $j \leftarrow 1, \dots, i - 1, i + 1, \dots, N$ **do**
8              **if** $position[i] \leq position[j]$ **then**
9                  $velocity[i] = -velocity[i]$;
10                  $velocity[j] = -velocity[j]$;
11          **end**
12      **end**
13      *update_positions*();
14 **end**

---

**Algorithm 8.2:** Parallel particle algorithm.

---

1 **Data:** *wind*() function used to update each particle velocity
2 **Data:** *t_delta* is set appropriately small
3 **Result:** *update_positions*() updates (x,y) position for each updated velocity
4 $velocity\_old = velocity$;
5 **for** $t \leftarrow t\_start, t\_start + t\_delta, \dots, t\_final$ **do**
6      **for** $i \leftarrow 1, \dots, N$ **do**
7          $velocity[i]+ = wind(x, y, t)$;
8          **for** $j \leftarrow 1, \dots, i - 1, i + 1, \dots, N$ **do**
9              **if** $position[i] \leq position[j]$ **then**
10                  $velocity[i] = -velocity\_old[i]$;
11                  $velocity[j] = -velocity\_old[j]$;
12          **end**
13      **end**
14      *update_positions*();
15 **end**

---

applied where the updates are *not* independent, then totally incorrect results may occur (due to "race conditions" updating a shared variable, for example).

Let us take a quick look at how the choice of "parallel processing elements" based on the hardware available and what parallel programming approaches we can use.

### 8.3.1 Making the Most of a Single Node

On a single node (such as a laptop and desktop), we may have one or perhaps two processors, each of which would one or more cores. In this case we would make use of two or more cores for parallel processing. For the main programming languages, we would recommend using OpenMP to parallelize your code. OpenMP works at the (CPU) thread level, where you typically have one OpenMP thread running per physical processor core and we consider the thread to be the parallel processing element. The official OpenMP website at https://www.openmp.org has many useful resources and gives the specification for C, C++, and Fortran bindings. It is also possible to use OpenMP from Java or Python. OpenMP is a portable thread-based standard for **shared memory programming** where each and every thread has access to the memory (RAM), as is the case on a single node. Other threading models exist such as pThreads (for C, C++) and Java threads.

OpenMP is based on a fork-join model which means the code starts running on one thread (e.g. for serial initialization) and when it encounters a user-defined parallel region, the operating system will spawn more threads. The work can then be shared over these threads such that the computational effort within that parallel region is split over concurrent threads to reduce the wall clock time taken. The programmer controls whether each thread shares a variable (so all threads can read but also write to the memory location of that variable) or has its own private copy of a variable (so each thread can independently update the value of a variable, but cannot share that value of the variable with other threads).

OpenMP can be used fairly quickly to parallelize independent loops to get reasonable speed-up factors (for those loops). Note that the maximum speed-up factor will be limited to the number of processor cores on the single node. So a 4 core, single processor, laptop could be used to get results 4 times more quickly (than a single core) but no faster.

### 8.3.2 Making Use of Multiple Nodes

To scale parallelism further than a single node requires the use of multiple nodes and exchanging data between the nodes by use of their interconnect. The *de facto* standard used for programming multiple nodes is the "Message Passing Interface" usually known just by its acronym "MPI" [19]. As the name reveals, to make use of multiple nodes requires passing of messages between "MPI processes" each of which has a unique "MPI rank" (an integer identifier counting upward from zero). Unlike OpenMP, the MPI run-time model is that each MPI process runs its own copy of the code and each has its own private copy of each and every variable. This approach is known as **distributed memory programming** and we consider the MPI process to be the parallel processing element, typically with one MPI process running on a physical processor core.

The work can be divided by, for example, using the MPI rank to divide up a loop. Consider a simple example where we have a loop with $N = 10$ iterations $i = 0, 1, 2, ..., 9$ to be run in parallel on $P = 2$ MPI processes. We can divide this such that the process with rank value 0 handles iterations $i = 0, 1, 2, 3, 4$ and that with rank value 1 handles $i = 5, 6, 7, 8, 9$. Clearly, this is programmable with each MPI rank starting its iteration at $rank * N/P$ and doing $N/P$ such iterations, for all cases where $N$ is divisible by $P$. When there is a remainder, we cannot ignore those iterations (since they are part of the original, serial, implementation) and can handle by giving one of the ranks all the remaining iterations to process. The MPI model is to write a single code that is run on each and every MPI process, but that the control flow will vary by use of MPI ranks as shown simplistically here. MPI requires much more user intrusion to parallelize a code than the use of OpenMP directives for shared memory programming.

Since each MPI process is running its own copy of the code and has its own private copies of variable, it will be necessary at some point to exchange data between MPI processes. This is achieved by calling MPI functions to pass messages. MPI requires more major code changes than OpenMP, and debugging is notoriously harder, but has the key advantage that one can use 100s or 1000s of processor cores concurrently. A key challenge is how to use such high processor core counts efficiently; while OpenMP can be applied loop-by-loop, efficient MPI requires the full computational load of a simulation to be distributed fairly over all the MPI processes.

One common technique, is to have a hybrid approach of using MPI to handle inter-node messages but to use OpenMP across the processor cores on a per-node basis. For example you would have one MPI process per physical socket, exchange some information at the start of a time-stepping iteration, and then use OpenMP parallelization across all cores of each node to process a sub-domain of a grid for that time step.

### 8.3.3 Other Technologies

We have focused on CPU technologies, including today's multi-core systems and briefly outlining how to use the thousands of nodes found in the most powerful supercomputers. There are other computing elements and approaches available, some in common use (as can be seen by examining the computer architectures of the "Top 500" [10]) and some more of an "emerging tech" nature, as discussed by conferences such as EMiT [20]. One technology that has emerged, and in common use in most supercomputers is the "graphical processing unit" (GPU). The modern computational GPU has evolved from the older units that helped accelerate displaying pixels on a screen. The modern computational GPU is commonly used to accelerate "kernels" of a simulation. Such GPUs are known as "thread throughput machines" and fall under the SIMT category of Flynn's taxonomy [7, 8]; see above. Consider the NVIDIA Volta VG100 used in two of the top ten supercomputers listed in the Top 500. These have 5,120 "CUDA cores" (i.e. they can run 5,120 lightweight threads which we would consider as the parallel processing elements discussed in Section 8.3) and a theoretical peak performance of 7.5 TFLOPS/sec for double precision calculations. To make efficient use of such hardware requires using all the threads, which practically means having a computation kernel (a heavily used subroutine, for example) that is processing a very large number of independent updates. Ideal for GPUs is a code comprising vector arithmetic, whereas unsuitable code for GPU is that which has lots of logical branches (for example "if" statements). There are several GPU manufacturers (NVIDIA, AMD, Intel, ARM) and for code to run across any GPU, the use of OpenACC [21] or OpenMP [22] is recommended. These are both directives-based approaches. For NVIDIA GPUs the use of their proprietary language CUDA [23] is likely to give higher performance than a directives-based approach, albeit requiring more input from the user to write CUDA kernels. Various extensions are now available to support use of GPUs from Python and Fortran.

It is key to note that while modern computational GPUs may have (say) 5,000 cores it is not true that porting a code to a GPU will make it 5,000 times quicker than running on a CPU. Firstly, note that a core of a GPU is very lightweight (less clever logic) and runs at a slower clock-speed than a CPU processor core. Secondly, some of the computation will remain on the processor and for that on the GPU there will be some overhead of transferring data across a PCI-Express lane from the CPU to the GPU, and back. And finally, we note that today's processors may have 64 (not a single) core. As [24, 25] and [26] showed while it is possible to accelerate aerosol models such as WRF-Chem, GEOS-Chem, COSMO-ART, SAPRC99 et al. the speed-up varies between 1.5 and 7 times depending upon the code and which CPU technology it is compared to. The limit on the performance is typically bounded for these codes by the large amount of data transfer compared to the amount of computation on the data.

Some researchers are also exploiting the virtues of Field Programmable Gate Array (FPGA) technologies, particularly since the tools are now being mature in supporting the port to these technologies. Historically, a FPGA has been harder to program than a CPU or a GPU, requiring knowledge of low-level programming languages (e.g. Verilog or VHDL). Recently, the use of "high-level synthesis" (HLS) tools allow easier programming of FPGAs and the technology now forms part of UK exascale roadmap [27] and European Processor Initiative [28, 33]. The basis of FPGA technology is configuring a grid of programmable silicon to implement the required algorithm. In theory having logic gates to run the algorithm (and not to be more general as would be the case for CPU) leads to faster implementations that are also an order of magnitude more efficient that the likes of GPUs. The general approach is to implement the algorithm in a "data flow" fashion (e.g. by rewriting the high-level C code), to optimize the computation per data element (e.g by some use of lower precision variables), and to make use of HLS pragmas to maximize the depth of pipelines on the silicon.

Further out on the horizon, quantum computing is making waves. Whereas traditional computing has bits that can be "0" or "1," quantum computing employs **qubits** whose state is all values between zero and one (via superposition).[2] Quantum computing also uses **entanglement** which means operations on one qubit affect simultaneously all of its entangled partners. Together, in theory, these principles (see also [29]) will allow quantum computers to undertake various operations (e.g. searches and finding minima) although current working quantum computers are limited by only having O(50) qubits. However, it seems too soon to tell whether it will come toward us as a tsunami of exciting opportunities or a more limited opportunity for those who are able to ride a small number of certain waves.

## 8.4 Collaborative Software Engineering

This book has discussed a lot about aerosol physics and chemistry and how to transform an algorithm describing the physical world (or indeed the physics of another world) into a form that computers can process, efficiently and correctly. We have covered the importance of portability and performance in terms of implementing an algorithm in one or more chosen computer languages. The art of writing computing codes can be thought of as **software engineering** and covers several key principles, among these being choices at each level of the **software development life cycle** (SDLC). Space and

---

2 However, at the point of any measurement, the state will collapse to a single observed value.

time limits us from an infinite discussion on these, but we outline the importance of style sheets, version control, and unit testing. Further information on these and many other aspects of software engineering can be found in textbooks such as [30, 31].

Underlying all the technical discussion of software engineering is the need for clear communication – whether that is in the "requirements capture" (what is it that the clients (which may be yourself) *really* want from the software?) or collaboratively working together via a single version control repository (not only in meaningful commit comments but also to ensure different parties are not making conflicting changes on the same file at the same time) through to the publication and presentation of your work.

### 8.4.1   Coding Stylesheets

Programming languages may be **statically typed** (as is typical of compiled languages) where it is necessary to declare the variable and its **type** (e.g. whether it presents an integer or single precision or a double precision real number) before use, or **dynamically typed** where they are inferred e.g. at run time for interpretted languages. Some languages are case sensitive for their variables, such that *no2* and *NO2* would be different variables, whereas others are case insensitive (so you can refer to the same variable as any of many of *No2*, *nO2*, *no2* and *NO2*). As codes grow in size, and the number of contributors also grow, you can see how a mix of different approaches can lead to unreadable and perhaps buggy code. A typical case of different approaches is for variable names with two or more parts. For example, would you name a variable *carbonmonoxide*, *CarbonMonoxide*, *carbonMonoxide*, or *carbon_monoxide*? These are all equally valid but imagine you have been asked to update the code and you expect one form but the previous author used another form.

This is just one example of why using a **coding stylesheet** aids code portability and readability; see [32], and implicitly also improving user productivity and resilience to introducing bugs. A coding stylesheet is a basic "recipe" for how to write code in a consistent manner. In this example it would explain how to name variables. It could also describe expected methods (e.g. whether there should always be a "get" and "set" method for an object).

### 8.4.2   Modularity and Re-use

It is good practice to break down your code in to a number of functions (or methods or modules), where each function has a clearly defined purpose, and a defined set of inputs and outputs. Functions should do one thing and have no side effects. For example, if you have a function to determine the mean of single precision input numbers then you should return the mean without changing the value of the input numbers, and give the function a meaningful name that describes its purpose (in line with your coding stylesheet).

By breaking the code down into simple functions, it is easier to read and you can use "unit tests" (see Section 8.4.5) to ensure working functions. You will also find that you can re-use functions, and this is good practise. Having a library of known working functions saves you from having to write and re-write functions, and the less you have to write then the less likely you are to introduce a bug.

### 8.4.3 Version Control

Whether working alone or with other team members, it is good practise to use a **version control repository** such as "subversion" [33] or "git" [34]. The repository is a place to keep known working snapshots of your code. You pull a copy of the latest snapshot to your local working environment, make some improvements, and once you are happy (e.g. everything still works) you can push the new working improvement to the repository. This allows you to make mistakes since you can just throw away a non-working version and just pull again a copy of the latest snapshot so you can try again. The version control software requires you to comment each push – which aids in documentation and later understanding *why* you made the changes - and also to label releases. This later point supports releasing different versions to your colleagues and communities.

A version control repository sits in the cloud. This allows many users to access it. As long as people are working on different functionalities and thus different files, it is possible to have "branches" of work, that can be combined later.

### 8.4.4 Software Development Life Cycle

Historically, codes were written, then bugs fixed, and perhaps new functionality added, commenting out older functionality, and applying more bug fixes and workarounds. Such codes can become unwieldy and highly expensive to maintain. One theory for the various stages of the life cycle of software development (from inception, requirements capture, design and then implement, test and release) came to prominence. The **waterfall method** broke down the life cycle in to a number of stages, each of which had to be completed (perhaps by a team specialized in that area) before moving on to the next stage. This works well when there are expert teams for a given area (since why should a programmer be an expert in understanding user requirements?) but does mean nothing at all is delivered until the final stage of the pipeline.

More recently, **agile methods** for the software development life cycle (SDLC) have risen to the fore. Simply put, "agile" is about getting something (that works even if only doing part of the full requirement) out of the door more quickly. Typically, a small team will both get some subset of the full requirements, then design, implement, and test these before delivering to the client and agreeing the next subset of requirements. of course, the client can be yourself so your agile approach to a full weather model may be to first get the transport methods working and release that as version one. Then to include some atmospheric chemistry, and release as version two. Contrast to the waterfall approach which will only release a version once it has both the transport and chemistry methods working.

There is much more that can be said on the agile approach, but let us focus on how we can help ensure that the incremental approach works.

### 8.4.5 Continuous Integration and Unit Tests

In the agile approach to software development, Continuous Integration (CI) and Unit Tests are frequently used. **Continuous Integration** is based upon incrementally improving the given software suite, making positive changes and checking these in regularly to the version control repository. In order to ensure that nothing has broken since the last check in, we can make use of **unit tests**. These are a set of tests based on user requirements that every code update has to pass before it is committed (pushed) to the version control repository. This ensures that only working code is held within the repository. Each test should test one basic aspect of the code, and may required associated data to be constructed (as part of the test). For example, for a stock control software system they may be a test to ensure one cannot take out more goods than exist. While computers may be happy with negative numbers of items on the shelf, this would not be acceptable for a storekeeper.

In some arenas, it is possible to automate calling the unit tests, as part of the process of checking in updates to a version control repository. It is also possible in some cases to automate the delivery or deployed of successful updates to production services. These would be known as **Continuous Delivery** or **Continuous Deployment**, but are just beyond the scope of this book.

## 8.5 In Conclusion

We discussed a little earlier about modularity and re-use of functional code, perhaps via a software library. We have also discussed optimizing parallelizing codes. It may not surprise you, that many giants have come before us and written optimized, parallel libraries of usual functions and that we can stand on their shoulders by using their efforts to help our own codes go faster or to expand the memory available. There may already be a code that does everything you want, as quick as you want, so why not use that? Or if there's a code that you can extend, that may be a good place to start. Even if you have to code up the algorithm of your physics and chemistry from scratch, there is most likely a relevant maths library that you can call.

For example, you would need to apply a lot of effort to code in C the most efficient implementation of a matrix-matrix multiplication routine, let alone handling when it is sparse or banded. Most systems now come with a set of "Basic Linear Alegra Subprograms" (BLAS) routines tuned (or even tunable!) for a system and these include efficient implementations of matrix-matrix multiplication routines, and you would call these rather than attempt to write your own. Much has been discussed over the BLAS, and extending them, notably within a recent Journal issue dedicated to discussion of BLAS [35]. In a similar vein, there are many linear algebra packages including LAPACK [36] and Trilinos [37], and software libraries for solving PDES such as PETSc [38].

On IBM machines the software libraries are encapsulated within the "Engineering and Scientific Subroutine Library" (ESSL) [39]. For AMD chip sets, it is the "AMD Optimizing CPU Libraries" (AOCL) [40]. For Intel chip sets, it is the "Maths Kernel Library" (MKL) [41]. For vendor-independent maths libraries you can consider "GNU Scientific Library" (GSL) [42].

It is important that you carefully consider available libraries, bearing in mind whether they provide the required accuracy, their licencing or costs, and even portability across platforms. And if you do decide to write your own code, we hope the guidance in this chapter will help you produce quality software that runs fast and can scale to bigger problem sizes efficiently – good luck!

## Bibliography

[1] Lecture 2: Instruction Set Architectures and Compilers. http://hpca23.cse.tamu.edu/taco/utsa-www/cs5513-fall07/lecture2.html.

[2] Compilers. https://highendcompute.co.uk/compilers.

[3] David Goldberg. What every computer scientist should know about floating-point arithmetic. *ACM Computing Surveys (CSUR)*, 23(1):5–48, 1991.

[4] Weather and climate science and services in a changing world, research and innovation strategy v.1. MetOffice. https://www.metoffice.gov.uk/binaries/content/assets/metofficegovuk/pdf/research/r-i_strategy_full_version_sml.pdf, April 2020.

[5] The cray xc40 supercomputing system. MetOffice. https://www.metoffice.gov.uk/about-us/what/technology/supercomputer.

[6] V. Pelletier. https://github.com/vpelletier/pprofile.

[7] Michael J Flynn. Very high-speed computing systems. *Proceedings of the IEEE*, 54(12):1901–1909, 1966.

[8] Michael J Flynn. Some computer organizations and their effectiveness. *IEEE Transactions on Computers*, 100(9):948–960, 1972.

[9] Performance Tools. https://highendcompute.co.uk/Performance.

[10] Top500 List – June 2021. https://www.top500.org/lists/top500/list/2021/06/.

[11] Antoine Petitet. HPL - a portable implementation of the high-performance linpack benchmark for distributed-memory computers. http://www.netlib.org/benchmark/hpl/, 2004.

[12] Power Management. The Linux Kernel. https://www.kernel.org/doc/html/v4.12/admin-guide/pm/index.html.

[13] Can't use 'userspace' cpufreq governor and set cpu frequency. StackExchange. https://unix.stackexchange.com/questions/153693/cant-use-userspace-cpufreq-governor-and-set-cpu-frequency.

[14] Standing on the shoulders of giants. Wikipedia. https://en.wikipedia.org/wiki/Standing_on_the_shoulders_of_giants.

[15] R. Rew and G. Davis. Netcdf: an interface for scientific data access. *IEEE Computer Graphics and Applications*, 10(4):76–82, 1990.

[16] https://github.com/compdyn/partmc.

[17] D.E. Knuth. *The Art of Computer Programming 2nd Edition: Seminumerical Algorithms*, vol. 2, Addison-Wesley Professional, 1981.

[18] Numba. http://numba.pydata.org/.

[19] MPI Forum. MPI Forum. https://www.mpi-forum.org/.

[20] EMiT News. EMiT: Emerging Technology Conference. https://emit.tech/.

[21] OpenACC. https://www.openacc.org/.

[22] OpenMP. https://www.openmp.org/specifications/.

[23] Cuda Toolkit Documentation v11.6.0. nvidia. https://docs.nvidia.com/cuda/index.html.

[24] John C Linford, John Michalakes, Manish Vachharajani, and Adrian Sandu. Automatic generation of multicore chemical kernels. *IEEE Transactions on Parallel and Distributed Systems*, 22(1):119–131, 2010.

[25] John C. Linford. Kppa: A high performance source code generator for chemical kinetics. EMiT15, Manchester, July 2015. https://emit.tech/wp-content/uploads/2015/10/EMiT2015_Linford.pdf.

[26] William Sawyer. Overview of GPU-enabled atmospheric models. ENES Workshop on Exascale Technologies & Innovation in HPC for Climate Models. Mar. 18, 2014, DKRZ, Hamburg. https://is.enes.org/archive-1/phase-2/documents/Talks/WS3HH/session-4-hpc-software-challenges-solutions-for-the-climate-community/william-sawyer-gpu-models, 2014.

[27] https://excalibur.ac.uk/projects/fpga-testbed/.

[28] https://www.european-processor-initiative.eu/wp-content/uploads/2019/12/EPI-Technology-eFPGA.pdf.

[29] Understanding quantum computing. Microsoft. https://docs.microsoft.com/en-us/azure/quantum/overview-understanding-quantum-computing.

[30] Roger S Pressman and Bruce R Maxim. *Software Engineering: a Practitioner's Approach*. McGraw-Hill, 2015.

[31] Robert C. Martin. *Clean Code: A Handbook of Agile Software Craftsmanship*. Prentice Hall, 2008.

[32] Titus Winters, Tom Manshreck, and Hyrum Wright. *Software Engineering at Google: Lessons Learned from Programming over Time*. O'Reilly Media, 2020.

[33] Apache Subversion. The Apache Software Foundation. https://subversion.apache.org/.

[34] Scott Chacon and Ben Straub. *Pro Git*. Springer Nature, 2014.

[35] *ACM Transactions on Mathematical Software*, 28(2), 2002. https://dl.acm.org/toc/toms/2002/28/2.

[36] Edward Anderson, Zhaojun Bai, Christian Bischof, L Susan Blackford, James Demmel, Jack Dongarra et al. *LAPACK Users' guide*. SIAM, 1999.

[37] Packages. Trilinos. https://trilinos.github.io/packages.html.

[38] PETSc. https://petsc.org/release/.

[39] Calling the BLAS and ESSL libraries. IBM. https://www.ibm.com/docs/en/aix/7.2?topic=techniques-calling-blas-essl-libraries.

[40] AMD Optimizing CPU Libraries (AOCL). AMD. https://developer.amd.com/amd-aocl/.

[41] Intel-Optimized Math Libary for Numerical Computing. Intel. https://software.intel.com/content/www/us/en/develop/tools/oneapi/components/onemkl.html.

[42] GSL – GNU Scientific Library. https://www.gnu.org/software/gsl/.

## Appendix A

## A.1 Exercises

1. Chapter 1

   In these exercises we dive straight in to implementing sectional and model representations of size distributions. Using Python in both instances, you are asked to implement the numerical frameworks presented in Sections 1.3.2 and 1.3.3. As with all exercises provided in this book, a complete solution is provided with additional material given with the book and the full code can be found in the book repository and archive (see Chapter 1.4).

   (a) In this exercise we need to go through a series of steps to create a discrete lognormal distribution that has a mean diameter of 150nm, a standard deviation of 1.7, and total number density of particles of 600 per cubic centimeter. In code snippet A.1 we have given you an unfinished template *.py* file with the

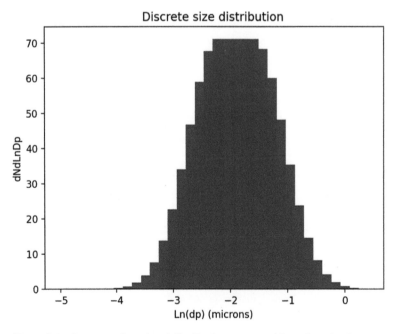

**Figure A.1** Log-normal sectional distribution generated from Exercise 1.

*Introduction to Aerosol Modelling: From Theory to Code.*
First Edition. Edited by David Topping and Michael Bane.
© 2022 John Wiley & Sons Ltd. Published 2022 by John Wiley & Sons Ltd.

lower and upper bin boundary and desired number of size bins set. We have also defined the name of each required variable and array before providing the plotting routine. Please note we define our diameter in microns and the number concentration is given in number per cubic centimeter. Can you fill in the blanks and run the resulting code? Your plot should match Figure A.1.

```python
import numpy as np
import matplotlib.pyplot as plt

# Volume ratio discrete distribution
d1 = 0.01 # Lowest size diameter
d_Nb = 1.0 # Diameter of largest bin [microns]
Nb = 30 # Number of size bins

V_rat =  # Volume ratio between bins
d_rat =  # Diameter ratio between bins

# Use the volume ratio to create an array of diameters as follows
# Create an empty diameter array
d_i=np.zeros((Nb), dtype=float) # Diameter array
d_i[0]=d1
# Populate values in array d_i
for step in range(Nb):
    if step > 0:
        d_i[step]=

# Log of Diameter array
log_di = np.log(d_i) # Log of Diameter array

# Create an array of lower bin boundaries
vi = (4.0/3.0)*np.pi*np.power(d_i/2.0,3.0)
vi_low = 2.0*vi/(1.0+V_rat)
di_low = 2.0*(np.power(vi_low/((4.0/3.0)*np.pi),(1.0/3.0)))

# Create a diameter width array
d_width =

# Set parameters of log-normal distribution
sigmag1 =  # Geometric standard deviation
mean1 =  # Mean particle size [150nm]

Ntot = 600.0 # Total number of particles [per cm-3]
# Implement equation 1.16
N_dist =
#- - - - - - - - - - - - - - - - - - - - - - - - - - - -

plt.bar(log_di, N_dist)
plt.ylabel(r'dNdLogDp')
plt.xlabel(r'LogDp')
plt.title(r'Discrete size distribution')
plt.show()
plt.show()
```

**Listing A.1** Template for creating a discrete log-normal in Python

(b) You have measured a number of bulk properties of a population of particles including a total number concentration of 2300 particles $cm^{-3}$, a surface area concentration of $2.14 \times 10^{-4}$ $m^2$ $m^{-3}$, and a mass concentration of 27.2 $\mu g$ $m^{-3}$. Assuming the size distribution follows a log-normal function, calculate the geometric mean diameter and standard deviation of the population. Assume the average particle density is 1400 kg $m^{-3}$.

(c) Using the parameters calculated in part (b), calculate the number, surface area, and mass concentration of particles between 50 nm and 400 nm in diameter.

2. Chapter 2

(a) In Section 2.2 we used the Volatility Basis Set (VBS) approach to simulate the partitioning of gas phase compounds to a pre-existing absorptive mass. In that example we manually set the concentration of pre-existing mass to $5\ \mu\mathrm{g\,m^{-3}}$. In this exercise you need to calculate the concentration of pre-existing mass using the properties of a mono disperse population. Specifically, using the same volatility profile and ambient conditions used in Section 2.2, set the number density of particles of our mono-disperse population to $300\ \mathrm{cm^{-3}}$ and assume a density of $1400\ \mathrm{kg\,m^{-3}}$. What is the final estimated secondary mass?

(b) If the density of the core material is now assumed to be $1200\ \mathrm{kg\,m^{-3}}$, how does this impact on the concentration of predicted secondary mass?

(c) What fraction of the highest volatility bin has condensed under equilibrium conditions for the second scenario?

(d) In Section 2.4.1 we constructed a sectional model of condensational growth for a poly-disperse population of particles. We assumed ideal solution thermodynamics, such that the activity coefficient $\gamma_i$ is equal to 1. In this exercise you are tasked with implementing a simple extension to our existing framework to account for changes in $\gamma_i$ as a function of mole fraction in solution. This hypothetical model follows the formula:

$$\gamma_i = 1.0 + x_i f_i \tag{A.1}$$

where $x_i$ is the mole fraction of compound $i$ in solution and $f_i$ is a scaling factor for each volatility bin listed in the table below:

Table A.1   Factors used in generating hypothetical activity coefficients.

| Partitioning parameters | |
| --- | --- |
| $\mathrm{Log10}(C_i^*)$ | $f_i$ |
| | 3.0 |
| -6 | |
| -5 | 3.0 |
| -4 | 3.0 |
| -3 | 4.0 |
| -2 | 5.5 |
| -1 | 6.5 |
| 0 | 7.0 |
| 1 | 10.0 |
| 2 | 15.0 |
| 3 | 20.0 |

When simulating growth for 5000 seconds, what is the percentage difference in predicted secondary mass when including this new parameterization of activity coefficients compared with the scenario assuming an ideal solution?

3. Chapter 3

(a) The BAT (light) model is a relatively simple predictive activity coefficient model for binary aqueous organic solutions. As discussed in Section 3.4.1, for certain organic compound properties an LLPS is detected to occur in some range of the binary composition space. Let us consider the case of 1,6-hexanediol as organic component in BAT (as well as molecules of similar properties).

Use the BAT program Fortran code from Example 2, specifically subroutine BAT_example_2 (from file chap3_alg6_BAT_Example_2.f90), as template to work on the following questions (introduce your modifications to the code where necessary).

(i) The molecular properties of 1,6-hexanediol ($C_6H_{14}O_2$) serving as inputs to the BAT model are: $M_{org} = 118.18 \text{ g mol}^{-1}$, $O:C = \frac{2}{6}$, $H:C = \frac{14}{6}$. Use np = 101 points to produce the BAT predictions of the water and organic activities from low to high organic mole fraction with associated plots (steps 2.1 – 2.6 in Example 2). Based on the plotted curves, is LLPS expected to occur in this binary system? Check your answer by running the phase separation detection code (step 2.7).

(ii) Consider a hypothetical molecule that shares some of the properties with 1,6-hexanediol, while differing in others. Keeping the set $M_{org}$ and H:C values, modify the O:C ratio by +15 % or by −15 % relative to its original value and run the BAT predictions in each case to determine whether phase separation occurs. In these cases, also update the estimated liquid-state density (code step 2.1, use of function density_est) based on the modified inputs. If LLPS is detected, report the computed extent of the miscibility gaps in terms of the $x_{org}$ values of phases $\alpha$ and $\beta$.

(iii) Similar to (ii), consider two cases where O:C remains fixed at the original value but instead $M_{org}$ is varied by ± 15 %. Do variations of molar mass affect the LLPS behavior in this case? What are the extents of miscibility gaps?

(iv) Compare your calculations for the cases in (ii) and (iii). For these cases, which input perturbation leads to the widest miscibility gap, +15 % or −15 % variation in O:C or in $M_{org}$? For each case (O:C or $M_{org}$ perturbation), qualitatively, what explains the observed trend in LLPS formation behavior and extent of LLPS or absence thereof? Hint: think of physicochemical properties and their influences on non-ideal mixing.

(b) Numerical determination of the onset O:C ratio causing LLPS. Related to Exercise (a) above, we will consider the BAT model and use 1,6-hexanediol as example for an organic component. Given the initial LLPS detection procedure outlined by code steps 2.1 to 2.7 (of BAT_example_2), write a program that numerically determines the critical O:C value of a hypothetical molecule for which LLPS is just detected (while absent at a higher value). Use the other organic molecule properties of 1,6-hexanediol (but update the density as above). Also use np = 15 as the number of points for initial activity curve characterizations. Determine the critical O:C value to a precision of at least $10^{-5}$. You can make use of the numerical methods provided in modules chap3_alg7_Mod_NumMethods and/or `chap3_alg8_Mod_MINPACK`. Report the value and the extent of the LLPS $x_{org}$ range for this critical O:C ratio.

(c) Methyltetrols are compounds found in aerosol particles as a product of isoprene oxidation in the atmosphere; some form oligomers [44]. Let us use a 2-methyltetrol dimer compound (SMILES: CC(O)C(C)(O)C(O)OCC(O) C(C)(O)CO; $C_{10}H_{22}O_7$) as example. Modify the BAT example program to compute the following properties of a binary water + 2-methyltetrol dimer system at $T = 20\,°$ C.

   (i) Compute and plot the activity coefficients of water and the 2-methyltetrol dimer as function of organic mole fraction. Approximately in which $x_{org}$ composition range are the largest deviations from ideal mixing found? Also, approximately over what value ranges do the two activity coefficients vary?

   (ii) Using the modified Raoult's law on mass concentration basis (Eq. 3.10), compute the equilibrium gas phase mass concentrations of water and the organic for $x_{org} = 0.3$. State assumptions made. Positive deviations from Raoult's law are characterized by a higher total equilibrium vapor pressure over a liquid solution than expected from ideal mixing. Is the calculated case an example of a positive or negative deviation from Raoult's law?

(d) Acidity of mixed aqueous organic–inorganic solutions. In Section 3.6.1, we have discussed the use of the AIOMFAC-LLE model for the purpose of activity coefficient and LLPS computations in multicomponent mixtures. Acidity in form of the pH value, $pH = -\log_{10}(a_{H^+}^{(m)})$, depends on the single-ion molal activity of the $H^+$ ion in solution. The acidity of aerosols and cloud droplets can impact a number of chemical reactions and gas–liquid exchange processes. Let us consider the concentration dependence of the pH in different solutions.

   (i) Consider a system consisting of four input components plus water: levoglucosan, 3-methylglutaric acid, ammonium nitrate, and ammonium bisulfate. Further, let us use a mixture in which the two electrolytes are mixed 3:1 by moles, the organics 1:1 by moles and the total organic to inorganic dry molar mixing ratio is 1:1. We can generate a few input compositions ranging from high ($x_w^{inp} = 0.9999$) to intermediate ($x_w^{inp} = 0.70$) water content; see the attached AIOMFAC-LLE input file, input_0606.txt. Note, this input file also defines the system components in terms of AIOMFAC subgroups.

   Using AIOMFAC-LLE, compute the equilibrium phase composition(s) and write code near the end of subroutine `OutputLLE_plots` to produce a plot of the pH value as a function of water activity for each liquid phase present. Discuss why the pH values of coexisting phases (in an LLPS case) are not identical.

   (ii) Similar to (i), simplify the system by removing the organic compounds, such that it only consists of water, ammonium nitrate, and ammonium bisulfate, with the electrolytes mixed 3:1 by moles. You can use file input_0606.txt as a template. Compute the equilibrium-state compositions and make again a plot of pH versus water activity. Compare the pH curves from the organic-containing system in (i) to those from the organic-free system in (ii). In which case will lower pH values result at the same water activity?

4. Chapter 4

   (a) In Chapter 2 we initialized a volatility distribution, or VBS, to represent a fixed concentration of material in the gas phase. In the VBS each volatility bin is linearly separated in log space. Using the concentration of condensed mass that would lead to 50% of the mass in each bin condensing as a volatility metric, these bins typically range from $10^{-6}$ to $10^3$ $\mu g\,m^{-3}$. In solving the droplet growth equation, we did not allow material to move between each bin. In this exercise, you are tasked with building a gas phase only simulation where material is re-distributed between each volatility bin through second-order reactions between material in each bin and an oxidant. In these simulations we specify a fixed abundance of our oxidant, in this example Ozone, in parts-per-billion [ppb] and use the same rate coefficient for each volatility bin. The table below outlines the starting conditions. Run the simulation for two hours and plot the change in abundance in each volatility bin. Use the Scipy ODE solver and the template provided in code listing 4.3 to write your own code. You do not need to provide a Jacobian.

   (b) Following on from the previous question, now modify your code to account for the loss of Ozone from the second-order reactions and an assumed constant emission rate of 10.0 $\mu g\,m^{-3}\,h^{-1}$ and an emission rate for the highest volatility bin of 1.5 $\mu g\,m^{-3}\,h^{-1}$. The molecular weight of Ozone is 47.998 $g\,mol^{-1}$ [https://pubchem.ncbi.nlm.nih.gov/compound/Ozone].

   (c) In this exercise you are tasked with plotting the ratio of the highest volatility bin from the different simulation structures outlined in the first and second exercise above. You could save the output from each simulation, but it may be useful practice to include two different functions that define the RHS of each structure, and thus output arrays, in one Python file so you can change initial conditions or model parameters and obtain a different answer.

   (d) In this exercise you are tasked with combining the gas phase only mechanism defined in Exercise 2, with variable loss/gains for ozone and the highest volatility bin, with the structure for solving the droplet growth equation for a poly-disperse sectional model as described in Section 4.1.2. As with all exercises, we provide you with a working answer should you wish to read through the code provided to understand the suggested mapping. Once again we use 10 volatility bins, each defined by a C* value and initial concentrations in the table below.

      In the gas phase, we assume that material is re-distributed between each volatility bin through second-order reactions between material in each bin and an oxidant. In this simulations we specify a starting concentration of our oxidant at 10 parts-per-billion [ppb]. The temperature is 298.15 K and the molecular weight of each representative compound, thus volatility bin, is 200 $g\,mol^{-1}$. We also assume the rate coefficient for each second-order reaction is $10^{-15}cm^3$ $molecule^{-1}\,s^{-1}$. You also need to initialize a sectional size distribution, with 8 size bins, assuming a total number density of 100 $cm^{-3}$, a mean particle diameter of 150 nm and standard deviation of 1.7. The core density is set to 1400 $kg\,m^{-3}$ with a molecular weight matching the condensing compounds (200 $g\,mol^{-1}$). Run the simulation for 2 hours and calculate the final concentration of condensed secondary mass in $\mu g\,m^{-3}$.

(e) Often the second moment of the modal approach is neglected in large-scale models due to numerical instability. Rewrite the modal condensation code in list 4.18 assuming only two moments and the standard deviation is fixed at 1.7. What differences in condensed mass are observed from the three-moment implementation?

(f) Now expand the three-moment modal code in snippet 4.18 to include an additional initial aerosol mode with 250 particles $cm^{-3}$, a geometric mean diameter equal to 700 nm, and a geometric standard deviation equal to 2.1. What is the ratio mass condensed to each of the two modes.

(g) In this exercise you are asked to partially recreate SIMPOL as it has been done within UManSysProp to estimate the vapor pressures of two carboxylic acids, acetic acid, and propanoic acid. To support this exercise the file SIMPOL_SMILES.prop contains the SMILES strings for both acetic and propanoic acid and SIMPOL.data contains the parameters needed for estimating the vapor pressures of carboxylic acids using SIMPOL. We provide you with a template file to complete (*SIMPOL_template.py*) which contains a Python script with certain areas that have been left blank. You are directed to fill in each of these areas with code according to the directions listed below. To complete the unfinished template, please follow the steps outlined below:

   (i) The first step requires SMARTS for each of the required functions to be written. For this exercise you will require the SMARTS to identify: (i) The total number of carbons present, (ii) the total number of hydroxyl groups bonded to non-aromatic carbon present, and (iii) the total number of carboxylic acid

**Table A.2** Initializing the abundance of condensable material in each volatility bin.

| Partitioning parameters | |
| --- | --- |
| Log10($C_i^*$) | $C_{t,i}$ ($\mu g\,m^{-3}$) |
| -6 | 0.1 |
| -5 | 0.1 |
| -4 | 0.15 |
| -3 | 0.22 |
| -2 | 0.36 |
| -1 | 0.47 |
| 0 | 0.58 |
| 1 | 0.69 |
| 2 | 0.84 |
| 3 | 1.0 |

groups present. An example of the format is shown in code listing 4.24. The website https://smarts.plus is a useful tool for visually representing SMARTS and might be useful for this exercise.

(ii) Where directed to in *SIMPOL_template.py* file, you need to specfiy the code that will read in the SMARTS from your .smarts file. An example of the format is shown in code listing 4.28.

(iii) Next, you need to define a function that will use SMARTS to search SMILES to identify which groups are present. An example of the format is shown in code listing 4.29.

(iv) Then you will need to define variables to read in the parameter values from the data file, SIMPOL.data. An example of the format is shown in code listing 4.27.

(v) Following this, define a function to estimate vapor pressure based on the identified groups. An example of the format is shown in code listing 4.26.

(vi) Finally, call the function to calculate the vapor pressures of acetic acid and propanoic acid in $\log_{10}$ atm. The expected output should be -2.514726286046664 for acetic acid and -2.938920416921228 for propanoic acid.

5. **Chapter 5**

(a) (i) A scanning mobility particle sizer spectrometer sampled 100 000 particles and determined their sizes. The diameter samples are recorded in datafile particle_samples.txt. Use this data to graph the number concentration density distributions $n(d_p)$ and $\tilde{n}(\log d_p)$, the corresponding surface concentration density distributions $S(d_p)$, $\tilde{S}(\log d_p)$, and volume concentration density distributions $V(d_p)$, $\tilde{V}(\log d_p)$.

(ii) Figure A.2 shows an aerosol size distribution $\tilde{n}(\log d_p)$ that consists of three lognormal modes. Estimate the three lognormal parameters that define each mode (total number concentration, geometric mean diameter, and geometric standard deviation).

(iii) An aerosol size distribution has a total number concentration of 3000 $cm^{-3}$, a geometric mean diameter of 200 nm, and a geometric standard deviation of 1.3. Generate 100 000 particle samples from this distribution. Use your code from Question (a)(i). above to check your result.

(b) Consider a monodisperse aerosol population with $N = 10^{11}$ particles and a particle size of 500 nm in a chamber of volume 4 $m^3$.

(i) How many coagulation events per second occur right at the start of the experiment?

(ii) Assuming that the coagulation rate remains constant at the value calculated in part ((b)(i)), how long will it take for the original particles to be depleted by coagulation?

(iii) In reality, will it take longer or shorter? Why?

(iv) If the chamber is twice as big and we had the same number of particles ($10^{11}$), what would change?

(v) If the chamber is half as big and the number concentration was the same, what would change?

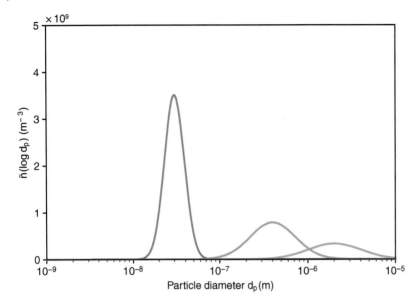

**Figure A.2** Size distribution $\tilde{n}(\log d_p)$ needed for Question a.

(c) (i) Section 5.4.2.2 introduced the sedimentation kernel. Make a figure like Figure 5.1, using the sedimentation kernel instead of the Brownian kernel. For simplicity, assume that the collision efficiency $E$ is 1, and that the terminal velocity can be approximated by the Stokes' formula:

$$v_t = \frac{d_p^2 g \rho_p}{18\mu}, \tag{A.2}$$

where $g$ is the gravitational constant, $\rho_p$ is the particle's density and $\mu$ is the dynamic viscosity of air.

(ii) Consider the case of a particle with diameter $d_{p1} = 1$ μm. Find the diameter range $d_{p2}$ for a second particle for which Brownian coagulation is more important and the diameter range where gravitational settling is more important. For which size range are the two processes of approximately similar importance?

(d) Figure 5.5 shows how Algorithms 5.1, 5.2, and 5.3 scale. Modify the code provided in *chap5_alg1_basic_method.py*, *chap5_alg2_binned_method.py*, and *chap5_alg3_tau_leaping_method.py* to generate suitable data to confirm these expected relationships by varying the number of particles and number of bins.

(e) Consider particle interaction events that are occurring at an average rate $\lambda = 6\,\text{s}^{-1}$ (that is 6 events per second).

(i) Generate 100 000 different event times $\tau$ corresponding to this rate. What is the average value of $\tau$?

(ii) Plot a histogram of the $\tau$ values and show that this histogram looks like $\exp(-6\tau)$. This is a reflection of the fact that interevent times for Markov processes are exponentially distributed according to the rate parameter.

(f) Algorithm 5.1 frequently reevaluates the kernel. With this question we want to investigate how this can be optimized. Set up Algorithm 5.1, using 100 initial computational particles and a simulation time of 3600 s. Set the computational volume so that the initial number concentration is $5 \times 10^{12}$ m$^{-3}$. Compare the runtime of this base case with the following two alternative methods:

(i) Rather than recalculating the complete kernel for each event, precompute the kernel once for all possible particle size combinations at the beginning.

(ii) Another technique is called "memoization." Rather than precomputing the kernel for all possible particle size combinations, we compute the combinations as needed as they occur and store them in a dictionary. The memoization algorithm looks like the following:

---

1 **Function** MemoizedKernel $(p_1, p_2)$
2    **if** $(p_1, p_2)$ *in* memo **then**
3        |  return memo$(p_1, p_2)$
4    **else**
5        |  memo$(p_1, p_2) = $ **Kernel** $(p_1 m_b, p_2 m_b)$
6        |  return memo$(p_1, p_2)$
7    **end**

---

Determine which other algorithms the technique of memoization applies to and implement this change for these algorithms.

(g) Algorithm 5.2 shows that for $k = \ell$, the number of pairs is $N_{pairs} = \frac{1}{2}N[k]$ $(N[l] - 1)$. Explain why this is indeed the case.

(h) Simulating coagulation using Algorithm 5.2, we would like to figure out how many coagulation events occur between pairs of particles. To do this, augment the code by introducing a matrix **M**, so that **M**$(i, j)$ is the number of coagulation events that occur for particles that belong in bins $i$ and $j$, respectively. When plotting this matrix for different simulation times, it may be beneficial to use a logarithmic colorscale.

(i) Algorithm 5.2 as written requires the recomputing of all values of $\tau$ after each coagulation event. In actuality, this is unnecessary for any values of $\tau$ where the $N[k]N[\ell]$ remains unchanged from the previous iteration. To reduce computational cost, keep all values of $\tau$, except those who were impacted by coagulation events, bin $k$, bin $\ell$, and bin $k + \ell$. Evaluate the decrease in computational time due to this adjustment. Construct a time series plot of mean and variance and verify this method against the mean and variance of the original method of Algorithm 5.2.

(j) Algorithm 5.3 has a timestep parameter and it is making the approximation that the rate $\lambda$ doesn't change during the timestep.

(i) Consider a monodisperse aerosol with $N = 10^{12}$ particles in a chamber volume $V = 1$ m$^3$ and assume that after particles coagulate they no longer interact. Write a version of Algorithm 5.3 to predict the number of particles $N_1(t)$, using a fixed timestep.

(ii) Run your new algorithm with different timesteps. Plot $N_1(t)$ for different timesteps ($\Delta t = 1$ second, 10 seconds, 100 seconds). How small does the timestep need to be to get reasonable results?

(k) Algorithms 5.1–5.5 are not only applicable for the process of coagulation, but for any process that consists of a series of stochastic events. Rewrite Algorithms 5.1

and 5.5 for the example of radioactive decay using a constant decay rate $\lambda$. (Hint: Algorithm 5.5 becomes an ordinary differential equation that has an analytical solution.)

(l) Setup a PartMC run as described in Section 5.5. Add emissions of a lognormal soot mode (representing emissions from traffic) to the scenario with the following specifications:

- Geometric mean diameter 80 nm, geometric standard deviation 1.5, total emission flux or particles $1.5 \times 10^8 \, \mathrm{m^{-2} \, s^{-1}}$
- Composition: 100% BC
- Add emissions for 6 hours of simulation time.

After 8 hours of simulation, determine the number fraction of particles that (1) are purely soot, (2) contain ammonium sulfate and soot, and (3) are soot-free.

The templates for the three input files related to emissions (aero_emit.dat, aero_emit_dist.dat, and aero_emit_comp.dat) are shown below.

The file aero_emit.dat specifies the time period of emission and the filename of the file with the size distribution information:

```
# time (s)
# rate (s^{-1})
# aerosol distribution filename
time    0    21600
rate    1    0
dist    aero_emit_dist.dat aero_emit_dist.dat
```

The file aero_emit_dist.dat specifies the emission size distribution parameters and the emission flux:

```
mode_name traffic
mass_frac aero_emit_comp_traffic.dat
# composition proportions of species
diam_type geometric              # type of diameter
                                   specified
mode_type log_normal             # type of distribution
num_conc 1.5e8                   # particle number
                                   concentration (#/m^2)
geom_mean_diam 8e-8              # geometric mean
                                   diameter (m)
log10_geom_std_dev 0.176        # log_10 of geometric
                                   std dev of diameter
```

The file aero_emit_comp_traffic.dat defines the composition of the emitted particles:

```
#          mass fractions
BC                 1.0
```

(m) Rewrite Eq. 5.11 in terms of particle volume as independent variable for the number density distribution function.

(n) (i) Create an ensemble of 10 runs using Algorithm 5.6. You can use the file *chap5_alg6_particle_resolved.py* as a starting point. Graph the number concentration of particles as a function of time, averaged over the 10 ensemble members, and indicate the standard deviation with errorbars.

(ii) Repeat the simulations from part (i), but use 10 times as many particles. How does the standard deviation change compared to part (i)?

(iii) Repeat the simulations from part (i), but change the number of bins, e.g., use 10 or 100 instead of 50. Observe that the accuracy does not change at all, i.e., the ensemble average looks the same.

6. Chapter 6

   (a) Model the steady-state cluster distribution in a one-component system using the codes in Section 6.2 as a starting point. As a simple (but quantitatively unrealistic) test case, start by assuming classical droplet model evaporation rates (cf. Eq. 6.7):

$$\gamma_{(i+1)\to i,1} = \beta_{i,1} \frac{P_{\text{sat}}}{k_{\text{B}}T} \exp\left[\frac{(A_{(i+1)} - A_i)\sigma}{k_{\text{B}}T}\right], \tag{A.3}$$

   where $P_{\text{sat}}$ is the saturation vapor pressure (use e.g. $10^{-10}$ Pa), $\sigma$ is the surface tension (e.g. $5 \times 10^{-2}$ N m$^{-1}$), $A_i$ is the surface area of cluster $i$, and only monomers evaporate. First, include in the cluster GDE only cluster–monomer collisions and evaporations. Set the monomer concentration to a constant value of e.g. $5 \times 10^6$ cm$^{-3}$, and ensure that the system is in a steady-state.

      (i) Increase the system size, starting from e.g. including clusters consisting of up to 5 molecules. What happens to the cluster concentrations? Why?

      ii. Include also cluster–cluster coagulation and scavenging, and explain the changes.

      (iii) Modify the evaporation rate of e.g. the 3-mer by a factor of 5. What happens? Do the exact evaporation rates of individual small clusters matter?

   (b) Plot the cluster–monomer collision and evaporation rate constants for the one-component system as a function of cluster size.

      (i) Compare the size-dependent rates to the steady-state cluster concentration simulated considering only cluster–monomer collisions in the GDE. How does the evaporation rate affect the concentration?

      (ii) How do the rate constant and concentration profiles relate to the cluster flux between consecutive sizes, shown in Figure 6.3? Which sizes have the highest concentrations and why?

   (c) Apply ACDC to simulate a 2-component system of electrically neutral $H_2SO_4$–$NH_3$ clusters using the example input in Section 6.4.1. Utilize the tools in the code repository.

      (i) Is the cluster set large enough for performing cluster formation simulations at atmospheric conditions? Why/why not?

      (ii) Test different criteria for allowing clusters to grow out of the simulation system. Which choices are reasonable? Why?

   (d) Simulate the formation rate for the $H_2SO_4$–$NH_3$ system as a function of

      (i) $H_2SO_4$ concentration,

      (ii) $NH_3$ concentration, and

      (iii) temperature.

      Explain the behavior of the formation rate and its slope.

7. Chapter 7

   Exercises in this chapter will be done using the Python routine *box_model.py*. Settings for Exercise (a)-(b) are given below in Listing A.2 to be used in *parameters.py*. For Exercises (c)-(d), the settings are identical except that the integration time is increased to 3600, the number of bins is decreased to 30, and the updraft velocity is set to 0.5. Note that these simulations are computationally fairly expensive and take up to minutes to run depending on the processor speed of the computer. If the simulation times on your computer are excessively long, you can reduce the number of bins to speed up the simulations. You can also speed up the simulation by commenting out or deleting the last line in *differential_equations.py* which indicates the elapsed time of the simulation. In addition, further ways to make the box model computationally more effective are presented in the next chapter.

```
# Initial aerosol size distribution
# geometric standard deviation of the mode
sigma_g = 1.7
# geometric mean diameter of the mode [m]
mu = 150.0 * 1.0e-9
# total number of particles in the mode [m-3]
N_T = 1.0e8

# initial ambient conditions
# relative humidity [%]
RH0 = 95.0
# temperature [K]
T0 = 298.15
# pressure [Pa]
p0 = 101325.0
# updraft velocity for cloud parcel runs [m]
w = 0.
# integration time for the simulation
integration_time = 500
# Number of size bins
Nb = 50
```

**Listing A.2** Model setting for exercises (*parameters.py*).

Use the box model to investigate how different remapping techniques will influence the aerosol size distribution.

(a) Run the box model with `model_setup=1` using full-moving, quasi-stationary, moving-center, and hybrid bin remapping (both linear and flat-top sub-grid size distribution assumptions). Plot the number size distribution $dN/d\log d_p$ at the end of the run. You can use Function `plot_dN_dlogDp` in Routine *plotting_scripts.py* for plotting the size distributions.

   (i) Assuming that the full moving method is numerically most accurate, which of the other methods reproduces the size distribution of the full moving method the best?

   (ii) What causes the variability in $dN/d\log d_p$ when the moving center method is used?

   (iii) Compare the difference between the hybrid bin method using linear sub-grid size distribution and flat-top sub-grid size distribution.

(b) Run the box model with `model_setup=2` using full-moving and hybrid bin remapping. Plot the number size distribution $dN/d\log D_p$ as a function of time. You can use Function *plot_N_R_t.py* in Routine *plotting_scripts.py* for plotting the size distributions as a function of time. Compare the size distributions.

(c) Simulate cloud formation for a case in which only water condenses on droplets. To do this, run the box model with `model_setup=3`.

(i) Plot the wet radii as a function of time. You can use the routine `plot_R_t` in *plotting_scripts.py* to do this. Explain the behavior of the radii of the droplet population.

(ii) Find the two size bins with the highest ratio of their wet radii ($R_{wet,i+1}/R_{wet,i}$) which are the smallest activated bin and the largest non-activated bin. Plot the saturation ratio of water at the surface of these two droplet sizes. In the same figure, plot also the saturation ratio of water at the surface of the largest and the smallest droplet, and the saturation ratio of water in the air. Explain the behavior of the curves. (Zoom in to the peak of saturation ratio of water in air to see the details of the behavior of the smallest activated and the largest non-activated bin.)

(d) Simulate cloud formation for a case in which both water and VBS species condense on the droplets. To do this, run the box model with `model_setup=4`.

(i) Plot the wet radii as a function of time.

(ii) What is the difference between the number of activated droplets compared to a case where only water condenses on droplets?

(iii) In which size bin, the concentration of VBS species is highest?

## A.2 Physical Constants

| Constants | | |
| --- | --- | --- |
| **Symbol** | **Definition** | **Unit** |
| $N_A$ | Avogadro's number | $6.02214076 \ 10^{23}$ molecules mol$^{-1}$ [1] |
| $R_{gas}$ | Universal gas constant | $8.31451$ J mol$^{-1}$ K$^{-1}$ [2] |
| $R_{gas}^*$ | —— | $8.20573 \ 10^{-5}$ m$^3$ atm K$^{-1}$ mol$^{-1}$ [3] |
| $R_{g,w}^*$ | Gas constant for water vapor $R_{gas}/M_w$ | $0.46140$ J g$^{-1}$ K$^{-1}$ [2] |
| $k_B$ | Boltzmann constant | $1.380649 \ 10^{-23}$ J K$^{-1}$ [4] |
| $\kappa_{w,air}$ | Thermal conductivity of moist air [299.9 K] | $2.60$ J cm$^{-1}$ s$^{-1}$ K$^{-1}$ [5] |
| $M_w$ | Molecular weight of water | $18.015$ g mol$^{-1}$ [6] |
| $M_{ozone}$ | Molecular weight of Ozone | $47.998$ g mol$^{-1}$ [6] |
| $\sigma_w$ | Surface tension of pure water [21.5 °C] | $0.07275$ N m$^{-1}$ [7] |
| $\rho_w$ | Density of liquid water | $995$ g cm$^{-3}$ [6] |

## A.3 Conversion Factor

| | Conversion factors | |
|---|---|---|
| **Unit 1** | **Example** | **Reference** |
| 1 ppm* | $2.463 \cdot 10^{13}$ molecules cm$^{-3}$ | [8] [p. 20] |
| 1 ppb* | $2.463 \cdot 10^{10}$ molecules cm$^{-3}$ | [8] [p. 20] |
| 1 ppt* | $2.463 \cdot 10^{7}$ molecules cm$^{-3}$ | [8] [p. 20] |
| 1 atm* | $2.46 \cdot 10^{19}$ molecules cm$^{-3}$ | [9], p. 34. |

* T=298K, $p_{amb}$=1 atm.

## A.4 Variable Definitions

### A.4.1 Chapter 1

| | List of variables, constants, and acronyms | |
|---|---|---|
| **Variable** | **Definition** | **Typical unit** |
| $n_{v,t}$ | Number concentration of particles with volume $v$ at time $t$ | m$^{-3}$ |
| $n_{v,t+\Delta t}$ | Number concentration of particles with volume $v$ at time $t + \Delta t$ | cm$^{-3}$ |
| $n_{v-\Delta v,t+\Delta t}$ | Number concentration of particles with volume $v - \Delta v$ at time $t + \Delta t$ | cm$^{-3}$ |
| $n_{v+\Delta v,t+\Delta t}$ | Number concentration of particles with volume $v + \Delta v$ at time $t + \Delta t$ | cm$^{-3}$ |
| $\dfrac{dn_{v,t}}{dt}$ | Total rate of change of concentration of particles with volume $v$ at a time $t$ | cm$^{-3}$ s$^{-1}$ |
| $\left(\dfrac{dn_{v,t}}{dt}\right)_{nucleation}$ | Rate of change of concentration of particles with volume $v$ at a time $t$ due to nucleation | cm$^{-3}$ s$^{-1}$ |
| $\left(\dfrac{dn_{v,t}}{dt}\right)_{coagulation}$ | Rate of change of concentration of particles with volume $v$ at a time $t$ due to coagulation | cm$^{-3}$ s$^{-1}$ |

*(Continued)*

*(Continued)*

**List of variables, constants, and acronyms**

| Variable | Definition | Typical unit |
|---|---|---|
| $\left(\dfrac{dn_{v,t}}{dt}\right)_{condensation}$ | Rate of change of concentration of particles with volume $v$ at a time $t$ due to condensation | $cm^{-3}\,s^{-1}$ |
| $\left(\dfrac{dn_{v,t}}{dt}\right)_{emission}$ | Rate of change of concentration of particles with volume $v$ at a time $t$ due to primary emissions | $cm^{-3}\,s^{-1}$ |
| $\left(\dfrac{dn_{v,t}}{dt}\right)_{removal}$ | Rate of change of concentration of particles with volume $v$ at a time $t$ due to removal mechanisms | $cm^{-3}\,s^{-1}$ |
| $p_{norm}(x)$ | Probability density represented by a normal distribution | – |
| $\sigma_p$ | Standard deviation [or scale parameter] used within a normal distribution | – |
| $\mu_p$ | Mean [or location parameter] used within a normal distribution | – |
| $p_{lognorm}(x)$ | Probability density represented by a log-normal distribution | – |
| $\sigma_g$ | Geometric standard deviation of log-normal distribution | |
| $\mu_{ln}$ | Geometric mean of log-normal distribution | |
| $V_{rat}$ | Volume ratio between bins in a sectional approach | – |
| $D_{rat}$ | Diameter ratio between bins in a sectional approach | – |
| $d_{sect,Nb}$ | Diameter of largest bin in a sectional approach | m |
| $d_{sect,1}$ | Diameter of smallest bin in a sectional approach | m |
| $d_{sect,i}$ | Diameter of particle in bin $i$ in a sectional approach | m |
| $N_{sect,b}$ | Number of size bins in a sectional model | – |
| $v_{sect,i}$ | Bin center volume of a particles in size bin $i$ in a sectional model | $m^3$ |
| $v_{sect,i,hi}$ | Upper volume of a particle in size bin $i$ in a sectional model | $m^3$ |

*(Continued)*

*(Continued)*

**List of variables, constants, and acronyms**

| Variable | Definition | Typical unit |
|---|---|---|
| $v_{sect,i,lo}$ | Lower volume of a particle in size bin $i$ in a sectional model | $m^3$ |
| $d_{sect,i,hi}$ | Upper diameter of a particle in size bin $i$ in a sectional model | m |
| $d_{sect,i,lo}$ | Lower diameter of a particle in size bin $i$ in a sectional model | m |
| $\Delta d_{sect,i}$ | Diameter width of size bin $i$ in a sectional model | m |
| $n_{sect,i}$ | Number concentration in size bin $i$ in a sectional model | $m^{-3}$ |
| $\Delta d_{sect,i}$ | Diameter width of size bin $i$ in a sectional model | m |
| $\Delta d_{sect,i}$ | Diameter width of size bin $i$ in a sectional model | m |
| $n_{ln}(ln(x))$ | Log-normal number probability distribution | $cm^{-3}$ |
| $d_p$ | Particle diameter | m |
| $ln(d_p)$ | Log of particle diameter | m |
| $N_L$ | Total number of aerosol particles per unit volume | $cm^{-3}$ |

## A.4.2  Chapter 2

**List of variables, constants, and acronyms**

| Variable | Definition | Typical unit |
|---|---|---|
| $\theta_{s,ads}$ | Number of monolayers adsorbed to the surface of a particle | – |
| $V_{s,ads}$ | Volume of gas adsorbed to the surface of a particle at a partial pressure $p$ | – |
| $V_{s,m}$ | Volume of gas adsorbed to the surface of a particle that would form a monolayer | – |
| $S_{sat,ads}$ | Saturation ratio of a gas involved in the process of adsorption | – |

*(Continued)*

*(Continued)*

**List of variables, constants, and acronyms**

| Variable | Definition | Typical unit |
|---|---|---|
| $C_{OA}^*$ | Total condensed mass of absorptive particulate matter | $\mu g\,m^{-3}$ |
| $C_{g,i}^*$ | Gas phase abundance of compound set $i$ in a volatility basis set context | $\mu g\,m^{-3}$ |
| $\varepsilon_i$ | Partitioning coefficient of compound set $i$ in a volatility basis set context | – |
| $core$ | abundance of an assumed in-volatile core in a volatility basis set context | $\mu g\,m^{-3}$ |
| $C_i^*$ | Representative volatility of compound set $i$ in a volatility basis set context | $\mu g\,m^{-3}$ |
| $M_i$ | Molecular weight of compound $i$ | $g\,mol^{-1}$ |
| $\gamma_i$ | Activity coefficient of compound $i$ in a liquid solution | – |
| $P_{sat,i}$ | Pure component vapor pressure of compound $i$ | $atm$ |
| $C_{a,i}^*$ | Condensed phase abundance of compound set $i$ in a volatility basis set context | $\mu g\,m^{-3}$ |
| $C_{a,i}^*$ | Condensed phase abundance of compound set $i$ in a volatility basis set context | $\mu g\,m^{-3}$ |
| $f(x)$ | Generic function of variable $x$ | NA |
| $f'(x)$ | Derivative of generic function $f(x)$ with respect to $x$ | NA |
| $T$ | Ambient temperature | K |
| $K_{p,i}$ | Equilibrium coefficient between gaseous and condensed phases | – |
| $M_{om}$ | Mass-weighted averaged molar mass of condensed material | $g\,mol^{-1}$ |
| $C_{i,mol}^*$ | Molar-based representative volatility of compound set $i$ in a volatility basis set context | $\mu mol\,m^{-3}$ |
| $Kn_i$ | Knudsen number for compound $i$ | – |
| $\lambda_i$ | Mean free path for compound $i$ in the gas phase | m |

*(Continued)*

(Continued)

### List of variables, constants, and acronyms

| Variable | Definition | Typical unit |
|---|---|---|
| $\alpha_i$ | Mass accommodation coefficient for compound $i$ | – |
| $f(Kn_i, \alpha_i)$ | Transition regime correction factor as a function of Knudsen number $Kn_i$ and mass accommodation coefficient $\alpha_i$ | – |
| $D_{g,i}$ | Gas phase diffusion coefficient of compound $i$ | $cm^2\,s^{-1}$ |
| $c_{g,i}$ | Mean thermal velocity of compound $i$ in the gas phase | $cm\,s^{-1}$ |
| $p_i^{eq}$ | Equilibrium vapor pressure of compound $i$ above a liquid mixture | atm |
| $p_i^o$ | Pure component saturation vapor pressure of compound $i$ | atm |
| $x_i$ | Mole fraction of compound $i$ in solution | – |
| $K_{surf}$ | Kelvin factor | – |
| $\sigma_{sol}$ | Surface tension of the air–surface interface | $N\,m^{-1}$ |
| $v$ | Volume occupied by compound $i$ in the liquid phase | $m^3\,mol^{-1}$ |
| $M_i$ | Molecular weight of compound $i$ | $kg\,mol^{-1}$ |
| $\rho_{sol}$ | Density of the solution | $kg\,m^{-3}$ |
| $S_i$ | Saturation ratio of compound $i$ | – |
| $\frac{dm}{dt}$ | Rate of change of mass of a single homogeneous liquid water droplet $i$ | $g\,s^{-1}$ |
| $R_{wet}$ | Radius of single homogeneous water droplet | cm |
| $D_{g,w}$ | Molecular diffusion coefficient of water vapor | $cm^2\,s^{-1}$ |
| $\rho_{g,w}$ | Density of water vapor | $g\,cm^{-3}$ |
| $n_w$ | Water droplets per cubic centimeter | $cm^{-3}$ |
| $\frac{dm_{a,w}}{dt}$ | Rate of change of liquid water in the condensed phase | $g\,s^{-1}$ |
| $C_{g,w}$ | Molecular abundance of in the gas phase | $molecules\,cm^{-3}$ |
| $C_{a,w}^*$ | Molecular abundance of water vapor above a droplet surface | $molecules\,cm^{-3}$ |

*(Continued)*

*(Continued)*

**List of variables, constants, and acronyms**

| Variable | Definition | Typical unit |
|---|---|---|
| $C_{a,w}$ | Molecular abundance of water vapour in the particulate phase | molecules $cm^{-3}$ |
| $\kappa_{w,air}$ | Thermal conductivity of moist air | $J\,cm^{-1}\,s^{-1}\,K^{-1}$ |
| $\frac{dQ_r}{dt}$ | Cooling rate at the droplet surface | $J\,s^{-1}$ |
| $Tr$ | Temperature of the droplet surface | K |
| $Tr$ | Temperature of the droplet surface | K |
| $p_{g,w}$ | Partial pressure of water vapor in the gas phase | – |
| $p_{g,w(R_{wet})}$ | Equilibrium pressure of water vapor at the droplet surface | – |
| $R_{g,w}$ | Gas constant for water vapor | $J\,g^{-1}\,K^{-1}$ |
| $R_{g,w}$ | Gas constant for water vapor | $J\,g^{-1}\,K^{-1}$ |
| $L_{e,w}$ | Latent heat of evaporation of water vapor | $J\,g^{-1}$ |
| $dm_{a,i}$ | Change in mass of compound $i$ to the particulate phase | g |
| $dm_{a,i}$ | Change in mass of compound $i$ to the particulate phase | g |
| $p_{g,i}$ | Partial pressure of compound $i$ in the gas phase | – |
| $p_{g,i,r}$ | Equilibrium partial pressure of compound $i$ above the surface of a droplet | – |
| $\kappa_{i,air}$ | Thermal conductivity of compound $i$ | $J\,cm^{-1}\,s^{-1}\,K^{-1}$ |
| $L_{e,i}$ | Latent heat of evaporation of compound $i$ | $J\,g^{-1}$ |
| $\frac{dm_{a,i}}{dt}$ | Rate of change of compound $i$ in the condensed phase | $g\,s^{-1}$ |
| $L_{e,i}$ | Latent heat of evaporation of compound $i$ | $J\,g^{-1}$ |
| $\rho_{g,i}$ | Density of compound $i$ in the gas phase | $g\,cm^{-3}$ |
| $\rho_{g,i,r}$ | Density of compound $i$ in the gas phase above the surface of a droplet | $g\,cm^{-3}$ |
| $\frac{dC_{a,i,m}}{dt}$ | Rate of change of compound $i$ in the condensed phase, specifically in size "bin" $m$ | molecules $cm^{-3}\,s^{-1}$ |

*(Continued)*

*(Continued)*

---

### List of variables, constants, and acronyms

| Variable | Definition | Typical unit |
|---|---|---|
| $C_{g,i}$ | Abundance of compound $i$ in the gas phase | molecules cm$^{-3}$ |
| $C_{a,i,m}$ | Abundance of compound $i$ in the condensed phase, specifically in size "bin" $m$ | molecules cm$^{-3}$ |
| $C_{a,i,m}$ | Abundance of compound $i$ in the condensed phase, specifically in size "bin" $m$ | molecules cm$^{-3}$ |
| $C_{a,i,m}^{eq}$ | Required abundance of compound $i$ above the surface of the condensed phase, specifically in size "bin" $m$, to maintain current composition | molecules cm$^{-3}$ |
| $k_{i,m}$ | First-order mass transfer coefficient for species $i$ and bin $m$ | s$^{-1}$ |
| $p_{i,m}^{eq}$ | Required equilibrium vapor pressure of compound $i$ above the surface of the condensed phase, specifically in size "bin" $m$, to maintain current composition | atm |
| $x_{i,m}$ | Mole fraction of compound $i$ in size bin $m$ | – |
| $\gamma_{i,m}$ | Mole-fraction-based activity coefficient of compound $i$ in size bin $m$ | – |
| $K_{surf,i,m}$ | Kelvin factor for compound $i$ in size bin $m$ | – |
| $\rho_{i,l}$ | Liquid density of compound $i$ | kg m$^{-3}$ |
| $Kn_{i,m}$ | The Knudsen number for component $i$ relative to size bin $m$ | – |
| $n_m$ | The number of particles in size bin $m$ | cm$^{-3}$ |
| $n_{w,sol}$ | Number of moles of water in solution | – |
| $n_{sol}$ | Number of moles of solute | – |
| $N_{solutes}$ | Total number of individual compounds defined as solutes | – |
| $M_{tot,i}$ | The concentration (mass) of solute $i$ | g |
| $\phi_{sol}$ | The osmotic coefficient of the solution | – |
| $\vartheta$ | Number of ions per solute molecule | – |

*(Continued)*

*(Continued)*

**List of variables, constants, and acronyms**

| Variable | Definition | Typical unit |
|---|---|---|
| $\rho_w$ | Density of pure water | $\mathrm{kg\,m^{-3}}$ |
| $d_{pc}$ | Critical droplet diameter | m |
| $S_c$ | Critical saturation ratio for water vapor above a liquid solution | – |
| $d_d$ | Diameter of a dry aerosol particle | m |
| $GF$ | Hygroscopic growth factor of an aerosol particle | – |
| $\kappa_K$ | Kappa Köhler factor for hygroscopic growth calculations | – |
| $V_s$ | Total volumes of a dry particle (both soluble and insoluble) | $\mathrm{m^3}$ |
| $V_w$ | Total volume of condensed water | $\mathrm{m^3}$ |
| $V_i$ | "Dry" volume of compound $i$ | $\mathrm{m^3}$ |
| $V_T$ | Total volume of solution | $\mathrm{m^3}$ |
| $\kappa_{K,i}$ | Kappa Köhler factor for component $i$ as used in hygroscopic growth calculations | – |
| $\varepsilon_i$ | Volume fraction of component $i$ | – |

### A.4.3 Chapter 3

**List of variables, constants, and acronyms**

| Variable | Definition | Typical unit |
|---|---|---|
| AIOMFAC | Aerosol Inorganic–Organic Mixtures Functional groups Activity Coefficients | – |
| AIOMFAC-LLE | AIOMFAC-based liquid–liquid equilibrium model | – |
| BAT | Binary Activity Thermodynamics | – |
| LLE | Liquid–liquid equilibrium | – |
| LLPS | Liquid–liquid phase separation | – |
| VLE | Vapor–liquid equilibrium | – |
| SLE | Solid–liquid equilibrium | – |

*(Continued)*

(Continued)

**List of variables, constants, and acronyms**

| Variable | Definition | Typical unit |
|---|---|---|
| $G$ | Gibbs energy | J |
| $G^E$ | Gibbs excess energy | J |
| $S$ | Entropy | $J\,K^{-1}$ |
| $T$ | Temperature (absolute) | K |
| $V$ | Volume | $m^3$ |
| $p$ | Total system pressure | Pa |
| $\mu_j^\alpha$ | Chemical potential of component $j$ in a phase $\alpha$ | $J\,mol^{-1}$ |
| $n_j$ | Molar amount of component $j$ | mol |
| $n$ | Total molar amount ($\sum n_j$) | mol |
| $R$ | Universal gas constant | $J\,mol^{-1}\,K^{-1}$ |
| $p_j^\ominus$ | Reference pressure (for component $j$) | Pa |
| $p_j$ | Partial pressure of component $j$ | Pa |
| $a_j^\alpha$ | Activity of species $j$ in phase $\alpha$ (on mole fraction basis for water and organics; on molality basis for ions) | – |
| $a_w$ | Water activity (mole-fraction-based) | – |
| $\gamma_j^\alpha$ | Activity coefficient of $j$ in phase $\alpha$ (on mole fraction basis for water and organics; on molality basis for ions) | – |
| $x_j$ | Mole fraction of $j$ (in phase specified by a superscript, e.g. $\alpha$ or total system $t$) | – |
| $\mu_j^{\alpha,\ominus}$ | Standard state chemical potential of $j$ in phase $\alpha$ | $J\,mol^{-1}$ |
| $\nu_+$ | Stoichiometric number of cations in a neutral electrolyte unit | – |
| $\nu_-$ | Stoichiometric number of anions in a neutral electrolyte unit | – |
| $(m)$ | Superscript denoting molality basis | – |
| $(x)$ | Superscript denoting mole fraction basis | – |

(Continued)

*(Continued)*

**List of variables, constants, and acronyms**

| Variable | Definition | Typical unit |
|---|---|---|
| $m^\circ$ | Unit molality ($= 1\ \mathrm{kg\,mol^{-1}}$) | $\mathrm{kg\,mol^{-1}}$ |
| $C_j^\circ$ | Pure component liquid-state saturation vapor concentration of $j$ | $\mathrm{\mu g\,m^{-3}}$ |
| $C_j^g$ | Mass concentration of $j$ in gas phase | $\mathrm{\mu g\,m^{-3}}$ |
| $C_j^\alpha$ | Mass concentration of $j$ in (liquid) phase $\alpha$ | $\mathrm{\mu g\,m^{-3}}$ |
| $M_j$ | Molar mass of $j$ | $\mathrm{kg\,mol^{-1}}$ |
| $\phi_o$ | Scaled volume fraction of organic component $o$ | – |
| $\rho_o$ | Liquid-state density of organic component $o$ | $\mathrm{g\,cm^{-3}}$ |
| $\rho_w$ | Liquid-state density of water | $\mathrm{g\,cm^{-3}}$ |
| $\vartheta$ | Elemental oxygen-to-carbon ratio (O:C) of organic compound | – |
| $s_1, s_2$ | BAT model constant coefficients | – |
| $c_i$ | BAT model parameterized coefficient functions | – |
| $a_{n,i}$ | BAT model constant coefficients | – |
| $\Delta G_{\mathrm{mix}}$ | Change of Gibbs energy due to mixing | J |
| $\dfrac{\Delta G_{\mathrm{mix}}}{nRT}$ | Normalized (dimensionless) change of Gibbs energy due to mixing | – |
| $f^\alpha$ | Fraction of the cumulative molar amounts residing in phase $\alpha$ | – |
| $x_{\mathrm{org}}^{\mathrm{init}}$ | Initial (input) organic mole fraction (prior to phase separation) | – |
| $q_j^\alpha$ | Fraction of $j$ in liquid phase $\alpha$ (in LLPS case) | – |
| $\omega$ | Molar $\beta$-to-$\alpha$ phase amount ratio, $\sum_j n_j^\beta / \sum_j n_j^\alpha$ (in LLPS case) | – |
| pH | $-\mathrm{Log}_{10}(a_{\mathrm{H^+}}^{(m)})$; acidity/basicity characteristic of a solution | – |

## A.4.4 Chapter 4

**List of variables, constants, and acronyms**

| Variable | Definition | Typical unit |
|---|---|---|
| $C_{g,i}$ | Abundance of compound $i$ in the gas phase | molecules cm$^{-3}$ |
| $r_l$ | Rate of reaction $l$. A product of compound abundance, reaction rate, and stoichiometry | molecules cm$^{-3}$s$^{-1}$. |
| $k_l$ | Reaction rate coefficient of reaction $l$ | See main text |
| $S_{i,l}$ | The stoichiometry matrix for compound $i$ in reaction $l$ | – |
| $y$ | Array that captures the states of the ODE being solved | – |
| $\vec{p}$ | Vector of parameters of the ODE | – |
| $f(y)$ | Generic right-hand side (RHS) function describing chemical reactions | – |
| $n_{chem}$ | Number of chemicals involved in multiple chemical reactions | – |
| $J$ | Jacobian matrix that contains the partial derivatives of each RHS function with respect to all components | – |
| $\dfrac{\partial f(y_1)}{\partial y_2}$ | Partial derivative of state $y_1$ with respect to $y_2$ | – |
| $\dfrac{(f(y_1+\Delta y_1)-f(y_1))}{\Delta y_1}$ | Finite difference approximation to a partial derivative used to construct a Jacobian matrix | – |
| $C_{g,i}$ | Abundance of compound $i$ in the gas phase | molecules cm$^{-3}$ |
| $C_{a,i,m}$ | Abundance of compound $i$ in the condensed phase, specifically in size "bin" $m$ | molecules cm$^{-3}$ |
| $C_{a,i,m}^{eq}$ | Required abundance of compound $i$ above the surface of the condensed phase, specifically in size "bin" $m$, to maintain current composition | molecules cm$^{-3}$ |
| $k_{i,m}$ | First-order mass transfer coefficient for species $i$ and bin $m$ | s$^{-1}$ |
| $p_{l,i}^{o}$ | Sub-cooled, pure component liquid vapor pressure for compound $i$ | atm |
| $\nu_{k,i}$ | Number of groups of type k in compound $i$ | – |

## A.4.5 Chapter 5

| | **List of variables, constants, and acronyms** | |
|---|---|---|
| **Variable** | **Definition** | **Typical unit** |
| $\bar{c}_i$ | Mean thermal speed of particle $i$ | $\mathrm{m\,s^{-1}}$ |
| $C_c$ | Cunningham correction factor | 1 |
| $d_g$ | Geometric mean diameter | m |
| $D_i$ | Brownian diffusivity of particle $i$ | $\mathrm{m^2\,s^{-1}}$ |
| $\Delta t$ | Time step | s |
| $d_p$ | Particle diameter | m |
| $\ell_i$ | The mean free path of particle $i$ | m |
| $E\left(d_{p1}, d_{p2}\right)$ | Collision efficiency of particles with sizes $d_{p1}$ and $d_{p2}$ | 1 |
| $k_B$ | Boltzmann constant | $\mathrm{J\,K^{-1}}$ |
| $K$ | Coagulation kernel | $\mathrm{m^3\,s^{-1}}$ |
| $\lambda$ | Probability rate of coagulation event | $\mathrm{s^{-1}}$ |
| $\lambda_{air}$ | Mean free path of air molecules | m |
| $m_i$ | Mass of particle $i$ | kg |
| $m_b$ | Base mass | kg |
| $M_k$ | $k$-th moment of a distribution $n(d_p)$ | $\mathrm{m^{k-3}}$ |
| $N$ | Number of particles | 1 |
| $N_{event}$ | The number of coagulation events | 1 |
| $C_t$ | Total number concentration | $\mathrm{m^{-3}}$ |
| $n(m, t)$ | Continuous number distribution density function | $\mathrm{m^{-3}\,kg^{-1}}$ |
| $p$ | Probability | 1 |
| $Q$ | Mass of particle divided by base mass $m_b$ | 1 |
| $\rho_p$ | Particle (material) density | $\mathrm{kg\,m^{-3}}$ |
| $\sigma_g$ | Geometric standard deviation | 1 |
| $\tau$ | Time increment | s |
| $t$ | Time | s |
| $T$ | Temperature | K |
| $\upsilon$ | Particle volume | $\mathrm{m^3}$ |
| $\upsilon_{t1}$ and $\upsilon_{t2}$ | Terminal velocities of the particles | $\mathrm{m\,s^{-1}}$ |
| $V$ | Volume that the particles reside in | $\mathrm{m^3}$ |
| $\mu$ | Viscosity of air | $\mathrm{N\,s\,m^{-2}}$ |

## A.4.6 Chapter 6

| List of variables, constants, and acronyms | | |
| --- | --- | --- |
| **Variable** | **Definition** | **Typical unit** |
| $C$ | Number concentration of molecules or molecular clusters | $\text{cm}^{-3}$ |
| $\beta$ | Collision rate constant between molecules and/or clusters | $\text{cm}^3\,\text{s}^{-1}$ |
| $\gamma$ | Evaporation rate constant of molecules or clusters from clusters | $\text{s}^{-1}$ |
| $Q$ | Source rate of molecules or clusters | $\text{cm}^{-3}\,\text{s}^{-1}$ |
| $S$ | Sink rate constant of molecules or clusters | $\text{s}^{-1}$ |
| $J$ | Formation rate of clusters of given size | $\text{cm}^{-3}\,\text{s}^{-1}$ |
| $I$ | Net flux of clusters between given sizes or compositions | $\text{cm}^{-3}\,\text{s}^{-1}$ |
| $t$ | Time | s |
| $T$ | Temperature | K |
| $P$ | Pressure | Pa |
| $m$ | Mass of molecule or cluster | kg |
| $V$ | Volume of molecule or cluster | $\text{m}^3$ |
| $\Delta G$ | Gibbs free energy of cluster formation | $\text{kcal mol}^{-1}$ |
| $\Delta H$ | Enthalpy of cluster formation | $\text{kcal mol}^{-1}$ |
| $\Delta S$ | Entropy of cluster formation | $\text{cal mol}^{-1}\,\text{K}^{-1}$ |
| $k_B$ | Boltzmann constant | $\text{J K}^{-1}$ |
| GDE | General dynamic equation | – |
| ACDC | Atmospheric Cluster Dynamics Code software | – |
| cluster | Small nanoparticle consisting of a discrete number of molecules (typically up to $\sim$10–20 molecules) | – |
| monomer | Smallest entity in cluster population modeling, i.e. a single vapor molecule | – |

## A.4.7 Chapter 7

### List of variables, constants, and acronyms

| Variable | Definition | Typical unit |
|---|---|---|
| $n_{\text{sect},i}$ | Number concentration in size bin $i$ in a sectional model | $m^{-3}$ |
| $\Delta n_{\text{sect},i}$ | Change in number concentration in size bin $i$ in a sectional model | $m^{-3}$ |
| $d_{\text{sect},i}$ | Diameter of particle in bin $i$ in a sectional approach | m |
| $v_{sect,i}$ | Bin center volume of a particles in size bin $i$ in a sectional model | $m^3$ |
| $v_{sect,i,hi}$ | Upper volume of a particle in size bin $i$ in a sectional model | $m^3$ |
| $v_{sect,i,lo}$ | Lower volume of a particle in size bin $i$ in a sectional model | $m^3$ |
| $d_{sect,i,hi}$ | Upper diameter of a particle in size bin $i$ in a sectional model | m |
| $d_{sect,i,lo}$ | Lower diameter of a particle in size bin $i$ in a sectional model | m |
| $V_{\text{sect},i,j}$ | Volume concentration of species $i$ in size bin $j$ in a sectional model | $m^3 m^{-3}$ |
| $V_{\text{sect},j}$ | Volume concentration of all species in size bin $j$ in a sectional model | $m^3 m^{-3}$ |
| $\Delta V_{\text{sect},i,j}$ | Change in volume concentration of species $i$ in size bin $j$ in a sectional model | $m^3 m^{-3}$ |
| $\Delta V_{\text{sect},j}$ | Change in volume concentration all species in size bin $j$ in a sectional model | $m^3 m^{-3}$ |
| $y$ | Linear function representing sub-grid size distribution | $m^{-3}$, $kg\ kg^{-1}$, $kg\ m^{-3}$, $m^3 m^{-3}$ |
| $k, k_*$ | Slope of the linear function representing sub-grid size distribution | 1 |
| $x$ | First moment of the linear function representing sub-grid size distribution | m, kg, $m^3$, 1 |

*(Continued)*

*(Continued)*

| | List of variables, constants, and acronyms | |
| --- | --- | --- |
| **Variable** | **Definition** | **Typical unit** |
| $x_0, x_*$ | Mid-point of the linear function representing sub-grid size distribution | m, kg, m$^3$, 1 |
| $y_0$ | Value of $y$ at the center ($x0$ or $x_*$) of the size bin | m$^{-3}$ |
| $n$ | Linear function representing sub-grid number size distribution | m$^{-3}$ |
| $n_t$ | Value of $n$ at time $t$ | m$^{-3}$ |
| $n_{t+\Delta t}$ | Value of $n$ at time $t + \Delta t$ | m$^{-3}$ |
| $\Delta n$ | Change in the value of $n$ in the linear function representing sub-grid number size distribution | m$^{-3}$ |
| $v_0, v_*$ | Volume of a particle at the center of the size bin | m$^{-3}$ |
| $n_0$ | Value of $n$ at the center ($v_0$ or $v_*$) of the size bin | m$^{-3}$ |
| $v_1$ | Volume at the lower edge of a bin with sub-grid size distribution | m$^{-3}$ |
| $v_2$ | Volume at the upper edge of a bin with sub-grid size distribution | m$^{-3}$ |
| $v_{1,t}$ | Volume at the lower edge of a bin with sub-grid size distribution at time $t$ | m$^{-3}$ |
| $v_{1,t+\Delta t}$ | Volume at the lower edge of a bin with sub-grid size distribution at time $t + \Delta t$ | m$^{-3}$ |
| $v_{2,t}$ | Volume at the upper edge of a bin with sub-grid size distribution at time $t$ | m$^{-3}$ |
| $v_{2,t+\Delta t}$ | Volume at the upper edge of a bin with sub-grid size distribution at time $t + \Delta t$ | m$^{-3}$ |
| $a, b$ | Integration limit | m, kg, m$^3$, 1 |
| $A$ | New particle formation rate constant | s$^{-1}$ |
| $C_{g,H_2SO_4}$ | Gas phase concentration of sulfuric acid | mol m$^{-3}$ |

*(Continued)*

*(Continued)*

### List of variables, constants, and acronyms

| Variable | Definition | Typical unit |
|---|---|---|
| $C_{a,H_2SO_4,1}$ | Molar concentration of sulfate in the smallest size bin | mol m$^{-3}$ |
| $v_{\text{NPF}}$ | Volume of the individual particles formed in new particle formation | m$^3$ |
| $\rho_{SO_4}$ | Density of sulfate | kg m$^{-3}$ |
| $M_{SO4}$ | Molar mass of sulfate | kg mol |
| $g$ | Gravitational acceleration | m s$^{-2}$ |
| $\rho$ | Density of air | kg m$^{-3}$ |
| $L_{e,i}$ | Latent heat of evaporation for species $i$ | J kg$^{-1}$ |
| $c_{p,i}$ | Heat capacity of species $i$ | J K$^{-1}$ |
| $m_i$ | Total mass of species $i$ in an air parcel | kg |
| $M_{air}$ | Average molar mass of air | kg mol |
| $w$ | Horizontal updraft velocity of the air parcel | m s$^{-1}$ |
| $R$ | Gas constant | J mol$^{-1}$K$^{-1}$ |
| $T$ | Air temperature | K |
| $P$ | Air pressure | Pa |
| $N_L$ | Total number of aerosol particles per unit volume | m$^{-3}$ |

## Bibliography

[1] 2018 CODATA Value: Avogadro constant. *The NIST Reference on Constants, Units, and Uncertainty*, 2019.

[2] M. Z. Jacobson. *Fundamentals of Atmospheric Modeling, Second Edition*. Cambridge University Press, New York, 2005.

[3] M. Barley, D. O. Topping, M. E. Jenkin, and G. Mcfiggans. Sensitivities of the absorptive partitioning model of secondary organic aerosol formation to the inclusion of water. *Atmospheric Chemistry and Physics*, 9(9):2919–2932, 2009.

[4] 2018 CODATA Value: Boltzmann constant. *The NIST Reference on Constants, Units, and Uncertainty*, 2019.

[5] Helmet Grünewald. Handbook of Chemistry and Physics. Von R. C. Weast. The Chemical Rubber Co., Cleveland, Ohio/USA 1972. 52. Aufl., XXVII, 2313 S., geb. DM 99.80. *Angewandte Chemie*, 84(9):445–446, 1972.

[6] Sunghwan Kim, Jie Chen, Tiejun Cheng, Asta Gindulyte, Jia He, Siqian He et al. PubChem in 2021: New data content and improved web interfaces. *Nucleic Acids Research*, 49(D1):D1388–D1395, 2021.

[7] Arthur W. Adamson and Alice P. (Alice Petry) Gast. Physical chemistry of surfaces. page 784, 1997.

[8] J. H. Seinfeld and S. N. Pandis. *Atmospheric Chemistry and Physics: From Air Pollution to Climate Change.* J. Wiley & Sons, New York, USA, 1998.

[9] Barbara J. Finlayson-Pitts and James N. Pitts. *Chemistry of the upper and lower atmosphere : theory, experiments, and applications.* Academic Press, 2000.

# Index

*Introduction to Aerosol Modelling: From Theory to Code.*
First Edition. Edited by David Topping and Michael Bane.
© 2022 John Wiley & Sons Ltd. Published 2022 by John Wiley & Sons Ltd.